CONTEMPORARY'S

GED

PREPARATION FOR THE HIGH SCHOOL EQUIVALENCY EXAMINATION

TEST 5:

THE MATHEMATICS TEST

JERRY HOWETT

GAIL SHEA
Consultant
Curriculum Specialist
Chicago Urban Skills Institute

CAREN VAN SLYKE
Editorial Director

Contemporary Books, Inc.
Chicago

Library of Congress Cataloging in Publication Data

Howett, Jerry.
    Contemporary's GED.

    1. Mathematics--Examinations, questions, etc.
2. General educational development tests.   I. Shea,
Gail.   II. Van Slyke, Caren.   Contemporary Books, inc.
IV. Title.   V. Title: Contemporary's G.E.D.   VI. Title:
GED.   VII. Title: G.E.D.   VIII. Title: Mathematics test
QA43.H7   1984   510'.76   84-23237
ISBN 0-8092-5590-1 (pbk.)

Published by Contemporary Books, Inc.
180 North Michigan Avenue, Chicago, Illinois 60601
Manufactured in the United States of America
International Standard Book Number: 0-8092-5590-1

Published simultaneously in Canada by
Beaverbooks, Ltd.
195 Allstate Parkway
Valleywood Business Park
Markham, Ontario L3P 4T8
Canada

Editorial Assistant
Karin Evans

Production Editor
Gerry Lynch

Production and Art
Princess Louise El
Ophelia M. Chambliss-Jones

Table of Contents

Contents

Contents

ALGEBRA——————————————————————299

MATHEMATICS POST-TEST——————————————353

Introducing the Mathematics Test

Perhaps it has been a long time since you took an important test. Maybe you are not sure about what is required to pass the math test, or you are nervous about the test-taking situation.

This book has been designed to help you succeed on the GED mathematics test. It will provide you with instruction in the math skills you need to pass the test, plenty of GED-type practice exercises, and test-taking hints. If you work carefully through this book, you should do well on the mathematics test.

USING THIS BOOK

Contemporary's GED series is a program that you or a teacher can use to determine your individual strengths and weaknesses.

1. Start your work with the three pre-tests at the beginning of the book. These will help you to preview what the mathematics test includes, but more importantly, they will help you to diagnose what areas you need to concentrate on. Use the skills chart at the end of the pre-tests to see what areas need special work.

   It would be best for you to work your way carefully through the entire book because you will learn problem-solving skills that will help you on the entire test. However, if your time is limited and you need more work on another test, you can use the pre-tests and skills chart to decide what areas to focus on.

2. Work your way carefully through this book, paying special attention to the word problems throughout. Most, if not all, of the GED mathematics test consists of word problems. To a large extent, your ability to pass the test depends on how well you understand and solve problems.

3. One of the key features of this book is the answers and explanations section at the end of every chapter. If you make a mistake, you can learn from it by reading the solution that follows the answer and then going back to the problem and analyzing your error.

4. Finally, the post-test is a simulated GED test. It presents problems in GED-type format and represents the level of difficulty of the questions. It will give you a chance to determine if you are ready for the test and, if not, what areas of the book you still need to review.

THE MATHEMATICS TEST

Test 5: The Mathematics Test is one of the tests in the GED examination series.

There are fifty multiple-choice questions on the test, some requiring only one step, some requiring two or more steps. Each question gives you five possible responses or choices, only one of which is the correct answer. Since the four wrong answer choices are written to be plausible, do all your work carefully and do not make any hasty conclusions.

THE GED TEST

Q: What does GED stand for?

A: GED stands for the Tests of General Educational Development. The credential received for passing the test is widely recognized by colleges, training schools, and employers as equivalent to a high school diploma. The GED is a national examination developed by the GED Testing Service of the American Council on Education.

The GED tests consist of five examinations in the areas of writing skills, social studies, science, reading skills, and mathematics. While the GED measures skills and knowledge normally acquired in four years of high school, much that you have learned informally or through other types of training can help you pass the test.

The GED test is available in English, French, and Spanish, and on audiocassette, in braille, and in large-print editions.

Q: Can I take the test?

A: Each year, more than 800,000 people take the GED test. In the United States, Canada, and many territories, people who have not graduated from high school and who meet specific eligibility requirements (age, residency, etc.) may take the test. Since eligibility requirements vary, it would be useful to contact your local GED testing center or the director of adult education in your state, province, or territory for specific information

Q: What should I know to pass the test?

A: The chart below outlines the main content areas, the breakdown of questions, and the time generally allowed per test.

THE GED TEST			
Content	**Minutes**	**Number of Questions**	**Percentage of Test**
Test 1 Writing Skills	75	80	
Spelling		10	12.5%
Capitalization & Punctuation		10	12.5%
Grammar & Usage		24	30%
Diction & Style		12	15%
Sentence Structure		12	15%
Logic & Organization		12	15%
Test 2 Social Studies	90	60	
U.S. History		15	25%
Economics		12	20%
Geography		9	15%
Political Science		12	20%
Behavioral Science		12	20%
Test 3 Science	90	60	
Biology		30	50%
Earth Science		12	20%
Chemistry		9	15%
Physics		9	15%
Test 4 Reading Skills	60	40	
Practical Reading		6	15%
General Reading		12	30%
Prose Literature		12	30%
Poetry		5	12.5%
Drama		5	12.5%
Test 5 Mathematics	90	50	
Arithmetic		28	55%
Geometry		10	20%
Algebra		12	25%

On all five tests, you are expected to demonstrate the ability to read and to understand what you are reading. You are also tested on many skills you have acquired from life experiences, reading, television, radio, newspapers, contracts, consumer products, and advertising. By and large, you are not expected to recall facts. However, parts of the social studies and science tests require you to remember some information from your past reading and learning.

In addition to the above information, keep these facts in mind:

1. Three of the five tests—reading, science, and social studies—require that you answer questions based mainly on reading written material or interpreting pictorial material in these content areas. Developing strong reading and analysis skills is the key to succeeding on these tests.

2. The writing skills test does not require you to write, but rather to detect errors of grammar and usage, punctuation, capitalization, and spelling. Also, you must be able to recognize correct sentence and paragraph construction.

3. The math test consists mainly of word problems to be solved. Therefore, you must be able to combine your ability to perform computations with reading comprehension skills.

Someone once said that an education is what remains after you've forgotten everything else. In many ways, this is what the GED measures.

Q: What is a passing score on the GED?

A: Again, this varies from area to area. To find out what you need to pass the test, contact your local GED testing center. However, you must keep two scores in mind. One score represents the minimum score you must get on each test. For example, if your state requires minimum scores of 35, you must get at least 35 points on every test. Additionally, you must meet the requirements of a minimum average score on all five tests. For example, if your state requires a minimum average score of 45, you must get a total of 225 points to pass. The two scores combined —the minimum score and the minimum average score—determine whether you pass or fail the GED.

To understand this better, look at the scores of three people who took the test in a state that requires a minimum score of 35 and a minimum average score of 45 (225 total). Heino and Bob did not pass, but Maria did. See if you can tell why.

	Heino	**Bob**	**Maria**
Test 1	45	38	43
Test 2	47	43	48
Test 3	33	42	47
Test 4	50	40	52
Test 5	50	39	49
	225	202	239

Heino made the total of 225 points but fell below the minimum score on Test 3. Bob passed each test but failed to get the 225 points needed—just passing each test was not enough. Maria passed all the tests and exceeded the minimum average score.

Generally, to receive a GED credential, you must correctly answer half or a little more than half of the questions on each test.

Q: What happens if I don't pass the test?

A: You are allowed to retake some or all of the tests. Again, the number of times that you may retake the test and the time you must wait before retaking it depend on your state, province, or territory. Some states require you to take a review class or to study on your own for a certain amount of time before retesting.

Q: How can I best prepare for the test?

A: Many libraries, community colleges, adult education centers, churches, and other institutions offer GED preparation classes. Some television stations broadcast classes to prepare people for the test. If you cannot find a GED preparation class locally, contact the director of adult education in your state, province, or territory.

Q: I need to study for the other tests. Are there other materials available?

A: Contemporary Books publishes a wide range of materials to help you prepare for the test. These books are designed for home study or class use. Contemporary's GED preparation books are available through schools and bookstores or directly from the publisher.

Now, let's focus on some useful test-taking tips. As you read this section, you should feel more confident about your ability to succeed on The Mathematics Test.

TEST-TAKING TIPS

1. Get prepared physically. Eat a well-balanced meal and get plenty of rest the night before the test so that you will have energy and will be able to think clearly. Last-minute cramming will probably not help as much as a relaxed and rested mind.

2. Arrive early. Be at the testing center at least fifteen to twenty minutes before the starting time. Have time to find the room and to get situated. Keep in mind that many testing centers refuse to admit latecomers.

3. Think positively. Tell yourself you will do well. If you have studied and prepared for the test, you should succeed.

4. Relax during the test. Take a half-minute several times during the test to stretch and breathe deeply, especially if you are feeling anxious or confused.

5. Read the test directions carefully. Be sure you understand how to answer the questions. If you have any questions about the test or about filling in the answer form, ask before the test begins.

6. Know the time limit for each test. The Mathematics Test has a time limit of ninety minutes, or one and one-half hours. Some testing centers allow extra time, while others do not. You may be able to find out the policy of your testing center before you take the test, but always work according to the official time limit. If you have extra time, go back and check your answers.

   For this fifty-item test, you should allow about a minute and a half to read and answer each question. However, this is not a hard and fast rule. Use it only as a guide to keep yourself within the time limit.

7. Have a strategy for answering problems. If you are having difficulty figuring out how to solve a problem, take a minute to restate the problem in your own words or draw a picture to represent the problem. Some people skim the answer choices to get an idea of what to do.

8. Don't spend a lot of time on difficult questions. If you're not sure of the answers, go on to the next question. Answer easier questions first and then go back to the more difficult questions. If you skip a question, be sure that you have skipped the same number on your answer sheet. Although this is a good strategy for making the most of your time, it can be very easy to get confused and throw off your whole answer key.

   Lightly mark the numbers of the questions you did not answer in the margin of your answer sheet so that you know what to go back to. Be sure to erase these marks completely after you answer the questions. In this way, there won't be any confusion when your test is graded.

9. Answer every question on the test. If you're not sure of an answer, take an educated guess. When you leave a question unanswered, you will lose points, but you can gain points if you make a correct guess.

   If you must guess, try to eliminate one or more answers that you are sure are not correct. Then, choose from the remaining answers. Remember, you greatly increase your chances if you can eliminate one or two answers before guessing. Of course, guessing should be used only when all else has failed.

10. Clearly fill in the circle for each answer choice. If you erase something, erase it completely. Be sure that you give only one answer per question; otherwise, no answer will count.

Use the exercises, reviews, and especially the post-test in this book to better understand your test-taking habits and weaknesses. Use them to practice different stategies, such as skimming answer choices before starting problems or skipping hard questions until the end. Knowing your own personal test-taking style is important to success on the GED.

Mathematics Pre-Tests

The following three tests will help you to evaluate your strengths and weaknesses in math. You may choose to take them all at once, before you start work on this book, or you may choose to come back to this section to take one before starting a new major area of this book.

Take your time to work through the problems. Read each one carefully before you start. Don't give up if you find that some problems are hard; there may be some later in the section that you remember how to do.

When you have finished, check the answers and the solutions that follow the tests. Fill in the charts at the end of the tests and use them as a guideline that tells you what areas need special work.

Whatever the outcome of these pre-tests, we do suggest that you work your way through the entire book, reviewing those areas with which you are already familiar. These sections contain hundreds of GED-type problems, helpful hints, and shortcuts to prepare you for the GED math test.

PRE-TEST I: WHOLE NUMBERS, DECIMALS, FRACTIONS, AND PERCENTS

Solve each problem.

1. $259 \times 78 =$

2. $5,291 \div 13 =$

3. $40,300 - 2,567 =$

4. $376 + 2,019 + 18 + 128 =$

5. Find the next term in the series 2, 9, 16, 23. . . .

6. 48 people went to the September meeting of a union. 57 people went to the October meeting. 63 people went to the November meeting. Find the average number of people that attended the union meetings.

7. Franklin bought a car on sale for $3,950. This was $1,599 less than the original price. What was the original price?

8. Anna-Marie can type 85 words per minute. How long will it take her to type a letter that contains 2,040 words?

9. The Central County Memorial Gymnasium holds 4,590 people. The gym was full three nights in a row for a basketball tournament. To the nearest thousand, find the number of people who attended the tournament.

10. Mr. and Mrs. Victoria paid $4,500 down and $280 a month for 360 months for their house. What total amount did they pay for the house?

11. $2.049 + .37 + .0895 =$ 14. $19\overline{)39.14}$

12. $13 - 3.964 =$ 15. $2.85 \div .38 =$

13. $6.15 \times 3.4 =$ 16. $6.5\overline{)156}$

17. What is 4,361.725 rounded off to the nearest hundredth?

18. With his old car, Ernesto could drive 12.6 miles on one gallon of gasoline. With his new car, he can drive 2.9 miles more on one gallon of gas. Find the number of miles Ernesto can drive his new car on one gallon of gas.

19. A batting average is the number of hits a player gets divided by the number of times he was at bat. Jack was at bat 60 times one season, and he made 14 hits. To the nearest thousandth, what was Jack's batting average?

20. One day Eli worked 7 hours at his regular rate of $6.50 an hour and 3.5 hours at the rate of $9.75 an hour. How much did he make for the day?

21. $8\frac{4}{5} + 7\frac{3}{5} =$ 25. $\frac{3}{4} \times \frac{6}{7} =$

22. $2\frac{3}{4} + 3\frac{5}{8} + 4\frac{7}{12} =$ 26. $3\frac{3}{4} \times 2\frac{4}{5} =$

23. $6\frac{5}{8} - 2\frac{1}{4} =$ 27. $3\frac{1}{2} \div \frac{3}{4} =$

24. $9\frac{1}{2} - 3\frac{2}{3} =$ 28. $16 \div 2\frac{2}{3} =$

29. In Max's GED class there are 12 women and 9 men. What fraction of the total class are men?

30. $\frac{3}{4}$ of the striking utility workers at the Central Utilities Company voted to accept a new contract. There are 284 workers at the company. How many workers voted to accept the contract?

31. Margarita bought $2\frac{1}{4}$ pounds of fish at $1.40 a pound and $2\frac{1}{2}$ pounds of ground beef at $1.30 a pound. How much did she pay altogether for these purchases?

32. The Richardsons spend $\frac{1}{4}$ of their income on rent and $\frac{1}{3}$ on food. They make $1,200 a month. How much do the Richardsons spend on rent and food in a month?

33. 30 is what percent of 40?

34. Find 85% of 40.

35. 60 is 10% of what number?

36. Ivan's gross salary is $1,100 a month. His employer withholds 24% of Ivan's salary. How much is withheld each month from his salary?

37. On a winter night only 12 people came to a tenants' organization meeting. This represents 25% of the membership. How many members are there in the organization?

38. In 1970 the population of Centerville was 40,000. By 1980, the population had dropped to 32,000. By what percent did the population drop in 10 years?

39. Find the interest on $800 at 12.5% annual interest for one year.

40. Pieter borrowed $1,000 at 18% annual interest for one year and three months. How much did he owe altogether at the end of that time?

Answers and solutions are on page 15.

PRE-TEST II: RATIO AND PROPORTION, GRAPHS, AND MEASUREMENT

Solve each problem.

1. Simplify the ratio 36:44.

2. Simplify the ratio $\frac{4}{5}:\frac{9}{10}$.

3. Paco got 25 hits in the 60 times he was at bat. What was the ratio of the number of hits to the number of times at bat?

4. Solve for c in $\frac{6}{c} = \frac{5}{9}$.

5. Find the value of m in $3:7 = 9:m$.

6. The ratio of men to women in Jorge's math class is 4:3. There are 12 men in the class. How many women are in the class?

7. Selma paid $4.65 for 3 pounds of pork chops. At the same rate how much would she pay for 5 pounds of pork chops?

8. For every $10 that Julio makes, his employer deducts $3. Julio makes $945 a month. How much does his employer deduct each month?

Use the table below to answer questions 9-11.

Consumption of Major Food Items in the U.S. (in pounds per person)			
Food Item	1970	1980	1981
Beef	84.0	76.5	77.2
Eggs	39.1	34.6	33.6
Chicken	40.4	50.1	51.7
Cheese	11.5	17.6	18.0
Fresh fruit	78.9	85.7	87.3

9. In 1980, how many pounds of chicken did the average person consume?

10. Between 1970 and 1980, what was the difference in the per person consumption of chicken?

11. For which food items did the per person consumption drop from 1970 to 1981?

Use the graph at the right to answer questions 12-14.

FOREIGN BORN AS A PERCENT OF REGIONAL POPULATION

12. In which region of the country did the percent of foreign-born population change the most from 1970 to 1980?

13. In which regions of the country were the percent of foreign-born population higher than 8% in 1980?

14. Which of the following statements is true?
    (1) The percentage of the foreign-born population dropped in all of the regions shown on the graph.
    (2) The percentage of the foreign-born population in the Northeast nearly doubled from 1970 to 1980.
    (3) In 1970, the region with the highest percentage of foreign-born population was the West.
    (4) The region with the greatest change in the percentage of the foreign-born population was the West.
    (5) The percentage of foreign-born population in the South was cut nearly in half from 1970 to 1980.

Use the graph at the right to answer questions 15-17.

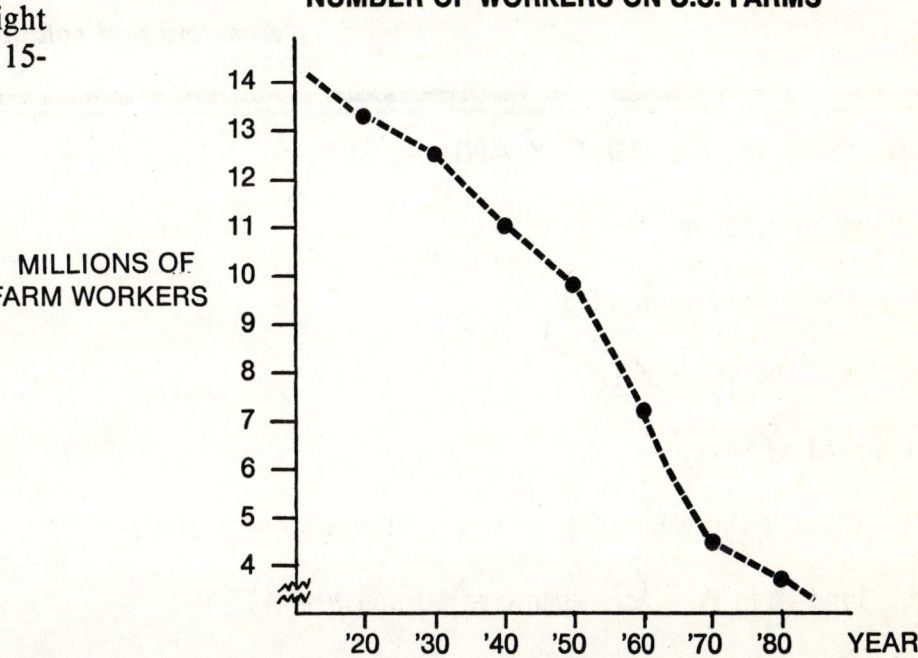

NUMBER OF WORKERS ON U.S. FARMS

15. In what year was the number of people employed on farms closest to 11 million?

16. From 1950 to 1970, by how much did the number of people employed on farms drop? (Round to the nearest million.)

17. If the trend shown on the graph continues, the number of people working on farms in 1990 will be closest to which of the following?
    (1) About twice the number as in 1980
    (2) About the same as in 1980
    (3) About $2\frac{1}{2}$ or 3 million
    (4) About the same as in 1970
    (5) About twice the number as in 1970

18. Change 72 ounces to pounds.

19. Change 150 inches to feet.

20. 6.4 kilograms is equal to how many grams?

21. Find the difference between 4 hr. 20 min. and 2 hr. 45 min.

22. From a 100-pound bag of cement Wilfredo used 66 lb. 12 oz. to mix concrete. How many pounds of cement were left in the bag?

23. What is the total weight of four packages each weighing 6 lb. 5 oz.?

24. In three days, Paul spent a total of 20 hrs. 45 min. driving. Find the average amount of time he drove each day.

25. Senta mixed 5 gal. 3 qt. of blue paint with 2 gal. 2 qt. of white paint. Find the total quantity of the mixture.

Answers and solutions start on page 17.

PRE-TEST III: GEOMETRY AND ALGEBRA

Solve each problem.

1. What is the value of 50^3?

2. Find the value of $12^2 - 3^4$.

3. Find $\sqrt{5041}$.

4. What is the value of $(s - t)(s + t)$ when $s = 8$ and $t = 5$?

5. Find A in $A = \frac{1}{2}bh$ when $b = 10$ and $h = 4\frac{1}{2}$.

6. The formula for changing Fahrenheit to Celsius temperature is $C = \frac{5}{9}(F - 32)$, F being the Fahrenheit temperature. Find the Celsius temperature that corresponds to $86°F$.

7. What is the perimeter of the rectangle shown at the right?

2.4 m

1.7 m

8. How many inches of weatherstripping are needed to go around a window measuring 52 inches on each side?

9. What is the perimeter of triangle MNO?

N

6½ in.   7½ in.

M   10 in.   O

10. A gallon of paint will cover about 400 square feet. The walls in Karin's house are $8\frac{1}{2}$ feet high. The total length of all the walls she wants to paint is 200 feet. How many gallons of paint should Karin buy?

11. Find the area of a square that measures $3\frac{1}{2}$ yards on each side.

12. The window in the attic of Udo's house is covered with a triangular piece of glass. The window measures 32 inches across the base and is 24 inches high. How many square inches of glass does Udo need to replace the old window?

13. To the nearest square inch, what is the area of a circle with a radius of 8 inches? Use $\pi = 3.14$.

14. The formula for the area of a trapezoid is $A = \frac{1}{2}h(b_1 + b_2)$ where h is the height and b_1 and b_2 are the two parallel sides. Find the area of the trapezoid shown at the right.

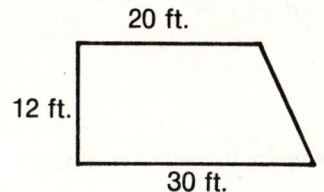

15. At the right is a floor diagram of Petra's living room and dining room. How many square feet of tiles does she need to cover the floors of the two rooms?

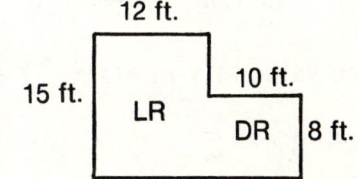

16. The picture at the right shows the length, width, and height of a large container. How many cubic feet can the container hold when it is full?

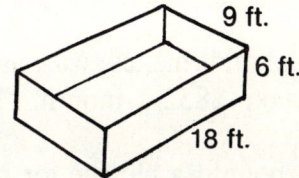

17. Find the measurement of angle x.

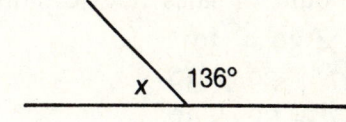

18. In triangle XYZ at the right, $\angle X = 35°$ $\angle Y = 55°$. What kind of triangle is $\triangle XYZ$?

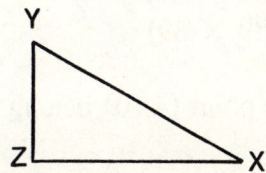

19. A six-foot-tall man casts a shadow 4 feet long. At the same time a telephone pole casts a shadow 42 feet long. How tall is the pole?

20. Find the distance from S to U in the picture at the right.

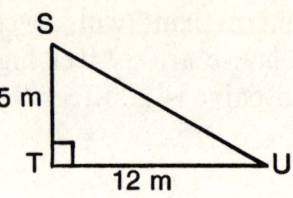

21. Which point on the line below corresponds to the number $-\frac{1}{2}$?

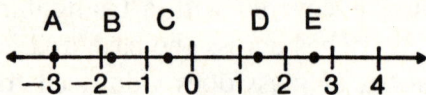

22. Find the sum of -13, $+15$, -6, and $+8$.

23. Solve the following: $(+9) - (-4) + (-6) =$

24. $(-20)(+\frac{3}{4})(+6) =$

25. $\dfrac{-48}{+60}$

26. Solve for c in $9c - 8 = 16$.

27. Solve for n in $12n - 4 = 7n + 36$.

28. Find the value of s in $6(s + 3) = 12$.

29. Find x in $8x - 2 > 5x + 10$.

30. Three times a number decreased by five is the same as that number increased by nine. Find the number.

31. Mrs. Schmidt makes twice as much as her daughter Monika. Together they make $852 a month. How much does Mrs. Schmidt make?

32. Utaka bought a bicycle for his son for $89 and a drill for himself for $39. He had to pay 7% sales tax on these items. Which expression tells the amount of sales tax he paid for the bicycle and drill?

(1) $7 \times 89 + 39$
(2) $.07 + 89 \times 39$
(3) $.07 + 89 + 39$
(4) $.07(89 + 39)$
(5) $.07(89 \times 39)$

33. Does the point $(4,10)$ belong to the graph of the equation $y = 3x - 2$?

34. What are the coordinates of the y-intercept of the equation $y = \frac{1}{2}x - 6$?

Answers and solutions are on page 18.

ANSWERS AND SOLUTIONS

Pre-Test I: Whole Numbers, Decimals, Fractions
and Percents

1. 20,202

$$
\begin{array}{r}
259 \\
\times 78 \\
\hline
2\,072 \\
18\,13 \\
\hline
20,202
\end{array}
$$

2. 407

$$
\begin{array}{r}
407 \\
13\overline{)5,291} \\
5\,2 \\
\hline
09 \\
0 \\
\hline
91 \\
91
\end{array}
$$

3. 37,733

$$
\begin{array}{r}
{}^{39}\,{}^{12\,9\,1} \\
4\cancel{0},\cancel{3}\cancel{0}\cancel{0} \\
-\ 2,567 \\
\hline
37,733
\end{array}
$$

4. 2,541

$$
\begin{array}{r}
376 \\
2,019 \\
18 \\
+\ 128 \\
\hline
2,541
\end{array}
$$

5. 30

$$
2 \nearrow 9 \nearrow 16 \nearrow 23 \nearrow \underline{30}
$$
$$
+7 \quad +7 \quad +7 \quad +7
$$

6. 56 people

$$
\begin{array}{r}
48 \\
57 \\
+\ 63 \\
\hline
168
\end{array}
\qquad
\begin{array}{r}
56 \\
3\overline{)168}
\end{array}
$$

7. $5,549

$$
\begin{array}{r}
3,950 \\
+1,599 \\
\hline
\$5,549
\end{array}
$$

8. 24 minutes

$$
\begin{array}{r}
24 \\
85\overline{)2,040} \\
1\,70 \\
\hline
340 \\
340
\end{array}
$$

9. 14,000 people

$$
\begin{array}{r}
4,590 \\
\times\ \ \ 3 \\
\hline
13,770 \text{ to the nearest thousand} = 14,000
\end{array}
$$

10. $105,300

$$
\begin{array}{r}
280 \\
\times\ \ 360 \\
\hline
16\,800 \\
84\,0 \\
\hline
100,800
\end{array}
\qquad
\begin{array}{r}
100,800 \\
+\ \ 4,500 \\
\hline
\$105,300
\end{array}
$$

11. 2.5085

$$
\begin{array}{r}
2.049 \\
.37 \\
+\ .0895 \\
\hline
2.5085
\end{array}
$$

12. 9.036

$$
\begin{array}{r}
{}^{2}\,{}^{9\,9\,1} \\
13.\cancel{0}\cancel{0}\cancel{0} \\
-\ 3.964 \\
\hline
9.036
\end{array}
$$

13. 20.91

$$
\begin{array}{r}
6.1\,5 \\
\times\ \ \ 3.4 \\
\hline
2\,4\,6\,0 \\
18\,4\,5 \\
\hline
20.9\,1\,0 = 20.91
\end{array}
$$

14. 2.06

$$
\begin{array}{r}
2.06 \\
19\overline{)39.14} \\
38 \\
\hline
1\,1 \\
0 \\
\hline
1\,14 \\
1\,14
\end{array}
$$

15. 7.5

$$
\begin{array}{r}
7.5 \\
.38.\overline{)2.85.0} \\
2\,66 \\
\hline
19\,0 \\
19\,0
\end{array}
$$

16. 24

$$
\begin{array}{r}
2\,4. \\
6.5.\overline{)156.0.} \\
130 \\
\hline
26\,0 \\
26\,0
\end{array}
$$

17. 4,361.73

18. 15.5 miles

$$\begin{array}{r} 12.6 \\ +\ 2.9 \\ \hline 15.5 \end{array}$$

19. .233

$$60\overline{)14.0000}$$ *to the nearest thousandth = .233*

$$\begin{array}{r} .2333 \\ 60\overline{)14.0000} \\ \underline{12\,0} \\ 2\,00 \\ \underline{1\,80} \\ 200 \\ \underline{180} \\ 200 \\ \underline{180} \end{array}$$

20. $79.63

$$\begin{array}{r} 6.50 \\ \times\quad 7 \\ \hline \$45.50 \end{array} \qquad \begin{array}{r} 9.7\,5 \\ \times\quad 3.5 \\ \hline 4\,8\,7\,5 \\ 29\,2\,5 \\ \hline \$34.1\,2\,5 \end{array}$$ *to the nearest penny = $34.13*

$$\begin{array}{r} 45.50 \\ +34.13 \\ \hline 79.63 \end{array}$$

21. $16\frac{2}{5}$

$$\begin{array}{r} 8\frac{4}{5} \\ +\ 7\frac{3}{5} \\ \hline 15\frac{7}{5} = 16\frac{2}{5} \end{array}$$

22. $10\frac{23}{24}$

$$\begin{array}{r} 2\frac{3}{4} = 2\frac{18}{24} \\ 3\frac{5}{8} = 3\frac{15}{24} \\ +4\frac{7}{12} = 4\frac{14}{24} \\ \hline 9\frac{47}{24} = 10\frac{23}{24} \end{array}$$

23. $4\frac{3}{8}$

$$\begin{array}{r} 6\frac{5}{8} = 6\frac{5}{8} \\ -2\frac{1}{4} = 2\frac{2}{8} \\ \hline 4\frac{3}{8} \end{array}$$

24. $5\frac{5}{6}$

$$9\frac{1}{2} = 9\frac{3}{6} = 8\frac{3}{6} + \frac{6}{6} = 8\frac{9}{6}$$
$$-3\frac{2}{3} = 3\frac{4}{6} = \qquad\qquad 3\frac{4}{6}$$
$$\overline{\qquad\qquad\qquad\qquad 5\frac{5}{6}}$$

25. $\frac{9}{14}$

$$\frac{3}{\cancel{4}_2} \times \frac{\cancel{6}^3}{7} = \frac{9}{14}$$

26. $10\frac{1}{2}$

$$3\frac{3}{4} \times 2\frac{4}{5} =$$
$$\frac{\cancel{15}^3}{\cancel{4}_2} \times \frac{\cancel{14}^7}{\cancel{5}_1} = \frac{21}{2} = 10\frac{1}{2}$$

27. $4\frac{2}{3}$

$$3\frac{1}{2} \div \frac{3}{4} = \frac{7}{2} \div \frac{3}{4} =$$
$$\frac{7}{\cancel{2}_1} \times \frac{\cancel{4}^2}{3} = \frac{14}{3} = 4\frac{2}{3}$$

28. 6

$$16 \div 2\frac{2}{3} = \frac{16}{1} \div \frac{8}{3} =$$
$$\frac{\cancel{16}^2}{1} \times \frac{3}{\cancel{8}_1} = \frac{6}{1} = 6$$

29. $\frac{3}{7}$

$$\begin{array}{ll} \text{women} = 12 & \\ \text{men} = \underline{+9} & \dfrac{men}{total} = \dfrac{9}{21} = \dfrac{3}{7} \\ \text{total} \quad\ 21 & \end{array}$$

30. 213

$$\frac{3}{\cancel{4}_1} \times \frac{\cancel{284}^{71}}{1} = \frac{213}{1} = 213$$

31. $6.40

$$2\frac{1}{4} \times 1.40 = \frac{9}{\cancel{4}_1} \times \frac{\cancel{1.40}^{.35}}{1} = \$3.15$$
$$2\frac{1}{2} \times 1.30 = \frac{5}{\cancel{2}_1} \times \frac{\cancel{1.30}^{.65}}{1} = \$3.25$$
$$\text{total} = \$6.40$$

32. $700

$$\begin{array}{l} \dfrac{1}{4} = \dfrac{3}{12} \\ +\dfrac{1}{3} = \dfrac{4}{12} \\ \hline \qquad \dfrac{7}{12} \end{array} \qquad \frac{7}{\cancel{12}_1} \times \frac{\cancel{1,200}^{100}}{1} = \$700$$

33. 75%

$$\frac{30}{40} = \frac{3}{4} = 75\%$$

34. 34

$$\begin{array}{r} 40 \\ \times\ .85 \\ \hline 2\,00 \\ 32\,0 \\ \hline 34.00 \end{array} = 34$$

35. 600

$$\begin{array}{r} 60 \\ \times\ 10 \\ \hline 600 \end{array}$$

36. $264

$$24\% = .24 \qquad \begin{array}{r} 1,100 \\ \times .24 \\ \hline 44\ 00 \\ 220\ 0 \\ \hline 264.00 \end{array}$$

37. 48

$$25\% = \frac{1}{4}$$

$$12 \div \frac{1}{4} =$$

$$\frac{12}{1} \times \frac{4}{1} = 48$$

38. 20%

$$\begin{array}{ll} 1970\ pop. & = 40,000 \\ 1980\ pop. & = \underline{32,000} \\ change & = 8,000 \end{array}$$

$$\frac{change}{original} = \frac{8,000}{40,000} = \frac{8}{40} = \frac{1}{5}$$

$$\frac{1}{5} = 20\%$$

39. $100

$$I = prt$$

$$I = \frac{\overset{8}{\cancel{800}}}{1} \times \frac{12.5}{\cancel{100}} \times 1 = 100$$

40. $1,225

$$1\ yr.\ 3\ mos. = 1\frac{3}{12} = 1\frac{1}{4} = \frac{5}{4}\ yr.$$

$$I = prt$$

$$I = \frac{\overset{5}{\cancel{\overset{10}{\cancel{1,000}}}}}{1} \times \frac{\overset{9}{\cancel{18}}}{\cancel{100}} \times \frac{5}{\cancel{4}} = \$225$$

$$\begin{array}{ll} amount\ borrowed & = \ \ \$1,000 \\ interest & = \underline{+\ \ \ 225} \\ total\ amount\ due & = \ \ \$1,225 \end{array}$$

Pre-Test II: Ratio and Proportion, Graphs, and Measurement

1. 9:11

$$36{:}44 = \frac{36}{44}$$

$$\frac{36 \div 4}{44 \div 4} = \frac{9}{11} = 9{:}11$$

2. 8:9

$$\frac{4}{5}{:}\frac{9}{10} = \frac{4}{5} \div \frac{9}{10} = \frac{4}{\cancel{5}} \times \frac{\cancel{10}^{2}}{9} = \frac{8}{9} = 8{:}9$$

3. 5:12

$$\frac{25}{60} = \frac{5}{12} = 5{:}12$$

4. $10\frac{4}{5}$

$$\frac{6}{c} = \frac{5}{9} \qquad \begin{array}{r} 6 \\ \times 9 \\ \hline 54 \end{array} \qquad \begin{array}{r} 10\frac{4}{5} \\ 5{\overline{)54}} \\ \underline{5} \\ 4 \end{array}$$

5. 21

$$3{:}7 = 9{:}m \qquad \frac{3}{7} = \frac{9}{m} \qquad \begin{array}{r} 7 \\ \times 9 \\ \hline 63 \end{array} \qquad \begin{array}{r} 21 \\ 3{\overline{)63}} \end{array}$$

6. 9 women

$$\frac{men}{women} \qquad \frac{4}{3} = \frac{12}{x} \qquad \begin{array}{r} 12 \\ \times 3 \\ \hline 36 \end{array} \qquad \begin{array}{r} 9 \\ 4{\overline{)36}} \end{array}$$

7. $7.75

$$\frac{lbs.}{\$} \qquad \frac{3}{4.65} = \frac{5}{x} \qquad \begin{array}{r} 4.65 \\ \times\ \ 5 \\ \hline 23.25 \end{array} \qquad \begin{array}{r} \$\ 7.75 \\ 3{\overline{)23.25}} \end{array}$$

8. $283.50

$$\frac{\$\ made}{\$\ deducted} \qquad \frac{10}{3} = \frac{945}{x}$$

$$\begin{array}{r} 945 \\ \times\ \ 3 \\ \hline 2,835 \end{array} \qquad \begin{array}{r} \$\ 283.50 \\ 10{\overline{)2,835.00}} \end{array}$$

9. 50.1

10. 9.7%

$$\begin{array}{l} 1980 = 50.1 \\ 1970 = \underline{40.4} \\ 9.7 \end{array}$$

11. Beef and eggs

12. West

13. Northeast and West

14. (4) The region with the greatest change in the percentage of the foreign-born population was the West.

15. 1940

16. 6 million

$$\begin{array}{l} 1950 = 10\ million \\ 1970 = \underline{4\ million} \\ 6\ million \end{array}$$

17. (3) About $2\frac{1}{2}$ or 3 million

18. $4\frac{1}{2}$ lb.

$$16\overline{)72}\,\,^{4\frac{8}{16}} = 4\frac{1}{2}$$
$$\underline{64}$$
$$8$$

19. $12\frac{1}{2}$ ft.

$$12\overline{)150}\,\,^{12\frac{6}{12}} = 12\frac{1}{2}$$
$$\underline{12}$$
$$30$$
$$\underline{24}$$
$$6$$

20. 6,400 g

$$6.4$$
$$\underline{\times 1,000}$$
$$6,400.0$$

21. 1 hr. 35 min.

$$4\ hr.\ 20\ min. = 3\ hr.\ 80\ min.$$
$$\underline{-2\ hr.\ 45\ min. = 2\ hr.\ 45\ min.}$$
$$1\ hr.\ 35\ min.$$

22. 33 lb. 4 oz.

$$100 \qquad\qquad = 99\ lb.\ 16\ oz.$$
$$\underline{-\ 66\ lb.\ 12\ oz. = 66\ lb.\ 12\ oz.}$$
$$33\ lb.\ \ \ 4\ oz.$$

23. 25 lb. 4 oz.

$$6\ lb.\ \ 5\ oz.$$
$$\underline{\times \qquad 4}$$
$$24\ lb.\ 20\ oz. = 25\ lb.\ 4\ oz.$$

24. 6 hr. 55 min.

$$3\overline{)20\ hr.\qquad 45\ min.}\ \ ^{6\ hr.\qquad 55\ min.}$$
$$\underline{18}$$
$$2\ hr. = 120\ min.$$
$$\underline{+45}$$
$$165\ min.$$
$$\underline{15}$$
$$15$$
$$\underline{15}$$

25. 8 gal. 1 qt.

$$5\ gal.\ 3\ qt.$$
$$\underline{+2\ gal.\ 2\ qt.}$$
$$7\ gal.\ 5\ qt. = 8\ gal.\ 1\ qt.$$

Pre-Test III: Geometry and Algebra

1. 125,000

$$50^3 = 50 \times 50 \times 50$$
$$= 125,000$$

2. 63

$$12^2 - 3^4$$
$$(12 \times 12) - (3 \times 3 \times 3 \times 3)$$
$$144 - 81$$
$$63$$

3. 71

Guess 70.

$$70\overline{)5041}\,\,^{72} \qquad \begin{array}{r}72\\ +70\\ \hline 142\end{array} \qquad 2\overline{)142}\,\,^{71}$$
$$\underline{490}$$
$$141$$
$$\underline{140}$$

4. 39

$$(s - t)(s + t)$$
$$(8 - 5)(8 + 5) =$$
$$(3)(13) =$$
$$39$$

5. $22\frac{1}{2}$

$$A = \frac{1}{2}\,bh$$
$$= \frac{1}{2} \times 10 \times 4\frac{1}{2}$$
$$= \frac{1}{2} \times \frac{\overset{5}{\cancel{10}}}{1} \times \frac{9}{2}$$
$$= \frac{45}{2} = 22\frac{1}{2}$$

6. 30° C

$$C = \frac{5}{9}(F - 32)$$
$$= \frac{5}{9}(86 - 32)$$
$$= \frac{5}{\cancel{9}} \times \frac{\overset{6}{\cancel{54}}}{1}$$
$$= 30$$

7. 8.2 m

$$P = 2l + 2w$$
$$= 2(2.4) + 2(1.7)$$
$$= 4.8 + 3.4$$
$$= 8.2$$

8. 208 in.

$$P = 4s$$
$$= 4 \times 52$$
$$= 208$$

9. 24 in.
$$P = s_1 + s_2 + s_3$$
$$= 6\frac{1}{2} + 7\frac{1}{2} + 10$$
$$= 24$$

10. 4.25 gal.
Find the area.
$$A = lw$$
$$= 200 \times 8\frac{1}{2} = \frac{\overset{100}{\cancel{200}}}{1} \times \frac{17}{\underset{1}{\cancel{2}}}$$
$$= 1{,}700 \text{ sq. ft.}$$
Divide by 400.

$$\begin{array}{r} 4.25 \\ 400{\overline{\smash{)}1{,}700.00}} \\ \underline{1{,}600} \\ 1{,}00\,0 \\ \underline{80\,0} \\ 2{,}000 \end{array}$$

11. $12\frac{1}{4}$ sq. yd.
$$A = s^2$$
$$= (3\frac{1}{2})^2 = 3\frac{1}{2} \times 3\frac{1}{2}$$
$$= \frac{7}{2} \times \frac{7}{2} = \frac{49}{4} = 12\frac{1}{4} \text{ sq. yd.}$$

12. 384 sq. in.
$$A = \frac{1}{2}bh$$
$$= \frac{1}{\underset{1}{\cancel{2}}} \times \frac{\overset{16}{\cancel{32}}}{1} \times \frac{24}{1}$$
$$= 384 \text{ sq. in.}$$

13. 201 sq. in.
$$A = \pi r^2$$
$$= 3.14 \times 8^2$$
$$= 3.14 \times 64$$
$$= 200.96 \text{ to the nearest}$$
$$\text{square inch} = 201.$$

14. 300 sq. in.
$$A = \frac{1}{2}h(b_1 + b_2)$$
$$= \frac{1}{2} \times 12(20 + 30)$$
$$= 6(50) = 300 \text{ sq. in.}$$

15. 260 sq. ft.
LR area: $A = lw$
$$= 15 \times 12 = 180 \text{ sq. ft.}$$
DR area: $A = lw$
$$= 8 \times 10 = \underline{+80} \text{ sq. ft.}$$
Total area: 260 sq. ft.

16. 972 cu. ft.
$$V = lwh$$
$$= 18 \times 9 \times 6$$
$$= 972 \text{ cu. ft.}$$

17. 44° $180° - 36° = 144°$

18. Right triangle
Find $\angle Z$.

$$\begin{array}{cc} 35° & 180° \\ \underline{+55°} & \underline{-90°} \\ 90° & 90° \end{array}$$

Since $\angle Z = 90°$, *XYZ is a right triangle.*

19. 63 ft.

$$\frac{height}{shadow} \quad \frac{6}{4} = \frac{x}{42}$$

$$\begin{array}{cc} 42 & \\ \underline{\times 6} & 4{\overline{\smash{)}252}} \\ 252 & \\ \end{array} \quad \begin{array}{r} 63 \\ \underline{24} \\ 12 \\ \underline{12} \end{array}$$

20. 13 m
$$c^2 = a^2 + b^2$$
$$c^2 = 5^2 + 12^2$$
$$c^2 = 25 + 144$$
$$c^2 = 169$$
$$c = \sqrt{169}$$
$$c = 13 \text{ m}$$

21. Point C

22. +4

$$\begin{array}{ccc} +15 & -13 & +23 \\ \underline{+\ 8} & \underline{-\ 6} & \underline{-19} \\ +23 & -19 & +\ 4 \end{array}$$

23. 7
$$(+9) - (-4) + (-6)$$
$$+9 + 4 - 6 =$$
$$+13 - 6 =$$
$$+7$$

24. −90
$$\frac{\overset{5}{\cancel{-20}}}{1} \cdot \frac{+3}{\underset{1}{\cancel{4}}} \cdot \frac{+6}{1} = -90$$

25. $-\frac{4}{5}$
$$\frac{-48}{+60} = -\frac{4}{5}$$

26. $2\frac{2}{3}$
$$\begin{array}{rcl} 9c - 8 &=& 16 \\ \underline{+\ 8} & & \underline{+8} \\ \frac{9c}{9} &=& \frac{24}{9} \\ c &=& 2\frac{6}{9} = 2\frac{2}{3} \end{array}$$

27. 8

$$12n - 4 = 7n + 36$$
$$\underline{-\ 7n \qquad\quad -7n}$$
$$5n - 4 = \qquad 36$$
$$\underline{\quad +4 \qquad\quad +\ 4}$$
$$\frac{5n}{5} = \frac{40}{5}$$
$$n = 8$$

28. -1

$$6(s + 3) = 12$$
$$6s + 18 = 12$$
$$\underline{\quad -18 \quad -18}$$
$$\frac{6s}{6} = \frac{-6}{6}$$
$$s = -1$$

29. $x > 4$

$$8x - 2 > 5x + 10$$
$$\underline{-5x \qquad\ -5x}$$
$$3x - 2 > \qquad 10$$
$$\underline{\quad +2 \qquad\quad +\ 2}$$
$$\frac{3x}{3} > \frac{12}{3}$$
$$x > 4$$

30. 7

$$3x - 5 = x + 9$$
$$\underline{-\ x \qquad\ -x}$$
$$2x - 5 = \qquad 9$$
$$\underline{\quad +5 \qquad +\ 5}$$
$$\frac{2x}{2} = \frac{14}{2}$$
$$x = 7$$

31. $568

Monika makes x.
Mrs. Schmidt makes 2x.
$$x + 2x = 852$$
$$\frac{3x}{3} = \frac{852}{3}$$
$$x = 284$$
Mrs. Schmidt makes 2(284) = $568

32. (4)   .07(89 + 39)

33. Yes

Substitute 4 for x in $y = 3x - 2$.
$$y = 3(4) - 2$$
$$y = 12 - 2$$
$$y = 10$$
The point (4,10) is on the graph.

34. $(0, -6)$

Substitute 0 for x in $y = \frac{1}{2}x - 6$.
$$y = \frac{1}{2}(0) - 6$$
$$y = 0 - 6$$
$$y = -6$$
The coordinates of the y-intercept are
$(0, -6)$.

EVALUATION CHARTS

Fill in the charts below. The results of these tests should tell you in which areas of the book you should do the most work.

PRE-TEST I EVALUATION CHART

Problems	Number Correct	Skill Area	Pages
1–10	_____ out of 10	Whole Numbers	23– 53
11–20	_____ out of 10	Decimals	54– 86
21–32	_____ out of 12	Fractions	87–137
33–40	_____ out of 8	Percents	138–174

PRE-TEST II EVALUATION CHART

Problem	Number Correct	Skill Area	Pages
1–8	_____ out of 8	Ratio and Proportion	176–183
9–17	_____ out of 9	Tables and Graphs	184–208
18–25	_____ out of 8	Measurements	209–232

PRE-TEST III EVALUATION CHART

Problem	Number Correct	Skill Area	Pages
1–6	_____ out of 6	Introduction to Algebra and Geometry	234–253
7–20	_____ out of 14	Geometry	254–298
21–34	_____ out of 14	Algebra	299–351

WHOLE NUMBERS DECIMALS FRACTIONS PERCENTS

Whole Numbers

PLACE VALUE

Digits (0, 1, 2, 3, 4, 5, 6, 7, 8, 9) are used to write **whole numbers**. 506 is a three-digit number. 5,600 is a four-digit number, even though the last two digits are the same.

Place value means that the position of each digit in a whole number determines its value. The diagram below shows the names of the first twelve whole number places. Notice that place value is based on counting in multiples of ten. The place values of whole numbers increase from right to left. Starting at the right (units or ones place), a comma is used to separate every group of three places.

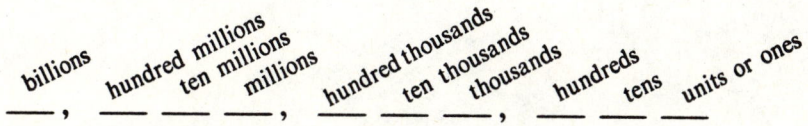

Place Value of Whole Numbers

Let's compare the digit 5 as it appears in two different numbers. The digit 5 in the number 506 is in the hundreds place. The value of the 5 in 506 is $5 \times 100 = 500$. The digit 5 in the number 5,600 is in the thousands place. The value of the 5 in 5,600 is $5 \times 1,000 = 5,000$.

EXAMPLE 1: What is the value of each digit in 7,403?

Step 1. 7 is in the thousands place.
$7 \times 1,000 = \mathbf{7,000}$

Step 2. 4 is in the hundreds place.
$4 \times 100 = \mathbf{400}$

Step 3. 0 is in the tens place.
$0 \times 10 = \mathbf{0}$
There are no tens.

Step 4. 3 is in the units (ones) place.
$3 \times 1 = \mathbf{3}$

23

EXAMPLE 2: What is the value of 6 in 5,600?

① 5,<u>6</u>00   ② <u>6</u> × 100 = **600**

Step 1. Tell what place the 6 is in. 6 is in the third place from the right, the hundreds place.

Step 2. Multiply 6 by 100.

WHOLE NUMBERS EXERCISE 1

What is the value of each underlined digit?

1. 1<u>4</u>6   20,0<u>1</u>9   167,28<u>3</u>

2. 9,<u>3</u>60,280   480,<u>9</u>23   <u>8</u>1,240

3. <u>2</u>7   6<u>7</u>,209,488   8,<u>4</u>43

4. 6,<u>4</u>09,255,108   342,774   21,<u>6</u>40,456

5. What is the value of each digit in 10,496?

    value of 1 =

    value of 0 =

    value of 4 =

    value of 9 =

    value of 6 =

Answers and solutions start on page 49.

ROUNDING OFF WHOLE NUMBERS

Is the number 362 closer to 300 or to 400? We could subtract to find out:

$$\begin{array}{r} 362 \\ -300 \\ \hline 62 \end{array} \qquad \begin{array}{r} 400 \\ -362 \\ \hline 38 \end{array}$$

362 is closer to 400. 400 is called a **round number** because it ends with zeros. 400 is 362 **rounded off** to the nearest hundred.

However, you do not have to subtract every time you want to round off a whole number.

Rules for Rounding Off Whole Numbers

1. Underline the digit in the place you are rounding off to.

2. a. If the digit to the right of the underlined digit is 5 or more, add 1 to the underlined digit.

b. If the digit to the right of the underlined digit is less than 5, leave the underlined digit as it is.

3. Put zeros in all the places to the right of the underlined digit.

EXAMPLE 1: Round off 26,489 to the nearest thousand.

① ‖ 26,489 ‖ ② ‖ **26,000** ‖

Step 1. Underline the digit in the thousands place, 6. Look at the digit to the right of 6. The digit is 4.

Step 2. Leave the underlined digit and all the numbers to the left of it. Put 0's to the right of 6.

EXAMPLE 2: Round off 196,274 to the nearest ten-thousand.

① ‖ 196,274 ‖ ② ‖ **200,000** ‖

Step 1. Underline the digit in the ten-thousands place, 9. Look at the digit to the right of 9. The digit is 6.

Step 2. Add 1 to the underlined digit. 9 + 1 = 10. When you add 1 to 9, carry 1 to the next column to the left. Add the carried 1 to the 1 already there. Then, put 0's to the right.

WHOLE NUMBERS EXERCISE 2

Round off each number to the place indicated.

1. 63 to the nearest ten.

2. 228 to the nearest ten.

3. 439 to the nearest hundred.

4. 5,620 to the nearest thousand.

5. 8,098 to the nearest thousand.

6. 6,982 to the nearest hundred.

7. 24,507 to the nearest thousand.

8. 38,496 to the nearest ten-thousand.

9. 1,063 to the nearest ten.

10. 28,092 to the nearest hundred.

11. 16,236 to the nearest hundred.

12. 1,475,290 to the nearest ten-thousand.

13. 2,951 to the nearest hundred.

14. 312 to the nearest ten.

15. 8,059 to the nearest hundred.

Answers and solutions start on page 49.

NUMBER FACTS

The mathematics test of the GED measures your ability to solve practical problems. To solve most problems, you need to know the arithmetic facts. If you cannot recall these facts quickly, you will waste time.

The Addition Facts

Here is a short review of the addition facts that you use in both addition and subtraction. Write down the answers that you know. Do not count to get any answers. Time yourself. This exercise should not take more than three minutes.

WHOLE NUMBERS EXERCISE 3

In each box, write the sum of the number on the top and the number on the side. Several have been done as examples.

	1	3	8	4	6	9	7	10	2	5
8	9									
1		4								
3										
6										
4					10					
9										
5										
7								17		
2										
10										

Answers and solutions start on page 49.

Multiplication Facts

Knowing the multiplication tables is also essential for your work in mathematics. Take the time now to make sure that you can remember them accurately and quickly.

WHOLE NUMBERS EXERCISE 4 ━━━━━━━━━━━━

For each box on the table below, write the product of the number on the top and on the side. A few answers have been filled in as examples.

	0	1	2	3	4	5	6	7	8	9	10	11	12
0			0										
1			2										
2			4		8							22	
3													
4				12									
5													
6												66	
7													
8													
9													
10												110	
11													
12													

Answers and solutions start on page 49.

You will also use these facts in your work with division. For example, $7 \times 3 = 21$ and $21 \div 3 = 7$. The following exercise gives you a chance to recall the multiplication facts in both multiplication and division.

WHOLE NUMBERS EXERCISE 5 ━━━━━━━━━━━━

Without looking at the multiplication table that you filled in, do the exercise below. Time yourself. This exercise should not take more than three minutes.

1. $5 \times 4 =$ 4. $5 \times 9 =$ 7. $9 \times 2 =$

2. $7 \times 8 =$ 5. $6 \times 5 =$ 8. $11 \times 11 =$

3. $3 \times 8 =$ 6. $7 \times 6 =$ 9. $12 \times 4 =$

10. 8 × 5 =   **17.** 42 ÷ 6 =   **24.** 56 ÷ 8 =

11. 6 × 3 =   **18.** 25 ÷ 5 =   **25.** 49 ÷ 7 =

12. 9 × 6 =   **19.** 24 ÷ 8 =   **26.** 45 ÷ 5 =

13. 8 × 9 =   **20.** 54 ÷ 9 =   **27.** 32 ÷ 4 =

14. 7 × 4 =   **21.** 36 ÷ 6 =   **28.** 63 ÷ 9 =

15. 9 × 4 =   **22.** 42 ÷ 7 =   **29.** 48 ÷ 6 =

16. 64 ÷ 8 =   **23.** 81 ÷ 9 =   **30.** 21 ÷ 3 =

Answers and solutions start on page 49.

Now is the time to review any addition or multiplication facts that you got wrong. Learn any that you missed before continuing to work with whole numbers.

THE BASIC OPERATIONS

This section gives you a chance to review the four basic whole number operations: addition, subtraction, multiplication, and division.

The following examples illustrate methods commonly used in the United States. You may use any method that works best for you, as long as you are careful and accurate.

Addition

Addition is the process of combining two or more numbers to find a total.

EXAMPLE 1:   2,723 + 8 + 700 + 925

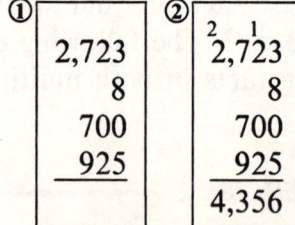

    Step 1. Line up the problem with units under units, tens under tens, and hundreds under hundreds.

    Step 2. Start with the units and add down each column. If the sum of any column is a two-digit number, put the digit on the right under the column you are adding and **carry** the number on the left to the next column. The total of the units column is 16, so you carry the 1. The hundreds column totals 23, so you carry 2.

To check an addition problem, add the numbers from the bottom to the top.

$$
\begin{array}{r}
34 \\
+28 \\
\hline
62
\end{array}
\qquad \textbf{\textit{Check:}} \qquad
\begin{array}{r}
28 \\
+34 \\
\hline
62
\end{array}
$$

Subtraction

When you subtract, you take one number away from another to find the **difference.** When you cannot subtract a larger number from a smaller number, you must **borrow.**

Borrowing doesn't change the value of a number, but it enables you to take from a higher place value and add to a lower place value number.

EXAMPLE 2: $6,039 - 1,982 =$

①
$$
\begin{array}{r}
6,039 \\
-1,982 \\
\hline
\end{array}
$$

②
$$
\begin{array}{r}
6,039 \\
-1,982 \\
\hline
7
\end{array}
$$

③
$$
\begin{array}{r}
{}^{5}\!\!\!\!\not{6},039 \\
-1,982 \\
\hline
7
\end{array}
$$

④
$$
\begin{array}{r}
{}^{5}\,{}^{9}\!\!\!\!\not{6},\!\not{0}39 \\
-1,982 \\
\hline
4,057
\end{array}
$$

Step 1. Put the larger number (6,039) on top. Line up the digits with units under units, tens under tens, etc.

Step 2. Starting with the units, subtract. Since you cannot take 8 away from 3, you must borrow.

Step 3. You cannot borrow from 0. Therefore, you must go one more place to the left. Borrow 1 from 6 to make 5 in the thousands place. Place 1 next to the 0 in the hundreds place to make 10.

Step 4. This still doesn't help the 3 in the tens place. Borrow again from the 10, put a 9 in the hundreds place, and put a 1 next to the 3 in the tens place. Subtract.

To check a subtraction problem, add the answer to the number being subtracted. The result should be the top number of the original problem.

$$
\begin{array}{r}
49 \\
-28 \\
\hline
21
\end{array}
\qquad \textbf{\textit{Check:}} \qquad
\begin{array}{r}
28 \\
+21 \\
\hline
49
\end{array}
$$

Multiplication

When you multiply two numbers, put the number with more digits on top. Start by multiplying the units.

EXAMPLE 3:   407 × 28 =

①	②	③	④
407 ×28	⁵ 407 ×28 3,256	¹ 407 ×28 3256 814	407 ×28 3256 814 **11,396**

Step 1.   Put 407 on top.

Step 2.   Multiply 407 by 8. Start this part of the answer under the units column.

Step 3.   Multiply 407 by 2. Start this part of the answer under the tens column.

Step 4.   Add the results from steps 2 and 3.

To check a multiplication problem, divide the answer by one of the numbers multiplied. The result should be the other number you multiplied.

$$\begin{array}{r} 18 \\ \times 6 \\ \hline \mathbf{108} \end{array} \qquad \textit{Check:} \quad \begin{array}{r} \mathbf{18} \\ 6\overline{)108} \end{array}$$

Division

The symbols ÷ and $\overline{)}$ both mean to divide. 56 divided by 8 is written as 56 ÷ 8 or 8$\overline{)56}$.

EXAMPLE 4:   798 ÷ 38

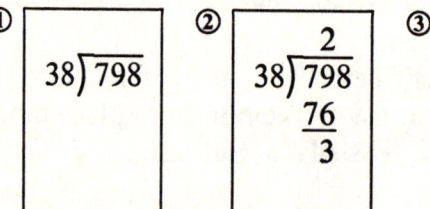

①	②	③
38$\overline{)798}$	2 38$\overline{)798}$ 76 3	**21** 38$\overline{)798}$ 76 38 38

Step 1.   Carefully set up your problem, putting the number being divided inside the frame. Here, 798 is being divided.

Step 2.   How many times does 38 divide into 7? None. Since 38 is too large to go into 7, you look at the first two digits together—79. Ask yourself, how many times does 38 go into 79? Two times. Put a 2 above the tens place and multiply 2 times 38, then subtract.

Step 3.   Bring down the 8. How many times does 38 go into 38? One time. Put the 1 above the units place. Multiply 1 times 38. Subtract. There is no remainder; 38 divides evenly into 798.

In division problems with a **remainder,** indicate the remainder as part of the answer.

To check a division problem, multiply the answer by the number you divided by. Then add the remainder if there is one. The result should be the number you divided into.

$$
\begin{array}{r}
17r3 \\
9\overline{)156} \\
\underline{9} \\
66 \\
\underline{63} \\
3
\end{array}
\qquad
\begin{array}{rl}
\textbf{\textit{Check:}} & 17 \\
& \underline{9} \\
& 153 \\
& \underline{+3}\text{ (remainder)} \\
& 156
\end{array}
$$

Division is the most difficult whole number operation. Often, you have to guess and test possible answers. Zeros in division problems can create extra trouble. Look at the following example.

$$
\begin{array}{r}
207 \\
13\overline{)2691} \\
\underline{26} \\
09 \\
\underline{0} \\
91 \\
\underline{91} \\
0
\end{array}
\qquad
\begin{array}{rl}
\textbf{\textit{Check:}} & 207 \\
& \underline{\times 13} \\
& 621 \\
& \underline{207} \\
& \textbf{2,691}
\end{array}
$$

The zero in the answer above holds the tens place. This zero means that 13 would not divide into 9.

WHOLE NUMBERS EXERCISE 6 ━━━━━━━━━━━━━━━━━━━━━

Solve and check each problem.

1. 25,624 + 92,183 =

2. 60,845 − 2,926 =

3. 48,005 − 6,774 =

4. 83 + 2,096 + 194 =

5. 6 × 5,708 =

6. 349 × 74 =

7. 65 × 50,000 =

8. 446 ÷ 17 =

9. 2,464 ÷ 8 =

10. 54,036 ÷ 6 =

11. 30,045 − 15,586 =

12. 194 + 8 + 2,366 + 850 =

13. 500 × 96 =

14. 3,000,000 − 816,000 =

15. 45,360 + 21,885 =

16. 7 × 29,058 =

17. 37,600 ÷ 800 =

18. 8,000 × 74 =

19. 5,040,000 − 264,500 =

20. 10,710 ÷ 35 =

Answers and solutions start on page 50.

MATHEMATICAL TERMS

Mathematics calls for careful reading. Mathematical terms are words that are sometimes used in ordinary English. In mathematics, these terms have special meanings. It is important to know these terms in order to understand questions and explanations about mathematics.

Here are four common mathematical terms:

    sum—the answer to an addition problem
    difference—the answer to a subtraction problem
    product—the answer to a multiplication problem
    quotient—the answer to a division problem

In the next exercise, there are no symbols for the basic operations. Read each problem carefully. Watch for the terms written above. Also watch for other words that tell you to add, subtract, multiply, or divide.

WHOLE NUMBERS EXERCISE 7 ━━━━━━━━━━━━━━━

Solve each problem.

1. Find the sum of 590, 2,041, 713, and 68.

2. Find the product of 59 and 603.

3. What is the difference between 10,230 and 8,907?

4. How much is 8 times 5,090?

5. How much greater is 103,460 than 98,567?

6. What is the quotient of 6,810 divided by 14?

7. 6,928 is how much less than 10,000?

8. Find the total of 467, 78, 896, and 2,043.

9. How much is 5,076 divided by 12?

10. What is the product of 346 and 450?

11. What is the difference between 88 and 7,000?

12. 400,300 is how much more than 9,216?

13. Find the quotient of 2,820 divided by 4.

14. Multiply the sum of 85 and 344 by 7.

15. The total of 857 and 1,604 is how much greater than their difference?

Answers and solutions start on page 50.

AVERAGE

To find the **average** of a group of numbers, add the numbers and then divide by how many numbers there are.

EXAMPLE 1: Find the average of 10, 15, 23, and 28.

①
```
  10
  15
  23
 +28
  76
```

②
$$
\begin{array}{r}
19 \\
4\overline{)76} \\
4 \\
\hline
36 \\
36 \\
\end{array}
$$

Step 1. Find the sum by adding the numbers.
Step 2. Divide the sum by the number of numbers, 4.

Finding an average is a common application of arithmetic skills.

EXAMPLE 2: Houses on this block sold for $40,000, $43,000, and $52,000. What was the average cost of a house?

①
```
$ 40,000
  43,000
  52,000
$135,000
```

②
$$
\begin{array}{r}
\$45,000 \\
3\overline{)135,000} \\
12 \\
\hline
15 \\
15 \\
\hline
0000 \\
\end{array}
$$

Step 1. Find the sum by adding the houses' costs.
Step 2. Divide by the number of houses, 3.

WHOLE NUMBERS EXERCISE 8 ━━━━━━━━━━━━━━━━━

Solve each problem.

1. Find the average of 353, 19, and 207.

2. What is the average of 2,043, 971, 3,116, and 1,850?

3. Find the average of 240, 313, 189, and 270.

4. José weighs 187 pounds. His brother Manny weighs 159 pounds. What is the average of their weight?

5. The noon temperature on Monday was 69°; on Tuesday, the noon temperature was 71°; on Wednesday, it was 56°; on Thursday, 63°; and on Friday, 66°. What was the average noon temperature for those days?

6. In 1980 Fran made $14,700. In 1981 she made $13,900. In 1982 she made $15,800. What was her average yearly salary for those years?

7. The McGlynn's phone bills were $25.66 in January, $33.27 in February, and $19.28 in March. What was the average of their bills for those months?

8. The total weekly payroll of the Central Electric Company is $108,000. 450 people work at Central Electric. What is the average pay of each worker?

9. A car salesman sold three used cars for $650 each, four used cars for $875 each, and two used cars for $1100 each. What was the average amount his customers paid for the used cars?

10. Sylvia received scores of 47, 42, 38 and 50 on four tests of the GED. What must she get on the fifth test in order to get an average score of 45? (**Hint:** What would be a five-test total for an average score of 45?)

Answers and solutions start on page 51.

NUMBER SERIES

A **number series** is a list of numbers in a special order or pattern. The numbers 1, 2, 3, 4, 5, 6 . . . form the most familiar series. The number 7 is the next **term** in the series.

To find a missing number of a number series, first find the pattern that changes the numbers from left to right.

EXAMPLE 1:   Find the next term in the series 5, 8, 11, 14 . . .

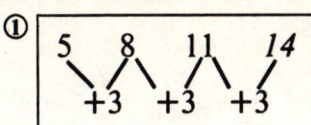   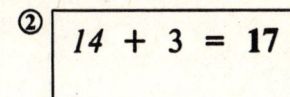

Step 1.   Find how the series is changing. This series is changing by adding 3.

Step 2.   Add 3 to the last term.

EXAMPLE 2:   Find the next term in the series 2, 7, 4, 9, 6, 11 . . .

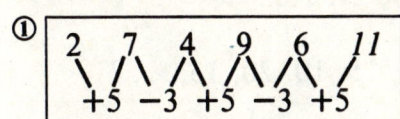   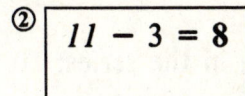

Step 1.   Find how the series is changing. This series first increases by 5 and then decreases by 3. At 11, the subtraction step is next.

Step 2.   Subtract 3 from the last term.

EXAMPLE 3:   Find the seventh term in the series 1, 3, 9, 27, 81 . . .

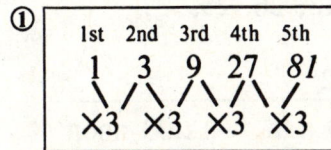   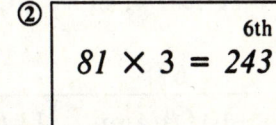   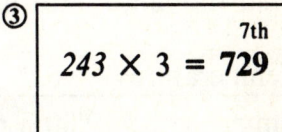

Step 1.   Find how this series is changing. This series is changing by multiplying by 3.

Step 2.   Find the 6th term. Multiply 81 by 3.

Step 3.   Find the 7th term. Multiply 243 by 3.

EXAMPLE 4:

Stock price per share	$33	$29	$25	—
Date	June 1	June 8	June 15	June 22

If the stock price shown in the table continued in the same pattern, what was the price of the stock on June 22?

① 33 29 25
      ↘₄↗ ↘₄↗

② $25 - 4 = \mathbf{21}$

Step 1.   Find how the series is changing. This series is changing by subtracting 4.

Step 2.   Subtract 4 from the last term. The June 22nd price was **$21**.

WHOLE NUMBERS EXERCISE 9 ───────────────

Solve each problem.

1. Find the next term in the series: 6, 12, 24, 48, . . .

2. What is the sixth term in the series: 81, 84, 87, 90, . . . ?

3. What is the next term in the series: 32, 29, 26, 23, . . . ?

4. Find the seventh term in the series: 13, 17, 21, 25, . . .

5. What is the next term in the series: 10, 5, 15, 10, 20, 15, . . . ?

6. Find the eighth term in the series: 100, 81, 64, 49, 36, 25, . . .

7. Find the next term in the series: 5, 9, 17, 33, . . .

8. What is the eighth term in the series: 2, 3, 5, 8, 12, 17, . . . ?

9. Find the next term in the series: 320, 160, 80, 40, 20, . . .

10. Find the ninth term in the series: 4, 8, 7, 14, 13, 26, 25, . . .

11.
Temperature	62°	67°	72°	—
Time	9:00 a.m.	10:00 a.m.	11:00 a.m.	12:00 noon

If the temperature continues in the same pattern, what will be the 12:00 noon temperature?

12.
Balance	Month
$29,000	March
23,000	April
17,000	May
11,000	June
?	July

Every month, a committee is withdrawing money from a building fund at the same rate. How much will be in the account in July?

13.
Year	Pop.
1900	1,500
1920	3,000
1940	6,000
1960	12,000
1980	—

This table shows the population of Bridge Creek. If the population continued in the same pattern, what was the 1980 population of Bridge Creek?

Answers and solutions start on page 52.

WHOLE NUMBER WORD PROBLEMS

Many students are confused by word problems. One goal of this book is to help you master such problems.

You have already solved some word problems in the number series and averages sections. Read through this section carefully. Think about the tips that will help you solve different kinds of problems.

The key to solving word problems is to organize your thinking about the problem. Take the time to do the following:

Step 1. Find the question — what is being asked for?

Step 2. Decide what information you need to answer the question. Be careful. Some problems may contain information you won't need to solve the problem.

Step 3. Decide what arithmetic operation to use (addition, subtraction, multiplication, or division). This requires a careful rereading of the problem in order to understand the situation. Some people draw diagrams or write down the information to help them see what to do.

Step 4. Do the arithmetic accurately and carefully. Check your math.

Step 5. Make sure that your answer is sensible.

Relate your answer to your understanding of the question (Step 1). You can often catch a mistake in your choice of math or in your work by making sure your answer is sensible.

One way of seeing whether your answer is sensible is to estimate your answer. To **estimate:** choose the math operation you will use, round off the numbers, and do the math. Look at your estimated answer in relationship to the question being asked. This will help you to see if the math you chose was right.

EXAMPLE 1: Laura works at a part-time job 20 hours a week. She makes $5 an hour. How much money does she make a week?

Step 1. *Question:* How much money does she make a week?

Step 2. *Information:* 20 hours at $5 an hour

Step 3. *Arithmetic:* Since she makes $5 each hour, and she works 20 hours, multiply.

Step 4. 20 hours × $5 = **$100** per week
Check: 100 ÷ 5 = 20.

Step 5. $100 is a sensible amount of money for a week's part-time salary.

If you had mistakenly decided to divide instead of to multiply, your answer would have been $4—not a sensible answer. How could she earn less in a week than she earns for one hour's work?

Estimate for a quick check. For example, if Laura had made $5.25 per hour for a 22-hour week, you could have rounded this off to $5 × 20 hours to know that your answer should be <u>close to</u> $100.

Addition and Subtraction

When you have to find a total amount, consider adding. Words such as *sum*, *total*, *altogether*, and *combined* often indicate addition.

> **EXAMPLE 2:** Mr. Swenson makes $260 a week, and Mrs. Swenson makes $240 a week. What is their combined income?
>
> **Solution:** Add their weekly incomes.
> $260 + $240 = **$500 per week**

The following phrases usually mean to subtract: *How much more? How much less? How much greater? How much smaller? Find the balance. Find the difference.*

> **EXAMPLE 3:** Mary weighs 120 pounds. Her daughter weighs 50 pounds. How much more does Mary weigh than her daughter?
>
> **Solution:** Subtract the daughter's weight from Mary's weight.
> 120 pounds − 50 pounds = **70 pounds**

The words <u>*net*</u> and <u>*gross*</u> also may suggest subtraction. Gross refers to an amount before something has been subtracted from it. Net refers to the rest after something has been subtracted from the gross.

> **EXAMPLE 4:** Mark's gross monthly salary is $1,000. His employer withholds $200 for taxes and social security. What is Mark's net salary for a month?
>
> **Solution:** Subtract the amount withheld from the gross amount.
> $1,000 − $200 = **$800**

Remember: When you have to combine, add; when you need to find a difference, subtract.

Multiplication and Division

Watch for situations that suggest multiplication. A problem may give you information for one thing. Then you will be asked to find information for several things.

EXAMPLE 5: There are 3 feet in one yard. How many feet are there in 20 yards?

Solution: Multiply the number of feet in one yard by 20.
3 feet × 20 yards = **60 feet**

EXAMPLE 6: On the open road, Jack can drive 24 miles on a gallon of gasoline. How far can he drive using 10 gallons of gasoline?

Solution: Multiply the number of miles Jack can drive on one gallon of gas by the number of gallons given in the problem.
24 miles per gallon × 10 gallons = **240 miles**

Some situations suggest division. A problem may give you information about several things. Then you will be asked to find information about one thing.

EXAMPLE 7: Last year, Ruth and Gordon together paid $3,000 for rent. How much rent did they pay each month?

Solution: Divide the amount of rent they paid in a year by the number of months in a year (12).
$3,000 ÷ 12 months = **$250 per month**

A problem may ask you to find how many of a certain item are in some bigger item. This situation also means to divide.

EXAMPLE 8: Soup is packed with 24 cans in a box. How many boxes are needed to ship 1,200 cans of soup?

Solution: Find out how many times 24 goes into 1,200. Divide the total number of cans by the number of cans one box can hold.
1,200 cans ÷ 24 per box = **50 boxes**

The ideas of sharing, cutting, or splitting also can indicate division.

EXAMPLE 9:   Fatima won $200 in a lottery. She decided to share the money equally among three friends and herself. How much did everyone get?

   Solution: Divide the amount Fatima won by the number of people who are going to share the money. Fatima and her three friends are four people.
$200 ÷ 4 people = **$50 per person**

These examples have all been one-step problems. Of course, some problems on the GED will require two or more steps.

Problems of Two or More Steps

EXAMPLE 10:   Celia bought two pounds of apples that cost 60¢ a pound and five pounds of potatoes at 40¢ a pound. How much did she pay for her purchases?

   Step 1.   Multiply the price of a pound of apples by 2.
$.60 × 2 = $1.20

   Step 2.   Multiply the price of a pound of potatoes by 5.
$.40 × 5 = $2.00

   Step 3.   Then add the results.
*$1.20 + $2.00 = **$3.20***

As the example shows, the problem does not give you enough information to answer the question immediately. You must find the total cost of the apples and the total cost of the potatoes before you add them in order to find out how much she paid for her purchases.

When you can see that a problem will involve more than one operation to solve, you must find out what information you will need to be able to perform the final operation.

EXAMPLE 11:   The charity drive raised $30,400. This was $3,000 less than last year's total. If last year's money had been divided among five agencies, how much did each agency receive?

   Step 1.   *Question:* How much did each agency receive (of last year's money)?

   Step 2.   *Information:* $30,400, $3,000 less, five agencies

   Step 3.   Find what you need to do your final operation. In this case, you need to know last year's total.
This year's total + the difference = last year's total
$30,400 + $3,000 = $33,400

   Step 4. $33,400 ÷ 5 agencies = **$6,680** for each agency.

The next exercise will help you think about the process of solving word problems.

WHOLE NUMBERS EXERCISE 10

Fill in the circle that corresponds to the best solution. Do not solve.

1. Peter is 37 years old. His son Chris is 8 years old. How much older is Peter than his son?   1 ① ② ③ ④ ⑤
   (1) Add 37 and 8.
   (2) Multiply 37 by 8.
   (3) Add 37 and 8. Then divide by 2.
   (4) Subtract 8 from 37.
   (5) Divide 8 into 37.

2. Caren works 40 hours a week for $260. How much does she make in an hour?   2 ① ② ③ ④ ⑤
   (1) Add 40 and $260.
   (2) Multiply $260 by 40.
   (3) Subtract 40 from $260.
   (4) Add 40 and $260. Divide by 2.
   (5) Divide 40 into $260.

3. In November, the Johnsons' bills for gas, electricity, and telephone were $12, $25, and $53 respectively. How much did the Johnsons pay that month for these three utilities?   3 ① ② ③ ④ ⑤
   (1) Add $12, $25, and $53. Then divide by 3.
   (2) Add $12, $25, and $53.
   (3) Add $12 and $25. Then subtract the total from $53.
   (4) Add $12, $25, and $53. Then divide by 12.
   (5) Add $12, $25, and $53. Then multiply the total by 12.

4. At the corner grocery, all but 5 dozen of the 20 dozen eggs in the cooler are white eggs. How many white eggs are in the cooler?   4 ① ② ③ ④ ⑤
   (1) Multiply 5 by 12. Multiply 20 by 12. Then find the difference by subtraction.
   (2) Subtract 5 from 20. Then multiply the result by 12.
   (3) Multiply 20 by 12.
   (4) Both solution (1) and solution (2).
   (5) None of the above.

5. Cheryl's gross income for the year was $18,296. Her total deductions were $4,680. What was her net income for the year?   5 ① ② ③ ④ ⑤
   (1) Add $18,296 and $4,680.
   (2) Multiply $18,296 by $4,680.
   (3) Subtract $4,680 from $18,296.
   (4) First subtract $4,680 from $18,296. Then multiply the difference by 12.
   (5) First add $4,680 and $18,296. Then divide the sum by 12.

6. Mrs. Hollis died in 1980 at the age of 92. Find the year in which   6 ① ② ③ ④ ⑤
she was born.
(1) Add 1980 and 92.
(2) Subtract 80 from 1980 and add 12.
(3) Subtract 90 from 1980 and add 2.
(4) Add 80 and 1980. Then subtract 12.
(5) Subtract 92 from 1980.

7. Of the 943 cars sold in Midvale last month, 387 of them were   7 ① ② ③ ④ ⑤
made in the U.S. and 198 of them were made in Japan. How
many of the cars sold in Midvale last month were made outside
the U.S.?
(1) 943 − 387
(2) 943 − 198
(3) 198 + 387
(4) 387 + 943
(5) 387 − 198

8. On Thursday, Pat worked for ten hours. She worked 8 hours at   8 ① ② ③ ④ ⑤
her regular rate of $6 an hour and 2 hours at the rate of $9 an
hour. How much did Pat earn altogether on Thursday?
(1) Add $6 and $9. Then multiply the result by 10.
(2) Add 8 and 2. Then multiply the result by $9.
(3) Multiply 8 by $6. Multiply 2 by $6. Then add the results.
(4) Multiply 8 by $6. Multiply 2 by $9. Then add the results.
(5) Add $6 and $9. Divide by 2. Then multiply the result by 10.

9. Allen's salary is $16,380 a year. How much does Allen make in   9 ① ② ③ ④ ⑤
one month?
(1) $16,380 × 12
(2) $16,380 ÷ 52
(3) $16,380 ÷ 12
(4) 12 ÷ $16,380
(5) $16,380 × 365

10. Grace can type 83 words per minute. About how many minutes   10 ① ② ③ ④ ⑤
will she need to type a 1,000-word letter?
(1) Divide 83 by 1,000.
(2) Divide 1,000 by 83.
(3) Multiply 83 by 60. Then divide by 1,000.
(4) Multiply 1,000 by 60. Then divide by 83.
(5) Multiply 83 by 1,000.

11. Middletown has 90 firemen and 158 policemen. The total 1982 payroll for firemen was $1,494,000. The total payroll for policemen was $2,717,600. What was the average income for a policeman in Middletown in 1982?

    (1) Divide $2,717,600 by 158.
    (2) Divide $1,494,000 by 90.
    (3) Add 90 and 158. Then divide the total into $2,717,600.
    (4) Divide $2,717,600 by 90.
    (5) Divide $1,494,000 by 158.

11 ① ② ③ ④ ⑤

12. California has 23,668,000 people, while New York has 17,558,000 and Texas has 14,229,000. How many more people live in the most populous state than in the second most populous?

    (1) Subtract 14,229,000 from 23,668,000.
    (2) Subtract 14,229,000 from 17,558,000.
    (3) Add 14,229,000 and 17,558,000. Then subtract the total from 23,668,000.
    (4) Subtract 23,668,000 from 17,558,000.
    (5) Subtract 17,558,000 from 23,668,000.

12 ① ② ③ ④ ⑤

Answers and solutions start on page 52.

Read Word Problems Carefully

You must read word problems carefully.

A word that is often used to suggest one operation may be used to mean another.

EXAMPLE 12: Deborah bought a pair of shoes on sale for $24. This was $10 less than the original price. What was the original price?

Solution: **$24 +10 = $34**

Here, <u>less than</u> is misleading. The sale price was $10 less than the original price. This means the original price was $10 <u>more</u> than the sale price. Add the savings to the sale price. The original price was $34.

As you saw in problems 7, 11, and 12 in Exercise 10, some problems contain unnecessary information.

EXAMPLE 13:   Adrienne makes $1,100 a month. She pays $215 a month for rent and $400 a month for food. How much more does Adrienne spend for food than for rent in a month?

Solution: $400 −215 = **$185**

The $1,100 Adrienne makes a month has nothing to do with the problem. Subtract the amount she spends on rent from the amount she spends on food. She spends $185 more for food than for rent in a month.

It is a good idea to think about the size of answers. When you multiply whole numbers you get a larger whole number as an answer. When you divide whole numbers, you get a smaller whole number. For example, an average is always less than the total of the numbers the average comes from.

For any word problem:
1. Read the problem carefully to get a sense of the situation.
2. Be sure you know what is being asked for.
3. Watch for any helpful words that tell you what operations to use, but be careful; some words are used misleadingly.

WHOLE NUMBERS EXERCISE 11 ———————————

Choose the best answer to each word problem.

1. A television was on sale for $198. This was $60 less than the original price. What was the original price?

   (1) $128
   (2) $138
   (3) $238
   (4) $258
   (5) $268

2. The average worker at the Central Power Plant makes $279 a week. There are 63 workers at the plant. What is the total weekly payroll for the workers?

   (1) $17,577
   (2) $16,577
   (3) $17,557
   (4) $16,557
   (5) $17,677

3. During a week in June 1982, factories produced 1,297,000 tons of steel. During the same period one year before, the production was 1,164,000 tons greater. How many tons of steel were produced in the corresponding week of June 1981?

   (1) 133,000
   (2) 1,330,000
   (3) 246,000
   (4) 1,451,000
   (5) 2,461,000

4. Each year, the employees of Schmidt Memorial Hospital share in the profits of the hospital. Last year, the total profit available for employees was $55,776. There are 83 employees at the hospital. What was the average amount of profit each employee received?

(1) $672
(2) $600
(3) $500
(4) $463
(5) $363

5. In one month, the leading U.S. car manufacturer sold 259,056 cars. The second largest manufacturer sold 120,947 cars. How many more cars did the leading manufacturer sell than the second largest manufacturer?

(1) 138,009
(2) 139,109
(3) 139,009
(4) 138,109
(5) 128,109

6. Last year, the Millhouse family took home $12,400. They spent $2,580 for rent, and they saved $1,560. After paying rent, how much did they have left for savings and other expenses?

(1) $9,820
(2) $10,840
(3) $8,260
(4) $11,380
(5) $13,420

7. When Mr. MacDonald died, he left his 825-acre farm to his three children. Each child got the same number of acres. How many acres did each child inherit?

(1) 175
(2) 275
(3) 247
(4) 225
(5) 2,475

8. Mrs. Seltzer paid $150 down and $45 a month for 18 months for new living room furniture. What total amount did she pay for the furniture?

(1) $660
(2) $740
(3) $810
(4) $960
(5) $1,040

9. The Upperville school district spends $7,800 a year to educate a child. The Central City school district spends $2,100 per child. Find the difference between the costs of educating 30 children in Upperville and 30 children in Central City.

(1) $171,000
(2) $5,700
(3) $9,900
(4) $297,000
(5) $330,000

10. In March 1982, 10,549,000 Americans were officially unemployed. In December 1981, 9,307,000 were unemployed. In March 1981, 8,248,000 were unemployed. By how many people did the unemployment count rise from March 1981 to March 1982?

(1) 1,242,000
(2) 1,059,000
(3) 2,301,000
(4) 1,879,000
(5) 1,985,000

11. Tina and her 3 sisters evenly split the cost of their parents' anniversary gift. If the present cost $72, how much did each girl spend?

(1) $18
(2) $24
(3) $32
(4) $36
(5) $69

12. Each week, Everett makes $240 and his wife makes $200. How much do they make in a year?

    (1) $2,080
    (2) $5,280
    (3) $10,560
    (4) $22,800
    (5) $22,880

13. In a recent election, Representative Sanders got 67,576 votes and his opponent got 62,881 votes. To the nearest hundred, how many more votes did Sanders get?

    (1) 4,600
    (2) 4,700
    (3) 5,000
    (4) 62,000
    (5) 67,000

14. For her store, Mrs. Rivera ordered 4 cases of green beans, 6 cases of canned corn, and 10 cases of beets. If there are 20 cans to a case, what is the total number of cans she ordered?

    (1) 20
    (2) 40
    (3) 120
    (4) 280
    (5) 400

15. Of the 2,500 adults in Cripple Creek, 1,250 are high school graduates and 750 finished college. How many adults did not finish high school?

    (1) 500
    (2) 750
    (3) 1,250
    (4) 1,750
    (5) 2,000

Answers and solutions start on page 52.

WHOLE NUMBERS REVIEW

Fill in the circle that corresponds to the correct answer.

1. What is the value of 8 in 3,286,925?
    (1) 8
    (2) 80,000
    (3) 800,000
    (4) 800
    (5) none of these

    1 ① ② ③ ④ ⑤

2. Round off 36,498 to the nearest thousand.
    (1) 36,500
    (2) 37,000
    (3) 36,000
    (4) 37,500
    (5) none of these

    2 ① ② ③ ④ ⑤

3. Find the difference between 100,000 and 79,028.
 (1) 21,072
 (2) 20,982
 (3) 21,972
 (4) 20,978
 (5) none of these

4. What is the quotient of 6,512 divided by 16?
 (1) 47
 (2) 407
 (3) 413¼
 (4) 470
 (5) none of these

5.

Year	1950	1955	1960	1965	1970	1975
No. of homes with TV's (in thousands)	1	3	7	15	31	—

The chart tells the number of homes in Central County with televisions. If the pattern continued, how many thousands of homes had televisions in 1975?
 (1) 63
 (2) 47
 (3) 62
 (4) 32
 (5) none of these

6. Everyday for a week, Sam timed his trip home. His trips took 27, 36, 28, 39, and 40 minutes. What was his average travel time?
 (1) 28 min.
 (2) 30 min.
 (3) 32 min.
 (4) 34 min.
 (5) 36 min.

7. Of the 420 employees of Apex, Inc., 63 regularly walk to work and 187 take public transportation. Choose the solution that tells the number of employees of Apex who do not regularly walk to work.
 (1) Subtract 187 from 420.
 (2) Subtract 63 from 187.
 (3) Subtract 63 from 420.
 (4) Add 63 and 187.
 (5) none of these.

8. Myron bought a suit that had been reduced by $57 to a sale price of $83. What had been the original price of the suit?   8 ① ② ③ ④ ⑤
   (1) $26
   (2) $57
   (3) $83
   (4) $140
   (5) none of these

9. Selma paid $200 down and $56 a month for 15 months for new bedroom furniture. What total amount did she pay for the furniture?   9 ① ② ③ ④ ⑤
   (1) $256
   (2) $840
   (3) $872
   (4) $940
   (5) none of these

10. A year's profit of $81,480 was split evenly among the 84 employees of the Central Electric Cooperative. How much did each employee receive?   10 ① ② ③ ④ ⑤
    (1) $814
    (2) $840
    (3) $920
    (4) $970
    (5) none of these

Answers and solutions are on page 53.

WHOLE NUMBERS REVIEW EVALUATION

Problem	Section	Starting Page
1	Place Value	23
2	Rounding Off Whole Numbers	24
3-4	Mathematical Terms	32
5	Number Series	34
6	Average	33
7-10	Whole Number Word Problems	37

<u>Passing score:</u>   8   right out of 10 problems.
<u>Your score:</u>   ___   right out of 10 problems.

   If you had less than a passing score, review the sections for those problems you missed. Then redo these questions before going on to the decimals section.
   If you had a passing score, correct any problem you got wrong. Then go on to the decimals section.

ANSWERS AND SOLUTIONS

Whole Numbers Exercise 1

1. 40   20,000   3
2. 300,000   900   1,000
3. 20   60,000,000   400
4. 400,000,000   40,000   600,000
5. $1 \times 10,000 = 10,000$
   $0 \times 1,000 = 0$ (no thousands)
   $4 \times 100 = 400$
   $9 \times 10 = 90$
   $6 \times 1 = 6$

Whole Numbers Exercise 2

1. 60
2. 230
3. 400
4. 6,000
5. 8,000
6. 7,000
7. 25,000
8. 40,000
9. 1,060
10. 28,100
11. 16,200
12. 1,480,000
13. 3,000
14. 310
15. 8,100

Whole Numbers Exercise 3

	1	3	8	4	6	9	7	10	2	5
8	9	11	16	12	14	17	15	18	10	13
1	2	4	9	5	7	10	8	11	3	6
3	4	6	11	7	9	12	10	13	5	8
6	7	9	14	10	12	15	13	16	8	11
4	5	7	12	8	10	13	11	14	6	9
9	10	12	17	13	15	18	16	19	11	14
5	6	8	13	9	11	14	12	15	7	10
7	8	10	15	11	13	16	14	17	9	12
2	3	5	10	6	8	11	9	12	4	7
10	11	13	18	14	16	19	17	20	12	15

Whole Numbers Exercise 5

1. $5 \times 4 = 20$
2. $7 \times 8 = 56$
3. $3 \times 8 = 24$
4. $5 \times 9 = 45$
5. $6 \times 5 = 30$
6. $7 \times 6 = 42$
7. $9 \times 2 = 18$
8. $11 \times 11 = 121$
9. $12 \times 4 = 48$
10. $8 \times 5 = 40$
11. $6 \times 3 = 18$
12. $9 \times 6 = 54$
13. $8 \times 9 = 72$
14. $7 \times 4 = 28$
15. $9 \times 4 = 36$
16. $64 \div 8 = 8$
17. $42 \div 6 = 7$
18. $25 \div 5 = 5$
19. $24 \div 8 = 3$
20. $54 \div 9 = 6$
21. $36 \div 6 = 6$
22. $42 \div 7 = 6$
23. $81 \div 9 = 9$
24. $56 \div 8 = 7$
25. $49 \div 7 = 7$
26. $45 \div 5 = 9$
27. $32 \div 4 = 8$
28. $63 \div 9 = 7$
29. $48 \div 6 = 8$
30. $21 \div 3 = 7$

Whole Numbers Exercise 4

	0	1	2	3	4	5	6	7	8	9	10	11	12
0	0	0	0	0	0	0	0	0	0	0	0	0	0
1	0	1	2	3	4	5	6	7	8	9	10	11	12
2	0	2	4	6	8	10	12	14	16	18	20	22	24
3	0	3	6	9	12	15	18	21	24	27	30	33	36
4	0	4	8	12	16	20	24	28	32	36	40	44	48
5	0	5	10	15	20	25	30	35	40	45	50	55	60
6	0	6	12	18	24	30	36	42	48	54	60	66	72
7	0	7	14	21	28	35	42	49	56	63	70	77	84
8	0	8	16	24	32	40	48	56	64	72	80	88	96
9	0	9	18	27	36	45	54	63	72	81	90	99	108
10	0	10	20	30	40	50	60	70	80	90	100	110	120
11	0	11	22	33	44	55	66	77	88	99	110	121	132
12	0	12	24	36	48	60	72	84	96	108	120	132	144

50   *Whole Numbers*

Whole Numbers Exercise 6

1. 117,807
$$\begin{array}{r} 25,624 \\ +92,183 \\ \hline 117,807 \end{array}$$

2. 57,919
$$\begin{array}{r} 60,845 \\ -2,926 \\ \hline 57,919 \end{array}$$

3. 41,231
$$\begin{array}{r} 48,005 \\ -6,774 \\ \hline 41,231 \end{array}$$

4. 2,373
$$\begin{array}{r} 83 \\ 2,096 \\ +194 \\ \hline 2,373 \end{array}$$

5. 34,248
$$\begin{array}{r} 5,708 \\ \times\ 6 \\ \hline 34,248 \end{array}$$

6. 25,826
$$\begin{array}{r} 349 \\ \times\ 74 \\ \hline 1\ 396 \\ 24\ 43 \\ \hline 25,826 \end{array}$$

7. 3,250,000
$$\begin{array}{r} 65 \\ \times\ 50,000 \\ \hline 3,250,000 \end{array}$$

Notice how this problem is lined up. Simply "bring down" the 0's and multiply 65 by 5.

8. 26r4
$$\begin{array}{r} 26r4 \\ 17\overline{)446} \\ \underline{34} \\ 106 \\ \underline{102} \\ 4 \end{array}$$

9. 308
$$8\overline{)2,464} \quad 308$$

10. 9,006
$$6\overline{)54,036} \quad 9,006$$

11. 14,459
$$\begin{array}{r} 30,045 \\ -15,586 \\ \hline 14,459 \end{array}$$

12. 3,418
$$\begin{array}{r} 194 \\ 8 \\ 2,366 \\ +\ 850 \\ \hline 3,418 \end{array}$$

13. 48,000
$$\begin{array}{r} 96 \\ \times\ 500 \\ \hline 48,000 \end{array}$$

Notice how this problem is lined up.

14. 2,184,000
$$\begin{array}{r} 3,000,000 \\ -\ 816,000 \\ \hline 2,184,000 \end{array}$$

15. 67,245
$$\begin{array}{r} 45,360 \\ +21,885 \\ \hline 67,245 \end{array}$$

16. 203,406
$$\begin{array}{r} 29,058 \\ \times 7 \\ \hline 203,406 \end{array}$$

17. 47
$$\begin{array}{r} 47 \\ 800\overline{)37,600} \\ \underline{32\ 00} \\ 5\ 600 \\ \underline{5\ 600} \\ 0 \end{array}$$

18. 592,000
$$\begin{array}{r} 74 \\ \times\ 8,000 \\ \hline 592,000 \end{array}$$

19. 4,775,500
$$\begin{array}{r} 5,040,000 \\ -\ 264,500 \\ \hline 4,775,500 \end{array}$$

20. 306
$$\begin{array}{r} 306 \\ 35\overline{)10,710} \\ \underline{10\ 5} \\ 21 \\ \underline{0} \\ 210 \\ \underline{210} \\ 0 \end{array}$$

Whole Numbers Exercise 7

1. 3,412
$$\begin{array}{r} 590 \\ 2,041 \\ 713 \\ +\ 68 \\ \hline 3,412 \end{array}$$

2. 35,577
$$\begin{array}{r} 603 \\ \times\ 59 \\ \hline 5,427 \\ 3,015 \\ \hline 35,577 \end{array}$$

3. 1,323
$$\begin{array}{r}10,230\\-8,907\\\hline 1,323\end{array}$$

4. 40,720
$$\begin{array}{r}5,090\\\times\quad 8\\\hline 40,720\end{array}$$

5. 4,893
$$\begin{array}{r}103,460\\-98,567\\\hline 4,893\end{array}$$

6. 486r6
$$\begin{array}{r}486r6\\14\overline{)6810}\\56\\\hline 121\\112\\\hline 90\\84\\\hline 6\end{array}$$

7. 3,072
$$\begin{array}{r}10,000\\-6,928\\\hline 3,072\end{array}$$

8. 3,484
$$\begin{array}{r}467\\78\\896\\+2,043\\\hline 3,484\end{array}$$

9. 423
$$\begin{array}{r}423\\12\overline{)5,076}\\48\\\hline 27\\24\\\hline 36\\36\\\hline 0\end{array}$$

10. 155,700
$$\begin{array}{r}346\\\times 450\\\hline 17\,300\\138\,4\\\hline 155,700\end{array}$$

11. 6,912
$$\begin{array}{r}7,000\\-\quad 88\\\hline 6,912\end{array}$$

12. 391,084
$$\begin{array}{r}400,300\\-9,216\\\hline 391,084\end{array}$$

13. 705
$$\begin{array}{r}705\\4\overline{)2,820}\\28\\\hline 02\\0\\\hline 20\\20\\\hline 0\end{array}$$

14. 3,003
$$\begin{array}{r}85\\+344\\\hline 429\end{array}\qquad\begin{array}{r}429\\\times\quad 7\\\hline 3,003\end{array}$$

15. 1,714
$$\begin{array}{r}1,604\\+\;857\\\hline 2,461\end{array}\qquad\begin{array}{r}1,604\\-\;857\\\hline 747\end{array}$$
$$\begin{array}{r}2,461\\-\;747\\\hline 1,714\end{array}$$

Whole Numbers Exercise 8

1. 193
$$\begin{array}{r}353\\19\\+207\\\hline 579\end{array}\qquad\begin{array}{r}193\\3\overline{)579}\end{array}$$

2. 1,995
$$\begin{array}{r}2,043\\971\\3,116\\+1,850\\\hline 7,980\end{array}\qquad\begin{array}{r}1,995\\4\overline{)7,980}\end{array}$$

3. 253
$$\begin{array}{r}240\\313\\189\\+270\\\hline 1,012\end{array}\qquad\begin{array}{r}253\\4\overline{)1012}\end{array}$$

4. 173 lbs.
$$\begin{array}{r}187\text{ lbs.}\\+159\\\hline 346\text{ lbs.}\end{array}\qquad\begin{array}{r}173\\2\overline{)346}\end{array}$$

5. 65°
$$\begin{array}{r}69°\\71°\\56°\\63°\\+66°\\\hline 325°\end{array}\qquad\begin{array}{r}65°\\5\overline{)325°}\end{array}$$

6. $14,800
$$\begin{array}{r}\$14,700\\13,900\\+15,800\\\hline \$44,400\end{array}\qquad\begin{array}{r}\$14,800\\3\overline{)\$44,400}\end{array}$$

7. $26.07
$$\begin{array}{r}\$25.66\\33.27\\+19.28\\\hline \$78.21\end{array}\qquad\begin{array}{r}\$26.07\\3\overline{)\$78.21}\end{array}$$

8. $240
$$\begin{array}{r}\$240\\450\overline{)\$108,000}\\90\,0\\\hline 18\,00\\18\,00\\\hline 00\end{array}$$

Notice that you already have the sum for this problem.

9. $850
$$3\times\;650=\$1,950$$
$$4\times\;875=\;3,500$$
$$2\times1100=+2,200$$
$$\overline{\qquad\quad\$7,650}$$
$$\begin{array}{r}\$850\\9\overline{)\$7,650}\end{array}$$

He sold 3 + 4 + 2 = 9 cars.

10. 48
$$\begin{array}{r}47\\42\\38\\+50\\\hline 177\end{array}\qquad\begin{array}{l}Total\\5\times45=225\\225\\-177\\\hline\;\;48\end{array}$$

Notice that this problem is different from the others. You must use the average of 45 to get the sum of 225 for five tests.

Whole Numbers Exercise 9

1. 96

$6 \xrightarrow{\times 2} 12 \xrightarrow{\times 2} 24 \xrightarrow{\times 2} 48 \xrightarrow{\times 2} \underline{96}$

2. 96

$81 \xrightarrow{+3} 84 \xrightarrow{+3} 87 \xrightarrow{+3} 90 \xrightarrow{+3} 93 \xrightarrow{+3} \underline{96}$

3. 20

$32 \xrightarrow{-3} 29 \xrightarrow{-3} 26 \xrightarrow{-3} 23 \xrightarrow{-3} \underline{20}$

4. 37

$13 \xrightarrow{+4} 17 \xrightarrow{+4} 21 \xrightarrow{+4} 25 \xrightarrow{+4} 29 \xrightarrow{+4} 33 \xrightarrow{+4} \underline{37}$

5. 25

$10 \xrightarrow{-5} 5 \xrightarrow{+10} 15 \xrightarrow{-5} 10 \xrightarrow{+10} 20 \xrightarrow{-5} 15 \xrightarrow{+10} \underline{25}$

6. 9

$100 \xrightarrow{-19} 81 \xrightarrow{-17} 64 \xrightarrow{-15} 49 \xrightarrow{-13} 36 \xrightarrow{-11} 25 \xrightarrow{-9} 16 \xrightarrow{-7} \underline{9}$

7. 65

$5 \xrightarrow{+4} 9 \xrightarrow{+8} 17 \xrightarrow{+16} 33 \xrightarrow{+32} \underline{65}$

8. 30

$2 \xrightarrow{+1} 3 \xrightarrow{+2} 5 \xrightarrow{+3} 8 \xrightarrow{+4} 12 \xrightarrow{+5} 17 \xrightarrow{+6} 23 \xrightarrow{+7} \underline{30}$

9. 10

$320 \xrightarrow{\div 2} 160 \xrightarrow{\div 2} 80 \xrightarrow{\div 2} 40 \xrightarrow{\div 2} 20 \xrightarrow{\div 2} \underline{10}$

10. 49

$4 \xrightarrow{\times 2} 8 \xrightarrow{-1} 7 \xrightarrow{\times 2} 14 \xrightarrow{-1} 13 \xrightarrow{\times 2} 26 \xrightarrow{-1} 25 \xrightarrow{\times 2} 50 \xrightarrow{-1} \underline{49}$

11. 77°

$62 \xrightarrow{+5} 67 \xrightarrow{+5} 72 \xrightarrow{+5} \underline{77}$

12. $5,000

$\$29,000 \xrightarrow{-\$6,000} \$23,000 \xrightarrow{-\$6,000} \$17,000 \xrightarrow{-\$6,000} \$11,000 \xrightarrow{-\$6,000} \underline{\$5,000}$

13. 24,000

$1,500 \xrightarrow{\times 2} 3,000 \xrightarrow{\times 2} 6,000 \xrightarrow{\times 2} 12,000 \xrightarrow{\times 2} \underline{24,000}$

Whole Numbers Exercise 10

1. (4) Subtract 8 from 37.
2. (5) Divide 40 into $260.
3. (2) Add $12, $25, and $53.
4. (4) Both solution (1) and solution (2).
5. (3) Subtract $4,680 from $18,296.
6. (5) Subtract 92 from 1980.
7. (1) 943 − 387. Although 198 cars were made in Japan, the question asks how many cars were made outside the U.S., which includes more than Japan.
8. (4) Multiply 8 by $6. Multiply 2 by $9. Then add the results.
9. (3) $16,380 ÷ 12.
10. (2) Divide 1,000 by 83.
11. (1) Divide $2,717,600 by 158. Notice you do not need the information about firemen.
12. (5) Subtract 17,558,000 from 23,668,000. Notice you do not need the population of Texas for this problem.

Whole Numbers Exercise 11

1. (4) $258

$$\begin{array}{r} \$198 \\ +\ 60 \\ \hline \$258 \end{array}$$

Notice the word <u>less</u> does not tell you to subtract in this problem. The original price was $60 <u>more</u> than the sale price. You must add.

2. (1) $17,577

$$\begin{array}{r} \$279 \\ \times\ 63 \\ \hline 837 \\ 1674 \\ \hline \$17,577 \end{array}$$

3. (5) 2,461,000

$$\begin{array}{r} 1,297,000 \\ +1,164,000 \\ \hline 2,461,000 \end{array}$$

4. (1) $672

$$\begin{array}{r} \$\ \ \ 672 \\ 83\overline{)\$55,776} \\ \underline{49\ 8} \\ 5\ 97 \\ \underline{5\ 81} \\ 166 \\ \underline{166} \end{array}$$

5. (4) 138,109

$$
\begin{array}{r}
259,056 \\
-120,947 \\
\hline
138,109
\end{array}
$$

6. (1) $9,820

$$
\begin{array}{r}
\$12,400 \\
-\ 2,580 \\
\hline
\$\ 9,820
\end{array}
$$

Notice you do not need to use the $1,560 savings to solve this problem.

7. (2) 275

$$
\begin{array}{r}
275 \\
3\overline{)825}
\end{array}
$$

8. (4) $960

$$
\begin{array}{r}
\$45 \\
\times 18 \\
\hline
360 \\
45 \\
\hline
\$810
\end{array}
\qquad
\begin{array}{r}
\$810 \\
+150 \\
\hline
\$960
\end{array}
$$

9. (1) $171,000

$$
\begin{array}{r}
\$7,800 \\
\times 30 \\
\hline
\$234,000
\end{array}
\qquad
\begin{array}{r}
\$2,100 \\
\times 30 \\
\hline
\$63,000
\end{array}
\qquad
\begin{array}{r}
\$234,000 \\
-\ 63,000 \\
\hline
\$171,000
\end{array}
$$

or

$$
\begin{array}{r}
\$7,800 \\
-2,100 \\
\hline
\$5,700
\end{array}
\qquad
\begin{array}{r}
\$5,700 \\
\times 30 \\
\hline
\$171,000
\end{array}
$$

10. (3) 2,301,000

$$
\begin{array}{r}
10,549,000 \\
-8,248,000 \\
\hline
2,301,000
\end{array}
$$

Notice you do not need the December 1981 unemployment figure of 9,307,000.

11. (1) $18

$$
\begin{array}{r}
18 \\
4\overline{)72} \\
4 \\
\hline
32 \\
32
\end{array}
$$

There is a total of 4 girls, Tina and 3 sisters.

12. (5) $22,880

$$
\begin{array}{r}
240 \\
+200 \\
\hline
440
\end{array}
\qquad
\begin{array}{r}
440 \\
\times 52 \\
\hline
880 \\
2200 \\
\hline
\$22,880
\end{array}
$$

13. (2) 4,700

$$
\begin{array}{r}
67,576 \\
-62,881 \\
\hline
4,695 \text{ to the nearest} \\
\text{hundred is 4,700}
\end{array}
$$

14. (5) 400

$$
\begin{array}{r}
4 \\
6 \\
+10 \\
\hline
20
\end{array}
\qquad
\begin{array}{r}
20 \\
\times 20 \\
\hline
400
\end{array}
$$

15. (3) 1,250

$$
\begin{array}{r}
2,500 \\
-1,250 \\
\hline
1,250
\end{array}
$$

The figure of 750 college graduates is not needed to solve the problem.

Whole Numbers Review

1. (2) 80,000

8 is in the ten-thousands place.
8 × 10,000 = 80,000

2. (3) 36,000

6 is in the thousands place. The 4 indicates that you should leave 6 and put zeros to the right of 6.

3. (5) none of these

$$
\begin{array}{r}
100,000 \\
-\ 79,028 \\
\hline
20,972
\end{array}
$$

4. (2) 407

$$
\begin{array}{r}
407 \\
16\overline{)6512} \\
64 \\
\hline
11 \\
0 \\
\hline
112 \\
112
\end{array}
$$

5. (1) 63

$$
1 \underset{+2}{\nearrow} 3 \underset{+4}{\nearrow} 7 \underset{+8}{\nearrow} 15 \underset{+16}{\nearrow} 31 \underset{+32}{\nearrow} 63
$$

6. (4) 34 min.

$$
\begin{array}{r}
27 \\
36 \\
28 \\
39 \\
40 \\
\hline
170
\end{array}
\qquad
\begin{array}{r}
34 \\
5\overline{)170} \\
15 \\
\hline
20 \\
20
\end{array}
$$

7. (3) Subtract 63 from 420.

8. (4) $140

$$
\begin{array}{r}
\$83 \text{ sale price} \\
+\$57 \text{ reduction} \\
\hline
\$140 \text{ original price}
\end{array}
$$

9. (5) none of these

$$
\begin{array}{r}
\$56 \\
\times 15 \\
\hline
280 \\
56 \\
\hline
\$840
\end{array}
\qquad
\begin{array}{r}
\$840 \\
+200 \\
\hline
\$1040
\end{array}
$$

10. (4) $970

$$
\begin{array}{r}
\$\ \ \ 970 \\
84\overline{)\$81,480} \\
75\,6 \\
\hline
5\,88 \\
5\,88 \\
\hline
00
\end{array}
$$

Decimals

WHAT ARE DECIMALS?

In the previous chapter, you worked with whole numbers. This chapter and the next concern two ways of writing parts of a whole—decimals and fractions. **Decimals** are parts that are expressed in tenths or multiples of tenths. Our money system is based on decimals, as is the metric system, a form of measurement that will be discussed later in the book.

In our money system, dollars are represented by whole numbers, and the cents, the parts of a dollar, are written as decimals.

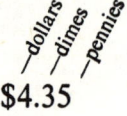

$4.35

The 4 in the amount $4.35 represents 4 whole dollars. The .35 represents 35 of 100 equal parts of a dollar (pennies) or 35 hundredths of a dollar.

The digit 3, one place to the right of the decimal point, is in the dimes place and has a value of 30 cents.

The digit 5 is in the pennies' place, two places to the right of the decimal point, and has a value of 5 cents.

You already know something about decimals from handling money every day.

EXAMPLE: What is the value of the 2 in $7.8<u>2</u>?
  Solution: Tell what place 2 is in. 2 is in the pennies' place. The 2 has a value of **2¢**.

DECIMALS EXERCISE 1

Find the value of each underlined digit.

1. $<u>4</u>.56 $.<u>2</u>8 $1.0<u>9</u>

2. $18.0<u>5</u> $1,922.<u>43</u> $87.3<u>1</u>

Answers and solutions start on page 81.

PLACE VALUE IN THE DECIMAL SYSTEM

As you saw above with dollars and cents, the first place to the right of the decimal point is the dimes' place. The second place to the right of the point is the pennies' place. <u>Without</u> the dollar sign, the first place to the right of the point is called the tenths' place. (A dime is one of ten equal parts of a dollar.) The second place is called the hundredths' place. (A penny is one of the hundred equal parts of a dollar.)

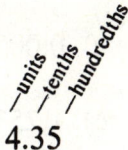

4.35

Below is a diagram of the first six whole number places and the first six decimal places.

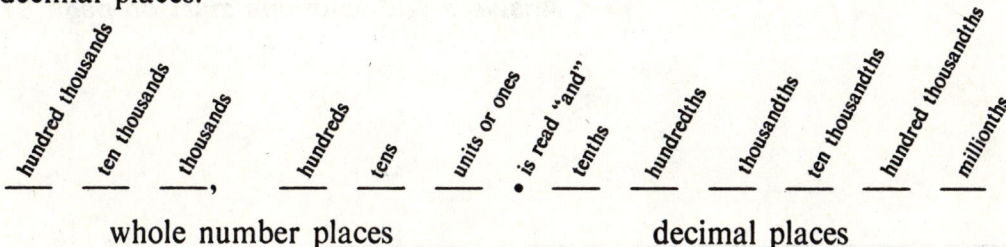

whole number places   decimal places

The **decimal point** separates whole numbers from decimal places. Notice that every decimal place ends with <u>ths</u>.

Hint: To learn the decimal place names, remember that 10 has one zero and tenths have one place. 100 has two zeros and hundredths have two places. 1,000 has three zeros and thousandths have three places.

EXAMPLE 1: Tell how many decimal places the number 18.456 has.
Solution: Count the number of digits to the right of the decimal point. 18.456 has three digits to the right of the point. 18.456 has **three** decimal places.

EXAMPLE 2: Tell what place the 6 in 18.456 is in.
Solution: Count how many places to the right of the decimal point the 6 is in. 6 is three places to the right of the point. The third place to the right of the point is called **thousandths.**

DECIMALS EXERCISE 2 ━━━━━━━━━━━━━━━━━━━━━━━━━━━━━

For items 1 to 4, tell how many decimal places each of the numbers has.

1. a) 4.907 b) 3,806.2 c) 314.26

2. a) 1,976,500 b) 18.04 c) 2.87654

3. a) 59 b) 9.3 c) 4.0035

4. a) 80.538 b) 1,264.7 c) 0.004502

For items 5 to 8, tell what place each underlined digit is in.

5. a) $8.4<u>7</u> b) 1.09<u>3</u>6 c) $267.8<u>1</u>

6. a) 0.073<u>8</u> b) $256.<u>3</u>0 c) 0.0004<u>5</u>

7. a) $198.1<u>3</u> b) <u>6</u>0.482 c) 1,<u>2</u>53.4

8. a) .00047<u>6</u> b) $93.<u>4</u>0 c) 7.81<u>5</u>

Answers and solutions start on page 81.

━━

READING DECIMALS

You have seen how place value works in the decimal system. Now, you will learn how to read a decimal. As you work ahead, remember this: the place of the last digit of a decimal names its value.

Look at the difference between .3 and .03. In the first case, the number 3 is in the tenths place and is read as three tenths. In the second case, the 3 is two decimal places over and is read as three hundredths.

How is .275 read? Read the number first (275) and the place that the last decimal place occupies (3 places over—thousandths). The number is read as 275 thousandths.

Finally, look at a **mixed decimal** such as 10.042. Read the whole number first (10). Read the decimal point as the word "and." Finally, read the decimal (42, with the last digit in the thousandths place). 10.042 is read as 10 and 42 thousandths. Don't let the zero in the tenths' place fool you. We'll discuss zeros in the next section.

DECIMALS EXERCISE 3 ─────────────────────────

Write out the following numbers.

1. .31   6. 403.1

2. .0780   7. 3.049

3. 2.12   8. .143

4. 100.03   9. 4.007

5. 13.13   10. 71.5

Answers and solutions start on page 81.

ZEROS

The digit 0 has no value, but some 0's hold other digits in their places. Look at the number below:

$$08.0060$$

The digit to the left of 8 has no use. The digit to the right of the 6 also has no use. However, the two 0's to the right of the point keep the digit 6 in the thousandths' place.

You can rewrite the number 08.0060, dropping two of the 0's, without changing its value.

$$08.0060 = 8.006$$

The 0's in the tenths' place and the hundredths' place mean that there are no tenths and no hundredths.

One exception to the rule of dropping unnecessary zeros is in our money system. We write $4.50 (four dollars and fifty cents) because the system is based on hundredths of a dollar.

Understanding the difference between necessary and unnecessary zeros will help you throughout your work with decimals.

DECIMALS EXERCISE 4 ─────────────────────────

Rewrite each number keeping only the necessary zeros.

1. a) .06700   b) 03.405   c) 8.0906

2. a) 80.0250   b) 0124.0090   c) 007.50

3. a) 5.0   b) .37080   c) 029.300

4. a) 06.3   b) .002300   c) 060.0502

Answers and solutions start on page 81.

WRITING DECIMALS

When you begin to write a decimal, first think about how many places you need.

EXAMPLE 1: Write fifty-two thousandths as a decimal.

Solution: The 52 is a two-digit number and the last word, thousandths, tells you the place value. Thousandths take three places, but 52 uses only two places. <u>Hold</u> the first place (tenths) with a zero. Fifty-two thousandths is written as **.052**.

Watch for the word <u>and</u>. Remember that <u>and</u> separates whole numbers from decimals. Whole numbers with a decimal are called mixed decimals.

EXAMPLE 2: Write twenty-three and nine tenths as a decimal.

Solution: First write the whole number (23) and the decimal point. Then think about how many decimal places you need. Tenths are in the first decimal place to the right of the point. Twenty-three and nine tenths can be written as **23.9**.

DECIMALS EXERCISE 5

Write each group of words as a decimal or mixed decimal.

1. four tenths

2. six and three tenths

3. eighteen hundredths

4. nine thousandths

5. two ten-thousandths

6. thirteen thousandths

7. ninety-six and four tenths

8. five and sixteen thousandths

9. seven thousand five hundred and eight tenths

10. one hundred twenty-five thousandths

11. eighty-four and nine thousandths

12. five thousand six hundred two and twenty-eight millionths

13. three hundred twelve ten-thousandths

14. two hundred four and three hundredths

15. seventy and three hundred forty-five thousandths

Answers and solutions start on page 81.

COMPARING DECIMALS

Which is larger, $.30 or $.04? You know that 30 cents is more than 4 cents. The decimals .30 and .04 have the same number of places. This makes them easy to compare.

Rules for Comparing Decimals

1. Give the decimals the same number of places. You can put zeros to the right of a decimal without changing its value.

2. Decide which decimal is larger.

EXAMPLE: Which decimal is larger, .06 or .052?

> .06 = .06<u>0</u>
> .052 = .052

Step 1: Give the decimals the same number of places. Put a zero to the right of .06 to give it three places. .06 = .060

Step 2: Decide which decimal is larger, .060 or .052? Sixty thousandths is larger than fifty-two thousandths. Therefore, **.06 is the larger number.**

DECIMALS EXERCISE 6 ━━━━━━━━━━━━━━━━━━━━━━━━━━━━━━━━

Circle the larger number in each pair.

1. a) .08 or .7   b) .62 or .062   c) .33 or .403

2. a) .0029 or .001   b) .01 or .101   c) .895 or .9

3. a) .8 or .098   b) 5.2 or 5.23   c) .4 or .0268

4. a) .31 or .295   b) 1.68 or 1.678   c) 3.004 or 3.04

Answers and solutions start on page 81.

━━━

ADDING DECIMALS

Think about adding money. Suppose you bought oranges for $1 and milk for 65¢. Your total bill would be $1.65. Automatically, you lined up the decimal points.

$$\begin{array}{r} \$1.00 \\ .65 \\ \hline \mathbf{\$1.65} \end{array}$$

Adding decimals is as easy as adding money or whole numbers.

Rules for Adding Decimals
1. Line up the decimals with <u>point under point</u>.
2. Add each column and bring the decimal point straight down into each answer. Carry if necessary. You can carry to the left of the decimal point.

Warning: Do not confuse the period at the end of a sentence with the decimal point.

EXAMPLE 1:   Add .8 and .4.

①
$$\begin{array}{r} .8 \\ +.4 \\ \hline \end{array}$$
②
$$\begin{array}{r} .8 \\ +.4 \\ \hline \mathbf{1.2} \end{array}$$

Step 1.   Line up the numbers with point under point. Notice that the point to the right of .4 is a period, not a decimal point.

Step 2.   Add the numbers and bring the decimal point straight down into the answer.

Notice in the example that eight tenths and four tenths give a sum of twelve tenths. Only one digit fits in the tenths' place. <u>Carry</u> the digit 1 to the units' place. This is just like the carrying you do when you add whole numbers.

EXAMPLE 2: Find the sum of 3.8, 47, .0092, and 1.83.

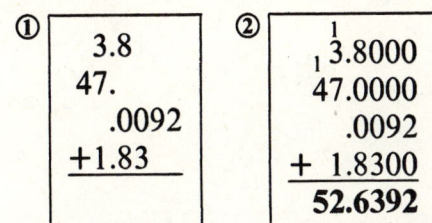

Step 1. Line up the numbers with decimal point under decimal point. Notice that the whole number 47 is understood to have a decimal point at the right even though it is not written in the problem.

Step 2. Add each column and bring the decimal point straight down into the answer. Carry where necessary.

Many students like to put zeros in decimal addition problems. The zeros help keep the other digits in the correct columns.

DECIMALS EXERCISE 7

Solve each problem.

1. .6, .4, + .9 =

2. .5 + 9 + 7.375 + 21.37 =

3. 16, 9.24, 170.3, + .369 =

4. 15.23 + 4 + 1.816 + 9.4 =

5. 3.6, 14.2, 10, + 7.05 =

6. 83 + 36.27 =

7. 4.036 + 23 + 2.19 + .084 =

8. Find the sum of 0.10905, 1.0687, and 2.00453.

9. Add 12.3, .016, 5, and 216.

10. What is the sum of 283.4, 87.49, and 107.3?

11. $84 + $19.65 + $.23 + $1.56 =

12. Find the total of .0075 + .00128 + .004 + .03806.

Answers and solutions start on page 81.

SUBTRACTING DECIMALS

Suppose you buy items for $1.65 at the grocery. Then you give the cashier a $5 bill. How much change should you receive? As in addition, you must line up the decimal points to subtract and find the answer.

$$\begin{array}{r} \overset{4\,9\,1}{\$\cancel{5}.\cancel{0}\cancel{0}} \\ -1.65 \\ \hline \$3.35 \end{array}$$

Rules for Subtracting Decimals

1. Use zeros to give each number the same number of decimal places. Compare the decimals to decide which is larger.
2. Line up the numbers with point under point. Be sure to put the larger number on top.
3. Subtract, borrowing if necessary. Bring the decimal point straight down to the answer.

In some problems, it is not immediately obvious which is the larger number. Since you must put the larger number on top, decide this first.

EXAMPLE: What is the difference between .254 and .7?

①
$$.254 = .254$$
$$.7 = 700$$

②
$$\begin{array}{r} \overset{6\,9\,1}{.\cancel{7}\cancel{0}0} \\ -.254 \\ \hline \mathbf{.446} \end{array}$$

Step 1.   Add zeros to give both numbers the same number of decimal places for comparison: .254 and .700.

Step 2.   Put the larger number on top. Borrow and subtract.

DECIMALS EXERCISE 8

Solve each problem.

1. 18 − .32

2. .09 − .075 =

3. 12 − 2.7 =

4. 16 − .04 =

5. 4.3 − 2.879 =

6. .3 − .094 =

7. From .008 take .0025.

8. Find the difference between .325 and .6.

9. What is the difference between 83 and 36.27?

10. How much more is $30 than $13.68?

11. Find the difference between 20 and 4.63.

12. Take .00925 from .03.

Answers and solutions start on page 82.

ADDITION AND SUBTRACTION OF DECIMALS: WORD PROBLEMS

The next problems will give you a chance to apply your problem solving skills to addition and subtraction of decimals. Solve each problem and select the correct answer from the five choices.

DECIMALS EXERCISE 9

Choose the best answer to each word problem.

1. In 1970, the population of New Mexico was 1.02 million people. By 1980, the population had increased by 0.28 million. What was the population of New Mexico in 1980?

   (1) .74 million
   (2) 1.48 million
   (3) 1.18 million
   (4) 1.08 million
   (5) 1.3 million

2. Mr. Rigby is 1.7 meters tall. His son Allen is 1.48 meters tall. How much taller is Mr. Rigby than his son?

   (1) 3.18 m
   (2) 0.22 m
   (3) 0.32 m
   (4) 0.38 m
   (5) 1.22 m

3. Martha's empty suitcase weighs 4.5 pounds. After she packed the suitcase, it weighed 16.25 pounds. How much did the contents of the suitcase weigh?

   (1) 20.75 lb.
   (2) 16.7 lb.
   (3) 11.75 lb.
   (4) 15.8 lb.
   (5) 12.25 lb.

4. Friday morning, Phil drove 85.4 miles. That afternoon he drove 98.2 miles. That night he drove another 63.7 miles. How many miles did Phil drive altogether on Friday?

   (1) 247.3 mi.
   (2) 248.2 mi.
   (3) 256.4 mi.
   (4) 257.3 mi.
   (5) 258.2 mi.

5. For Medicaid benefits, the U.S. government spent $17.1 million in 1981, $18.4 million in 1982, and $19.9 million in 1983. By how much did government spending on these benefits increase from 1981 to 1983?

   (1) $1.8 million
   (2) $2.8 million
   (3) $3.8 million
   (4) $1.3 million
   (5) $1.5 million

6. To pay for a new sports center, the town of Troy raised $1.35 million from the state government, $.85 million from the county government, and $1.05 million from local businesses and individuals. The total estimated cost of the project is $4 million. How much more money does the city need?

   (1) $0.3   million
   (2) $1.8   million
   (3) $2.65 million
   (4) $1.25 million
   (5) $0.75 million

Answers and solutions start on page 82.

MULTIPLYING DECIMALS

Lou bought three pairs of socks at a cost of $1.75 per pair. To find the price of the three pairs of socks, multiply the price of one pair by 3.

$$\begin{array}{r} {\scriptstyle 2\ 1} \\ \$1.75 \\ \times\ \ 3 \\ \hline \$5.25 \end{array}$$

To multiply a whole number by dollars and cents, mark off two decimal places for the cents in the answer. Notice the numbers in the problem are lined up at the right. This is not the same as in addition and subtraction.

Rules for Multiplying Decimals

1. Place one number under the other, lined up on the right side for easy multiplication. Multiply.
2. Count the number of decimal places in both numbers. Decimal places are to the right of the decimal point.
3. Counting from the right, put the total number of decimal places in your answer. Use zeros if you need more places than you have numbers.

Study each example below carefully. See how each problem is written. Notice where the decimal point is placed in each answer.

EXAMPLE 1: Find the product of 3.26 and .4.

      ^{1 2}

    3.26 two places

    × .4 one place

    1.304 three places

EXAMPLE 2: Multiply 27 by .4.

     ²

    27 no places

    × .4 one place

    10.8 one place

EXAMPLE 3: Find the product of 2.0413 and .006.

     ^{2 1}

    2.0413 four places

    ×.006 three places

    .0122478 add zero to make seven places

Notice that the answer had only six digits, yet the problem requires a seven-place answer. You need to add a zero on the left side—between the decimal point and the first number on the left.

There are shortcuts for multiplying a decimal or mixed decimal by 10, 100, or 1000.

To multiply by 10, move the decimal point one place to the right.

To multiply by 100, move the decimal point two places to the right.

To multiply by 1,000, move the decimal point three places to the right.

This will be useful when you work with metric measurement.

EXAMPLE 4: Find the product of 34.2 and 100.

    34.2 × 100 = 34.20. = **3,420.**

Put a zero to the right of 2 in order to move the decimal point two places to the right.

Solve each problem.

1. .8 × .7 =

2. Multiply 20.6 by .3.

3. 2.09 × .4 =

4. 8.3 × 2.7 =

5. .378 × .6 =

6. Multiply .0004 by 30.

7. .8 × .01 =

8. Multiply 1.439 by .8.

9. .076 × 1.5 =

10. 4.2 × $24.80 =

11. $250 × 2.3 =

12. $1.16 × 1.25 =

13. 1.432 × 10 =

14. Multiply .056 by 100.

15. .083 × 10 =

16. Multiply 19.7 by 1,000.

17. Find the product of 15 and .3.

18. What is the product of 143 and 2.5?

19. Find the product of .34 and 12.61.

20. Find the product of 1,000 and 7.95.

Answers and solutions start on page 82.

DIVISION WITH DECIMALS

Dividing Decimals by Whole Numbers

The bill for three people in a restaurant came to $14.55. If they shared the bill equally, how much did each person owe?

To solve this, you would divide the total bill by the number of people who shared it.

When you divide a decimal by a whole number, put the decimal point in the answer directly above its position in the problem.

$$\text{divisor} \longrightarrow 3 \overline{) \underset{\uparrow}{\$14.55}} \quad \begin{array}{l} \longleftarrow \text{quotient} \\ \longleftarrow \text{dividend} \end{array}$$

with $4.85 as the quotient.

Each person owes $4.85.

In the rules below, the term **dividend** means the number being divided. The **divisor** is the number being divided into the dividend. The term **quotient** is the answer to a division problem.

Rules for Dividing a Decimal by a Whole Number

1. Put the point in the quotient directly above its position in the dividend.

2. Divide as you would for whole numbers.

Study these examples carefully.

EXAMPLE 1: .168 ÷ 2 =

```
     .084
  2) .168
     16
     08
     08
```

Notice the 0 in the quotient. Since 2 does not divide into 1, you must put the 8 above the 6 in the hundredths' place. The 0 shows that there are no tenths and holds the 8 in the hundredths' place.

EXAMPLE 2: Divide 29.72 by 5.

```
      5.944
  5) 29.720
     25
      4 7
      4 5
       22
       20
        20
        20
```

Notice the extra zero that has been added to the dividend. This zero does not change the value of the dividend. Adding a zero may allow you to carry out the answer one more decimal place and eliminate a remainder.

In Example 2, the extra 0 in the dividend makes the problem come out evenly. In Example 1, the 0 in the quotient keeps the other digits in the correct places. These uses of zeros show some differences between dividing whole numbers and dividing decimals.

There are shortcuts for dividing a number by 10, 100, or 1,000. These shortcuts will be useful in your work with percents and with metric measurements.

To divide a number by 10, move the decimal point one place to the left.

$$14.3 \div 10 = 1.4.3 = \textbf{1.43}$$

To divide a number by 100, move the decimal point two places to the left.

$$14.3 \div 100 = .14.3 = \textbf{.143}$$

To divide a number by 1,000, move the decimal point three places to the left.

$$14.3 \div 1,000 = .014.3 = .0143$$

DECIMALS EXERCISE 11 ─────────────────────

Solve each problem.

1. Divide 52.608 by 8.

2. 9 into 758.7 =

3. $14\overline{)10.5}$

4. .039 split 3 ways =

5. Divide 2.375 by 25.

6. $7\overline{).0168}$

7. 9.6 ÷ 24 =

8. $24\overline{).96}$

9. $18\overline{)2.7}$

10. Divide 36.72 by 12.

11. 42.3 ÷ 100 =

12. Divide .04 by 10.

13. 1,000 into 19.5 =

14. .65 ÷ 100 =

Answers and solutions start on page 83.

Dividing by Decimals

To divide a number by a decimal, you must first change the divisor to a whole number.

Rules for Dividing by a Decimal

1. Make the divisor a whole number. Move the point to the right as far as it will go.

2. Move the point in the dividend to the right the <u>same</u> <u>number</u> <u>of</u> <u>places</u> you moved the point in the divisor. You may need to add zeros to the dividend.

3. Bring the point up in the quotient directly above its <u>new</u> position in the dividend and divide.

EXAMPLE 3:   Divide 1.842 by .06.

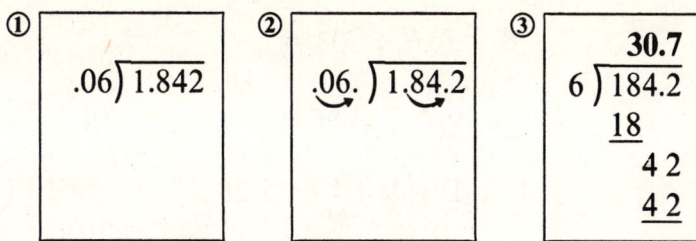

Step 1. Set the problem up for long division.

Step 2. Make the divisor a whole number. Move the decimal point two places to the right. Move the decimal point in the dividend two places to the right.

Step 3. Bring the decimal point up in the quotient directly above its new position in the dividend. Divide.

Study the next example carefully.

EXAMPLE 2. Divide .8 by .002.

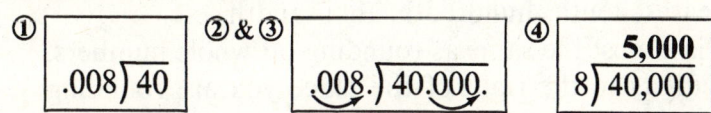 In this case, two zeros were added.

If you are dividing a decimal into a whole number, be sure to add zeros to the number being divided.

EXAMPLE 3: Divide 40 by .008.

① .008)40   ② & ③ .008.)40.000.   ④ 5,000
8)40,000

Step 1. Set the problem up for long division.

Step 2. Make the divisor a whole number. Move the decimal point three places to the right.

Step 3. Put a point at the right of 40. Put three zeros at the right of the point. Then move the point the same number of places to the right as you did in the divisor.

Step 4. Divide.

DECIMALS EXERCISE 12 —————————————————

Solve each problem.

1. 8.4 ÷ .2 =   **2.** Divide 5.25 by .5

3. $.13\overline{)\,.338}$

4. $.42 \div .028 =$

5. $1.2\overline{)\,.324}$

6. $.04\overline{)\,83.6}$

7. Divide 101.2 by 25.3

8. $90.6 \div .03 =$

9. $6.7 \div .134 =$

10. Divide 64 by 1.28.

11. $.12\overline{)\,96}$

12. $261 \div 4.35 =$

Answers and solutions start on page 84.

ROUNDING OFF DECIMALS

Suppose George works for $5.75 an hour. How much will he make in 3.5 hours? Multiply $5.75 by 3.5.

$$
\begin{array}{r}
\$5.75 \\
\times\ 3.5 \\
\hline
2875 \\
1725 \\
\hline
\$20.125
\end{array}
$$

Our money system has only two decimal places. In this problem, **round off** $20.125 to the nearest hundredth. In other problems, you may be asked to find the answer to the nearest tenth, hundredth, thousandth, etc.

Rounding off decimals is almost the same as rounding off whole numbers. With decimals, drop the digits to the right of the place you are rounding off to.

Rules for Rounding Off Decimals

1. Underline the digit in the place you are rounding off to.

2. If the digit to the right is 5 or more, add 1 to the underlined digit.

3. If the digit to the right is less than 5, leave the underlined digit as it is.

4. Drop the digits to the right of the underlined digit.

EXAMPLE: Round off 4.1942 to the nearest thousandth.

① $\boxed{4.19\underline{4}2}$ ② $\boxed{\textbf{4.194}}$

Step 1: Underline the digit in the thousandths' place.

Step 2: Look at the digit to the right of the underlined digit. Since the number is less than 5, keep the underlined digit and drop the number at the right.

DECIMALS EXERCISE 13 ━━━━━━━━━━━

Round off each of the following decimals to the place indicated.

1. $.328 to the nearest penny.

2. 6.527 to the nearest tenth.

3. 20.063 to the nearest tenth.

4. 8.0376 to the nearest thousandth.

5. .00825 to the nearest ten-thousandth.

6. $163.63 to the nearest dollar.

7. .27543 to the nearest thousandth.

8. 24.0273 to the nearest tenth.

9. $.5982 to the nearest penny.

10. .40365 to the nearest ten-thousandth.

Answers and solutions start on page 84.

REPEATING DECIMALS

In the previous sections, decimal division problems worked out evenly. However, some problems don't work out so smoothly. For instance, if you divide .20 by 3 you get a **repeating decimal**, no matter how many zeros you add to the dividend.

$$
\begin{array}{r}
.0666 \\
3\overline{)\,.2000} \\
\underline{18} \\
20 \\
\underline{18} \\
20 \\
\underline{18} \\
2
\end{array}
$$

Although you could write this with a fractional remainder (.0666⅔), it is more common to round off your answer. Generally, a problem will indicate to which decimal place you should round off.

EXAMPLE: Find .5 ÷ 8 to the nearest hundredth.

① 8 ⟌ .5

②
```
       .062
  8 ⟌ .500
       0
       50
       48
       20
       16
        4
```

③ .0̲62 to the nearest hundredth is **.06**

Step 1. Set up as a division problem.

Step 2. Place the decimal point in the quotient and work out the division. Carry out the division to one more place than the answer requires, so you can round off.

Step 3. To find the answer to the nearest hundredth, underline the number in that place, look at the number to the right, and round off.

DECIMALS EXERCISE 14

Divide and round off to the decimal place indicated.

1. .4 ÷ 15 to the nearest hundredth.

2. .02 ÷ 3 to the nearest hundredth.

3. .5 ÷ 6 to the nearest tenth.

4. .3 ÷ .8 to the nearest tenth.

Answers and solutions start on page 84.

MULTIPLICATION AND DIVISION OF DECIMALS: WORD PROBLEMS

These problems will give you a chance to apply your multiplication and division of decimals skills. Read each problem carefully. Remember in division problems to put the amount being divided inside the ⟌ sign. In some problems, you will have to round off your answers.

DECIMALS EXERCISE 15

Choose the correct answer to each word problem.

1. Fred's weekly salary is $240. He works 37.5 hours each week. How much does Fred make in one hour?

   (1) $2.40
   (2) $4.60
   (3) $5.80
   (4) $6.40
   (5) $8.10

2. The electricity to operate a color television costs $.034 an hour. The Walek family has its television turned on for an average of 52 hours each week. How much do the Waleks pay each week to run their television?

   (1) $1.77
   (2) $1.98
   (3) $2.60
   (4) $5.40
   (5) $8.60

3. Colin cut a piece of electric cable 3.5 meters long into four equal pieces. How long was each piece?

   (1) 1.15 m
   (2) .875 m
   (3) .915 m
   (4) 0.5 m
   (5) 3.1 m

4. On Monday, Grace worked 7 hours at her regular rate of $5.20 an hour and 2.5 hours at her overtime rate of $7.80 an hour. How much did Grace earn on Monday?

   (1) $55.90
   (2) $52.00
   (3) $57.20
   (4) $59.50
   (5) $54.70

5. The number of hits a baseball player gets divided by the number of times he comes to bat gives his batting average. Last season, Phil was at bat 75 times, and he made 22 hits. Find his batting average. Round off your answer to the nearest thousandth.

   (1) .356
   (2) .333
   (3) .320
   (4) .299
   (5) .293

6. Mrs. Goncalves talked to her sister in Portugal for twelve minutes on Tuesday night. The night rate for a call to Portugal is $3.15 for the first three minutes and $1.05 for each additional minute. How much did the call to Portugal cost Mrs. Goncalves?

   (1) $14.70
   (2) $12.60
   (3) $11.40
   (4) $10.50
   (5) $9.60

Answers and solutions start on page 84.

DECIMALS EXERCISE 16

Solve the decimal word problems below and choose the best answer.

1. In 1981–82, the U.S. government spent $874.5 million on special education programs. In 1982–83, the government spent $931 million on these programs. By how much did the budget for these programs increase from 1981–82 to 1982–83?

   (1) $54.5 million
   (2) $56.5 million
   (3) $57.5 million
   (4) $86 million
   (5) $105.5 million

2. Mark drove 344 miles on 16 gallons of gasoline. Find the average number of miles Mark drove on one gallon of gasoline.

   (1) 7.5 miles
   (2) 30 miles
   (3) 18.5 miles
   (4) 21.5 miles
   (5) 20 miles

3. Don weighs 210 pounds. One pound is the same as 0.45 kilogram. Find Don's weight in kilograms.

   (1) 467 kg
   (2) 165 kg
   (3) 94.5 kg
   (4) 92.3 kg
   (5) 46.7 kg

4. Frieda bought two packages of ground beef. One package weighed 2.65 pounds. The other weighed 1.8 pounds. What was the combined weight of the packages?

   (1) 0.85 lb.
   (2) 2.47 lb.
   (3) 2.83 lb.
   (4) 4.45 lb.
   (5) 3.65 lb.

5. Colette bought 4.3 pounds of tomatoes for $2.97. Find the price of one pound of tomatoes.

   (1) $0.84
   (2) $0.72
   (3) $0.69
   (4) $0.64
   (5) $0.58

6. Dolores bought 2.5 pounds of sausage at $1.90 per pound and 3 pounds of ground beef at $1.65 per pound. What was the total cost of her purchases?

   (1) $11.15
   (2) $9.70
   (3) $8.75
   (4) $6.85
   (5) $3.55

7. The area of Mexico is .762 million square miles. The area of the United States is 3.615 million square miles. The area of Canada is 3.852 million square miles. The combined area of Mexico and the U.S. is how much more than the area of Canada?

   (1) .425 million sq. mi.
   (2) .515 million sq. mi.
   (3) .525 million sq. mi.
   (4) 1.425 million sq. mi.
   (5) 1.515 million sq. mi.

8. In 1982, the average outlay per person for Medicaid benefits was $800. 23 million people received Medicaid in 1982. Find the total expenditures for Medicaid that year.

(1) $18.4 billion
(2) $82.3 billion
(3) $8.23 billion
(4) $1.84 billion
(5) $184 billion

9. Sam drove a truck from Phoenix to Dallas in 18 hours. The total distance from Phoenix to Dallas is 887 miles. Find Sam's average driving speed. Round off your answer to the nearest tenth.

(1) 62.4 mph
(2) 55.0 mph
(3) 49.2 mph
(4) 49.3 mph
(5) 46.5 mph

10. From a piece of copper tubing exactly 2 meters long, Pete cut one piece measuring .65 meter and another measuring .45 meter. How long was the remaining piece?

(1) .9 m
(2) 1.1 m
(3) 1.09 m
(4) 1.5 in
(5) 1.9 m

Answers and solutions start on page 85.

DECIMALS TEST ━━━━━━━━━━━━━━━━━━━━━━

Fill in the circle that corresponds to the correct answer.

1. Tell what place the 8 in 4,026.385 is in.   1 ① ② ③ ④ ⑤
   (1) tens
   (2) hundreds
   (3) tenths
   (4) hundredths
   (5) thousandths

2. Which of the following has the same value as 060.04300?   2 ① ② ③ ④ ⑤
   (1) 6.43
   (2) 60.043
   (3) 600.43
   (4) 60.0043
   (5) 64.03

3. Write the number four hundred eight and fifteen ten-thousandths as a decimal.   3 ① ② ③ ④ ⑤
   (1) 408.15000
   (2) 408.0015
   (3) .040815
   (4) 408.00015
   (5) 40,815

4. Which of the following has the greatest value?   4 ① ② ③ ④ ⑤
 (1) .086
 (2) .608
 (3) .0088
 (4) .8006
 (5) .806

5. Find the sum of 36, 2.93, .065, and 4.   5 ① ② ③ ④ ⑤
 (1) 3.359
 (2) 38.999
 (3) 42.995
 (4) 39.395
 (5) 39.8

6. Take .0063 from .08.   6 ① ② ③ ④ ⑤
 (1) .0055
 (2) .0017
 (3) .0737
 (4) .55
 (5) .017

7. In 1970, the population of Florida was 6.79 million. By 1980, the   7 ① ② ③ ④ ⑤
 population had increased by 3.39 million. What was the popula-
 tion of Florida in 1980?
 (1) 10.18 million
 (2) 11.28 million
 (3) 11.08 million
 (4) 10.08 million
 (5) 3.4 million

8. From a board 3 yards long, Allen cut a piece 1.75 yards long.   8 ① ② ③ ④ ⑤
 How long was the remaining piece?
 (1) 4.75 yd.
 (2) 1.75 yd.
 (3) 2.75 yd.
 (4) 1.25 yd.
 (5) 2.25 yd.

9. Find the product of 3.2 and .005.   9 ① ② ③ ④ ⑤
 (1) .016
 (2) .0016
 (3) 1.6
 (4) .1600
 (5) .160

10. What is .405 divided by 15?
    (1) .027
    (2) .27
    (3) .207
    (4) .37
    (5) .307

10 ① ② ③ ④ ⑤

11. What is the quotient of 9 divided by .045?
    (1) 2
    (2) .005
    (3) 200
    (4) .05
    (5) 20

11 ① ② ③ ④ ⑤

12. Find the cost of 4.3 pounds of tomatoes at $0.68 a pound.
    (1) $3.02
    (2) $2.92
    (3) $2.72
    (4) $2.04
    (5) $1.11

12 ① ② ③ ④ ⑤

13. David drove 223 miles on 12 gallons of gasoline. To the nearest tenth, find the average number of miles he drove on one gallon of gasoline.
    (1) 19.4 mi.
    (2) 18.6 mi.
    (3) 18.2 mi.
    (4) 16.8 mi.
    (5) 20.2 mi.

13 ① ② ③ ④ ⑤

14. All but 56.3 pounds of topsoil were sold at the sale. If the store began with 100 pounds of the topsoil and charged $.59 per pound, how much money did they make?
    (1) $59.00
    (2) $44.29
    (3) $25.00
    (4) $257.80
    (5) $25.78

14 ① ② ③ ④ ⑤

Answers and solutions start on page 86.

DECIMALS TEST EVALUATION

Passing score: 10 right out of 14 problems.

Your score: _____ right out of 14 problems.

If you had less than a passing score, review the sections for the problems you got wrong. Then repeat this test before you go on to the mixed review.

If you had a passing score, correct any problem you got wrong. Then go to the mixed review.

MIXED REVIEW

These problems give you a chance to practice the skills you have learned so far in this book. For each problem, fill in the circle that corresponds to the correct answer.

1. Round off 128,746.2413 to the nearest thousand.

    (1) 129,000
    (2) 130,000
    (3) 128,700
    (4) 100,000
    (5) 128,750

2. Round off 123.876 to the nearest hundredth.

    (1) 124
    (2) 123
    (3) 123.88
    (4) 123.87
    (5) 123.8

3. Margaret is a waitress. Thursday, she made $32.40 in tips. Friday, she made $47.75. Saturday, she made $42.10. What was her average amount of tips?

   (1) $47.75
   (2) $40.75
   (3) $41.25
   (4) $42.75
   (5) $32.40

4. Sam bought a stereo on sale for $279. This was $35 less than the original price. What was the original price?

   (1) $304
   (2) $314
   (3) $244
   (4) $254
   (5) $299

5. Write the number three hundred nine ten-thousandths as a decimal.

   (1) 300.9
   (2) 300.009
   (3) .309
   (4) .00309
   (5) .0309

6.

53.1°	55.3°	57.5°	59.7°	?
9 a.m.	10 a.m.	11 a.m.	12 p.m.	1 p.m.

   If the temperature continues to change at the same rate every hour, what will the temperature be at 1 p.m.?

   (1) 61.7°
   (2) 61.9°
   (3) 60°
   (4) 60.7°
   (5) 62.4°

7. Postage stamps cost $.20 a piece. Mrs. Ramos paid for 10 stamps with a $5 bill. How much change did she receive?

   (1) $15
   (2) $5.20
   (3) $2.00
   (4) $3.00
   (5) $15.20

8. In 1952, the budget for Westchester County was $26.3 million. By 1982, the budget was $445.5 million more than in 1952. Find the amount of the 1982 budget.

    (1) $419.2 million
    (2) $428.2 million
    (3) $480.8 million
    (4) $561.8 million
    (5) $471.8 million

9. 4,745 widgets were packed in boxes holding 65 widgets each. How many boxes were used?

    (1) 308,555
    (2) 4,810
    (3) 4,680
    (4) 4,745
    (5) 73

10. In 1952, an average monthly electricity bill was $7.80. In 1982, an average bill was $38.70. An average 1982 electricity bill was approximately how many times as big as a 1952 bill?

    (1) 2 times
    (2) 3 times
    (3) 4 times
    (4) 5 times
    (5) 1.5 times

Answers and solutions start on page 86.

ANSWERS AND SOLUTIONS

Decimals Exercise 1

1. $4   20¢   0¢   2. 5¢   40¢   1¢

Decimals Exercise 2

1. a) three   b) one   c) two
2. a) none   b) two   c) five
3. a) none   b) one   c) four
4. a) three   b) one   c) six
5. a) hundredths or pennies
   b) thousandths
   c) tenths or dimes

6. a) ten-thousandths
   b) tenths or dimes
   c) hundred-thousandths
7. a) hundredths or pennies
   b) ten<u>s</u>
   c) hundred<u>s</u>

8. a) millionths
   b) tenths or dimes
   c) thousandths

Decimals Exercise 3

1. thirty-one hundredths
2. seven hundred eighty ten-thousandths
3. two and twelve hundredths
4. one hundred and three hundredths
5. thirteen and thirteen hundredths

6. four hundred three and one tenth
7. three and forty-nine thousandths
8. one hundred forty-three thousandths
9. four and seven thousandths
10. seventy-one and five tenths

Decimals Exercise 4

1. a) .067   b) 3.405   c) 8.0906
2. a) 80.025   b) 124.009   c) 7.5
3. a) 5 Notice you do not need to write the decimal point with a whole number.
   b) .3708   c) 29.3
4. a) 6.3   b) .0023   c) 60.0502

Decimals Exercise 5

1. .4
2. 6.3
3. .18
4. .009
5. .0002
6. .013
7. 96.4
8. 5.016
9. 7,500.8
10. .125
11. 84.009
12. 5,602.000028
13. .0312
14. 204.03
15. 70.345

Decimals Exercise 6

1. a) .7   b) .62   c) .403
2. a) .0029   b) .101   c) .9

3. a) .8   b) 5.23   c) .4
4. a) .31   b) 1.68   c) 3.04

Decimals Exercise 7

1.
```
  .6
  .4
+ .9
 1.9
```

2.
```
   .5
  9.
  7.375
+21.37
 38.245
```

3.
```
  16.
   9.24
 170.3
+    .369
 195.909
```

4.
```
 15.23
  4.
  1.816
+ 9.4
 30.446
```

5.
```
  3.6
 14.2
 10.
+ 7.05
 34.85
```

6.
```
 83.
+36.27
119.27
```

7.
```
  4.036
 23.
  2.19
+  .084
 29.310 = 29.31
```

8.
```
 0.10905
 1.0687
+2.00453
 3.18228
```

9.
```
 12.3
   .016
  5.
+216.
 233.316
```

10.
```
 283.4
  87.49
+107.3
 478.19
```

Exercise 7 cont'd.

11.	$ 84.	12.	.0075
	19.65		.00128
	.23		.004
	+ 1.56		+.03806
	$ 105.44		.05084

Decimals Exercise 8

1.	18.00	4.	16.00	7.	.0080	10.	$30.00
	− .32		− .04		−.0025		− 13.68
	17.68		15.96		.0055		$16.32

2.	.090	5.	4.300	8.	.600	11.	20.00
	−.075		−2.879		−.325		− 4.63
	.015		1.421		.275		15.37

3.	12.0	6.	.300	9.	83.00	12.	.03000
	− 2.7		−.094		−36.27		−.00925
	9.3		.206		46.73		.02075

Decimals Exercise 9

1. (5) 1.3 million

 1.02 million
 +0.28
 1.30 = 1.3 million

2. (2) 0.22 m

 1.70 m
 −1.48
 0.22 m

3. (3) 11.75 lb.

 16.25 lb.
 −4.50
 11.75 lb.

4. (1) 247.3 mi.

 85.4 mi.
 98.2
 +63.7
 247.3 mi.

5. (2) $2.8 million

 $19.9 million
 −17.1
 $ 2.8 million

Notice you do not need the 1982 figure to solve this problem.

6. (5) $0.75 million

 $1.35 million
 .85
 +1.05
 $3.25 million

 $4.00 million
 −3.25
 $0.75 million

Decimals Exercise 10

1. .56

 .8 one place
 × .7 one place
 .56 two places

2. 6.18

 20.6 one place
 × .3 one place
 6.18 two places

3. .836

 2.09 two places
 × .4 one place
 .836 three places

4. 22.41

 8.3 one place
 × 2.7 one place
 581
 166
 22.41 two places

5. .2268

 .378 three places
 × .6 one place
 .2268 four places

6. .012

 .0004 four places
 × 30 no places
 .0120 four places

7. .008

 .8 one place
 × .01 two places
 .008 three places

8. 1.1512

 1.439 three places
 × .8 one place
 1.1512 four places

9. .114

$$\begin{array}{r} .076 \text{ three places} \\ \times\ 1.5 \text{ one place} \\ \hline 380 \\ 76 \\ \hline .1140 \text{ four places} \end{array}$$

10. $104.16

$$\begin{array}{r} \$24.80 \text{ two places} \\ \times 4.2 \text{ one place} \\ \hline 4960 \\ 9920 \\ \hline \$104.160 \text{ three places} = \$104.16 \end{array}$$

11. $575

$$\begin{array}{r} \$250 \text{ no places} \\ \times 2.3 \text{ one place} \\ \hline 750 \\ 500 \\ \hline \$575.0 \text{ one place } = \$575 \end{array}$$

12. $1.45

$$\begin{array}{r} \$1.16 \text{ two places} \\ \times 1.25 \text{ two places} \\ \hline 580 \\ 232 \\ 116 \\ \hline \$1.4500 \text{ four places} = \$1.45 \end{array}$$

13. 14.32

$1.432 \times 10 =$
$1.4.32 = 14.32$

14. 5.6

$.056 \times 100 =$
$.05.6 = 5.6$

15. .83

$.083 \times 10 =$
$.0.83 = .83$

16. 19,700

$19.7 \times 1,000 =$
$19.700. = 19,700$

17. 4.5

$$\begin{array}{r} 15 \text{ no places} \\ \times\ .3 \text{ one place} \\ \hline 4.5 \text{ one place} \end{array}$$

18. 357.5

$$\begin{array}{r} 143 \text{ no places} \\ \times\ 2.5 \text{ one place} \\ \hline 715 \\ 286 \\ \hline 357.5 \text{ one place} \end{array}$$

19. 4.2874

$$\begin{array}{r} 12.61 \text{ two places} \\ \times\ .34 \text{ two places} \\ \hline 5044 \\ 3783 \\ \hline 4.2874 \text{ four places} \end{array}$$

20. 7,950

$7.95 \times 1,000 =$
$7.950. = 7,950$

Decimals Exercise 11

1. 6.576

$$\begin{array}{r} 6.576 \\ 8\overline{)52.608} \\ \underline{48} \\ 4\,6 \\ \underline{4\,0} \\ 60 \\ \underline{56} \\ 48 \\ \underline{48} \end{array}$$

2. 84.3

$$\begin{array}{r} 84.3 \\ 9\overline{)758.7} \\ \underline{72} \\ 38 \\ \underline{36} \\ 27 \\ \underline{27} \end{array}$$

3. .75

$$\begin{array}{r} .75 \\ 14\overline{)10.50} \\ \underline{9\,8} \\ 70 \\ \underline{70} \end{array}$$

4. .013

$$\begin{array}{r} .013 \\ 3\overline{).039} \\ \underline{3} \\ 09 \\ \underline{9} \end{array}$$

5. .095

$$\begin{array}{r} .095 \\ 25\overline{)2.375} \\ \underline{2\,25} \\ 125 \\ \underline{125} \end{array}$$

6. .0024

$$\begin{array}{r} .0024 \\ 7\overline{)\,.0168} \\ \underline{14} \\ 28 \\ \underline{28} \end{array}$$

7. .4

$$\begin{array}{r} .4 \\ 24\overline{)9.6} \\ \underline{9\,6} \end{array}$$

8. .04

$$\begin{array}{r} .04 \\ 24\overline{)\,.96} \\ \underline{96} \end{array}$$

9. .15

$$\begin{array}{r} .15 \\ 18\overline{)2.70} \\ \underline{1\,8} \\ 90 \\ \underline{90} \end{array}$$

10. 3.06

$$\begin{array}{r} 3.06 \\ 12\overline{)36.72} \\ \underline{36} \\ 0\,7 \\ \underline{0} \\ 72 \\ \underline{72} \end{array}$$

11. .423 $42.3 \div 100 = .42.3 = .423$

12. .004 $.04 \div 10 = .0.04 = .004$

13. .0195 $19.5 \div 1000 = .019.5 = .0195$

14. .0065 $.65 \div 100 = .00.65 = .0065$

Decimals Exercise 12

1. 42

$$
\begin{array}{r}
42. \\
.2\overline{)8.4} \\
\underline{8} \\
4 \\
\underline{4}
\end{array}
$$

2. 10.5

$$
\begin{array}{r}
10.5 \\
.5\overline{)5.2\,5} \\
\underline{5} \\
25 \\
\underline{25} \\
0
\end{array}
$$

3. 2.6

$$
\begin{array}{r}
2.6 \\
.13\overline{)33.8} \\
\underline{26} \\
78 \\
\underline{78}
\end{array}
$$

4. 15

$$
\begin{array}{r}
15. \\
.028\overline{)420.} \\
\underline{28} \\
140 \\
\underline{140}
\end{array}
$$

5. .27

$$
\begin{array}{r}
.27 \\
1.2\overline{)3.24} \\
\underline{24} \\
84 \\
\underline{84}
\end{array}
$$

6. 2,090

$$
\begin{array}{r}
2090. \\
.04\overline{)83.60.} \\
\underline{8} \\
360 \\
\underline{36} \\
0
\end{array}
$$

7. 4

$$
\begin{array}{r}
4. \\
25.3\overline{)101.2} \\
\underline{101\,2} \\
0
\end{array}
$$

8. 3,020

$$
\begin{array}{r}
3020. \\
.03\overline{)90.60.} \\
\underline{9} \\
06 \\
\underline{06} \\
00
\end{array}
$$

9. 50

$$
\begin{array}{r}
50. \\
.134\overline{)6.700.} \\
\underline{6\,70} \\
00
\end{array}
$$

10. .50

$$
\begin{array}{r}
50. \\
1.28\overline{)64.00.} \\
\underline{64\,0} \\
0
\end{array}
$$

11. 800

$$
\begin{array}{r}
800. \\
.12\overline{)96.00.} \\
\underline{96} \\
0\,00
\end{array}
$$

12. 60

$$
\begin{array}{r}
60. \\
4.35\overline{)261.00.} \\
\underline{261\,0} \\
00
\end{array}
$$

Decimals Exercise 13

1. $ .33	**3.** 20.1	**5.** .0083	**7.** .275	**9.** $.60
2. 6.5	**4.** 8.038	**6.** $164	**8.** 24.0	**10.** .4037

Decimals Exercise 14

1. .03

$$
\begin{array}{r}
.026 \\
15\overline{).40} \\
\underline{30} \\
100 \\
\underline{90} \\
10
\end{array}
$$
to the nearest hundredth is .03

2. .01

$$
\begin{array}{r}
.006 \\
3\overline{).020} \\
\underline{18} \\
2
\end{array}
$$
to the nearest hundredth is .01

3. .1

$$
\begin{array}{r}
.08 \\
6\overline{).50} \\
\underline{48} \\
2
\end{array}
$$
to the nearest tenth is .1

4. .4

$$
\begin{array}{r}
.37 \\
8\overline{)3.00} \\
\underline{2\,4} \\
60 \\
\underline{56} \\
4
\end{array}
$$
to the nearest tenth is .4

Decimals Exercise 15

1. (4) $6.40

$$
\begin{array}{r}
\$6.40 \\
37.5\overline{)\$240.0.00} \\
\underline{225\,0} \\
15\,0\,0 \\
\underline{15\,0\,0} \\
00
\end{array}
$$

2. (1) $1.77

$$
\begin{array}{r}
\$.034 \\
\times \quad 52 \\
\hline
68 \\
1\,70 \\
\hline
1.768
\end{array}
$$
to the nearest hundredth = $1.77

3. (2) .875 m

$$4\overline{)3.500}$$ .875 m

$$\begin{array}{r} 3\,2 \\ \hline 30 \\ 28 \\ \hline 20 \\ 20 \end{array}$$

4. (1) $55.90

$$\begin{array}{r} \$5.20 \\ \times\ \ 7 \\ \hline \$36.40 \end{array}$$

$$\begin{array}{r} \$7.80 \\ \times\ 2.5 \\ \hline 3\,90\,0 \\ 15\,60 \\ \hline \$19.50\,0 \end{array} \qquad \begin{array}{r} \$19.50 \\ +36.40 \\ \hline \$55.90 \end{array}$$

5. (5) .293

$$75\overline{)22.0000}$$ .2933 to the nearest thousandth = .293

$$\begin{array}{r} 15\,0 \\ \hline 7\,00 \\ 6\,75 \\ \hline 250 \\ 225 \\ \hline 250 \\ 225 \\ \hline 25 \end{array}$$

6. (2) $12.60   *first 3 min.* = $3.15
next 9 min. = $1.05

$$\begin{array}{r} \times\ \ 9 \\ \hline \$9.45 \end{array}$$

$$total = \begin{array}{r} \$9.45 \\ +\ 3.15 \\ \hline \$12.60 \end{array}$$

Decimals Exercise 16

1. (2) $56.5 million

$$\begin{array}{r} \$931.0\ million \\ -874.5 \\ \hline \$\ 56.5\ million \end{array}$$

2. (4) 21.5 mi.

$$16\overline{)344.0}$$ 21.5

$$\begin{array}{r} 32 \\ \hline 24 \\ 16 \\ \hline 8\,0 \\ 8\,0 \end{array}$$

3. (3) 94.5 kg

$$\begin{array}{r} 210 \\ \times.45 \\ \hline 10\,50 \\ 84\,0 \\ \hline 94.50 = 94.5\ kg. \end{array}$$

4. (4) 4.45 lb.

$$\begin{array}{r} 2.65\ lb. \\ +1.8 \\ \hline 4.45\ lb. \end{array}$$

5. (3) $0.69

$$4.3\overline{)\$2.9.700}$$ $ 0.690 to the nearest penny = $0.69

$$\begin{array}{r} 2\,5\,8 \\ \hline 3\,90 \\ 3\,87 \\ \hline 30 \end{array}$$

6. (2) $9.70

$$\begin{array}{r} \$1.90 \\ \times\ 2.5 \\ \hline 95\,0 \\ 3\,80 \\ \hline \$4.75\,0 \end{array} \quad \begin{array}{r} \$1.65 \\ \times\ 3 \\ \hline \$4.95 \end{array} \quad \begin{array}{r} \$4.95 \\ +4.75 \\ \hline \$9.70 \end{array}$$

7. (3) .525 million sq. mi.

$$\begin{array}{r} .762\ million\ sq.\ mi. \\ +3.615 \\ \hline 4.377\ million\ sq.\ mi. \\ -3.852 \\ \hline .525\ million\ sq.\ mi. \end{array}$$

8. (1) $18.4 billion

$$\begin{array}{r} 23\ million \\ \times\$800 \\ \hline \$18,400\ million \end{array}$$

Billions are to the left of millions. Move the point in $18,400 million 3 places to the left: $18,400 million = $18.4 billion

9. (4) 49.3

$$18\overline{)887.00}$$ 49.27 to the nearest tenth = 49.3

$$\begin{array}{r} 72 \\ \hline 167 \\ 162 \\ \hline 5\,0 \\ 3\,6 \\ \hline 1\,40 \\ 1\,26 \end{array}$$

10. (1) .9 m

$$\begin{array}{r} 2.00\ m \\ -\ .65 \\ \hline 1.35\ m \\ -\ .45 \\ \hline .90\ m = .9\ m \end{array}$$

Decimals Test

1. (4) hundredths

2. (2) 60.043

3. (2) 408.0015

4. (5) .806

5. (3) 42.995

$$
\begin{array}{r}
36. \\
2.93 \\
.065 \\
+\ 4. \\
\hline
42.995
\end{array}
$$

6. (3) .0737

$$
\begin{array}{r}
.0800 \\
-.0063 \\
\hline
.0737
\end{array}
$$

7. (1) 10.18 million

$$
\begin{array}{r}
6.79\ million \\
+3.39 \\
\hline
10.18\ million
\end{array}
$$

8. (4) 1.25 yd.

$$
\begin{array}{r}
3.00\ yd. \\
-1.75 \\
\hline
1.25\ yd.
\end{array}
$$

9. (1) .016

$$
\begin{array}{r}
3.2 \\
\times .005 \\
\hline
.0160 = .016
\end{array}
$$

10. (1) .027

$$
\begin{array}{r}
.027 \\
15\overline{)\,.405} \\
\underline{30} \\
105 \\
\underline{105}
\end{array}
$$

11. (3) 200

$$
\begin{array}{r}
200. \\
.045\overline{)\,9.000.}
\end{array}
$$

12. (2) $2.92

$$
\begin{array}{r}
\$.68 \\
\times 4.3 \\
\hline
204 \\
272 \\
\hline
\$2.924
\end{array}
$$
to the
nearest penny = $2.92

13. (2) 18.6 mi.

$$
\begin{array}{r}
18.58 \\
12\overline{)\,223.00} \\
\underline{12} \\
103 \\
\underline{96} \\
7\,0 \\
\underline{6\,0} \\
1\,00 \\
\underline{96}
\end{array}
$$
to the nearest
tenth = 18.6

14. (5) $25.78

$$
\begin{array}{r}
100.0 \\
-\ 56.3 \\
\hline
43.7
\end{array}
\qquad
\begin{array}{r}
43.7 \\
\times\ \$.59 \\
\hline
3933 \\
2185 \\
\hline
25.783
\end{array}
$$

Mixed Review

1. (1) 129,000

2. (3) 123.88

3. (2) 40.75

$$
\begin{array}{r}
\$\ 32.40 \\
47.75 \\
42.10 \\
\hline
\$122.25
\end{array}
\qquad
\begin{array}{r}
\$40.75 \\
3\overline{)\,122.25} \\
\underline{12} \\
2\,2 \\
2\,1 \\
\hline
15
\end{array}
$$

4. (2) $314

$$
\begin{array}{r}
\$279 \\
+35 \\
\hline
\$314
\end{array}
$$

5. (5) .0309

6. (2) 61.9°

$$
\begin{array}{ccccc}
53.1 & 55.3 & 57.5 & 59.7 & 61.9 \\
& +2.2 & +2.2 & +2.2 & +2.2
\end{array}
$$

7. (4) $3.00

$$
\begin{array}{r}
.20 \\
\times 10 \\
\hline
\$2.00
\end{array}
\qquad
\begin{array}{r}
\$5.00 \\
-\ 2.00 \\
\hline
\$3.00
\end{array}
$$

8. (5) $471.8 million

$$
\begin{array}{r}
\$\ 26.3\ million \\
445.5 \\
\hline
\$471.8\ million
\end{array}
$$

9. (5) 73

$$
\begin{array}{r}
73 \\
65\overline{)\,4,745} \\
\underline{4\,55} \\
195 \\
\underline{195} \\
0
\end{array}
$$

10. (4) 5 times

$$
\begin{array}{r}
4.9 \\
7.80\overline{)\,38.70.0} \\
\underline{31\,20} \\
7\,50\,0 \\
\underline{7\,02\,0} \\
48\,0
\end{array}
$$
to the nearest
unit is 5

Fractions

WRITING FRACTIONS

A **fraction**, like a decimal, is a way of showing a part of a whole. A quarter is one of four equal parts of a dollar or $\frac{1}{4}$ of a dollar. A can of Coke is one of six equal parts of a six-pack, or $\frac{1}{6}$ of a six-pack. Seven eggs are seven of the twelve equal parts in a dozen, or $\frac{7}{12}$ of a dozen.

A fraction has a top number and a bottom number. These numbers are called the

$$\frac{\text{numerator}}{\text{denominator}}$$
- the number of parts you have
- the number of equal parts the whole is divided into

The 7 in the fraction $\frac{7}{12}$ is the **numerator**. The 7 means there are 7 parts. The 12 in $\frac{7}{12}$ is the **denominator**. The 12 means the whole is divided into twelve equal parts.

With fractions, any number (except 0) can be in the denominator. This differs from decimals, which are based on tenths or multiples of tenths.

EXAMPLE 1: Write a fraction that shows what part of the circle at the right is shaded.

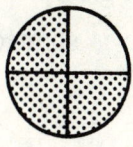

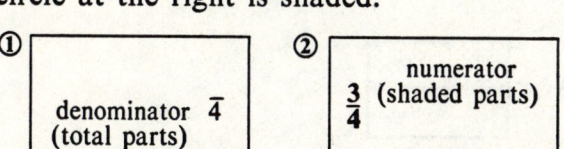

Step 1. Count the total number of parts. The circle is divided into 4 parts. 4 is the denominator.

Step 2. Count the number of shaded parts. 3 parts are shaded. 3 is the numerator.

EXAMPLE 2:   There are 24 students in Joe's math class. 13 of the students are women. Women make up what fraction of the class?

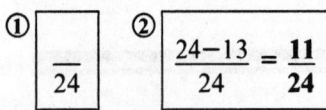

Step 1.   The denominator is the total number of students in the class. 24 is the denominator.

Step 2.   The numerator is the number of women in the class. 13 is the numerator.

EXAMPLE 3:   Based on the last example, men make up what fraction of the class?

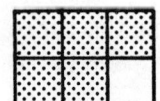 

Step 1.   The denominator again represents the total number of students in the class. 24 is the denominator.

Step 2.   The numerator is the number of men in the class. To find the number of men, subtract the number of women from the total number of students.

There are three uses of fractions with which you should become familiar:

1. As a part of a whole
$\frac{5}{6}$ of this box is shaded.

2. As a part of a group
$\frac{2}{3}$ of this group is shaded.

3. As a representation of division

This box is $\frac{1}{2}$ shaded. The fraction bar is also a division bar and can be read as "is divided by." $\frac{1}{2}$ also means 1 divided by 2 or 1 whole divided into 2 parts.

FRACTIONS EXERCISE 1 ───────────────────

Solve each problem.

1. Write a fraction that tells what part of each circle is shaded.

   a   b   c   d   e

   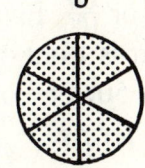 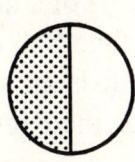  

2. A foot has 12 inches. What fraction of a foot is 5 inches?

3. A week has 7 days. What fraction of a week is the weekend?

4. Since there are 100 pennies in a dollar, how would you write 27¢ as a fraction?

5. Margaret's gross salary is $250 a week. She takes home $209 a week. Her take-home pay is what fraction of her salary?

6. In the previous problem, the amount withheld from Margaret's weekly check is what fraction of her salary?

7. A kilometer has 1,000 meters. What fraction of a kilometer is 333 meters?

8. Brian is running in a 24-mile marathon. In 2 hours, he ran 11 miles. What fraction of the marathon did he have left to run?

9. The Millers make $745 a month. They pay $164 a month for rent. Rent is what fraction of their monthly income?

10. Lois is trying to save $600 as a down payment on a car. So far, she has saved $517. What fraction of the down payment does she still have left to save?

Answers and solutions start on page 126.

REDUCING FRACTIONS

Reducing is one of the most important operations with fractions. Every answer to a fraction problem should be reduced to its lowest terms.

You know that fifty pennies are 50 out of 100 parts of a dollar or $\frac{50}{100}$. Five dimes represent 5 out of 10 parts of a dollar or $\frac{5}{10}$. One-half dollar is 1 out of 2 equal parts of a dollar or $\frac{1}{2}$. In every case, the value is the same: 50¢ or $\frac{1}{2}$ of a dollar. $\frac{1}{2}$ represents all of the values reduced to their lowest terms.

To reduce a fraction, divide the numerator and the denominator by the same number. When a fraction can be reduced no further, it is reduced to its **lowest terms.**

When you reduce a fraction, you have not changed its value. The circles below show that $\frac{3}{4}$ and $\frac{6}{8}$ have the same value ($\frac{6 \div 2}{8 \div 2} = \frac{3}{4}$). Fractions whose values are the same are called **equivalent fractions.**

$\frac{3}{4}$ shaded $\frac{6}{8}$ shaded

Often, more than one number can be divided evenly into both the numerator and the denominator. In the fraction $\frac{24}{36}$, both 24 and 36 can be divided evenly by 2, 3, 4, 6, or 12. Using the largest common number can save you several steps.

$$\frac{24 \div 12}{36 \div 12} = \frac{2}{3}$$

Suppose you had first reduced by 4:

$\frac{24 \div 4}{36 \div 4} = \frac{6}{9}$ This fraction can still be reduced by 3.

$\frac{6 \div 3}{9 \div 3} = \frac{2}{3}$ This is now the same as when $\frac{24}{36}$ was divided by 12.

Rules for Reducing Fractions

1. Divide both the numerator and the denominator by the largest number that will go evenly into both.
2. Check the result to see whether another number will go evenly into both.

The following are a few hints for reducing.

If the numerator and the denominator are both even numbers, you can divide by 2.

EXAMPLE 1: Reduce $\frac{18}{32}$ to lowest terms.

$$\frac{18 \div 2}{32 \div 2} = \frac{9}{16}$$

Step 1. Divide 18 and 32 by 2.
Step 2. Check to see whether this can be reduced further. No other number divides evenly into both 9 and 16. $\frac{9}{16}$ is reduced to lowest terms.

If one number in a fraction ends with a 5 and the other number ends with a 0, you can divide by 5.

EXAMPLE 2: Reduce $\frac{25}{70}$ to lowest terms.

$$\frac{25 \div 5}{70 \div 5} = \frac{5}{14}$$

If both the numerator and the denominator end with a zero, you can divide each of them by 10.

EXAMPLE 3: Reduce $\frac{70}{120}$ to lowest terms.

$$\frac{70 \div 10}{120 \div 10} = \frac{7}{12}$$

You may often have to reduce more than once.

EXAMPLE 4: Reduce $\frac{48}{64}$ to lowest terms.

①$\frac{48 \div 8}{64 \div 8} = \frac{6}{8}$ ②&③$\frac{6 \div 2}{8 \div 2} = \frac{3}{4}$

Step 1. Divide 48 and 64 by 8.
Step 2. Check to see whether another number can go evenly into both 6 and 8. 2 divides evenly into both. Divide 6 and 8 by 2.
Step 3. No other number divides evenly into both 3 and 4. $\frac{3}{4}$ is reduced to lowest terms.

The best tool for reducing is the multiplication table. The better you know the multiplication table, the easier you will find it to reduce fractions.

FRACTIONS EXERCISE 2

For problems 1 to 15, reduce each fraction to its lowest terms.

1. $\frac{25}{35}$ 　　　 2. $\frac{8}{24}$ 　　　 3. $\frac{12}{27}$

4. $\frac{50}{80}$ 8. $\frac{48}{60}$ 12. $\frac{15}{50}$

5. $\frac{30}{200}$ 9. $\frac{50}{250}$ 13. $\frac{18}{36}$

6. $\frac{90}{96}$ 10. $\frac{55}{77}$ 14. $\frac{81}{90}$

7. $\frac{36}{44}$ 11. $\frac{80}{160}$ 15. $\frac{45}{70}$

Solve problems 16 through 20. Remember that a fraction problem is not complete until it has been reduced to its lowest terms.

16. On a test with 50 problems, John got 45 of the problems right. What fraction of the problems did he get right?

17. There are 12 inches in a foot. What fraction of a foot is 8 inches?

18. Last season, the Consolidated Electric softball team played 21 games and won 15 of them. What fraction of the games did they lose?

19. There are 1,000 meters in a kilometer. What fraction of a kilometer is 650 meters?

20. In Jeanne's English class, there are eight women and six men. Women make up what fraction of the class?

Answers and solutions start on page 126.

WRITING MIXED NUMBERS

$1.25 is one whole dollar and one quarter. In fractional form, $1.25 can be written as $1\frac{1}{4}$. $1\frac{1}{4}$ is a **mixed number.** A mixed number combines a whole number and a fraction.

EXAMPLE: Write a mixed number that tells how many circles are shaded.

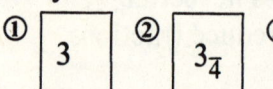

① $\boxed{3}$ ② $\boxed{3\frac{}{4}}$ ③ $\boxed{3\frac{1}{4}}$

Step 1. Count the number of completely shaded circles. 3 circles are completely shaded. 3 is the whole number.

Step 2. Count the total number of parts in the partially shaded circle. The circle is divided into 4 parts. 4 is the denominator.

Step 3. Count the number of shaded parts in the partially shaded circle. 1 part is shaded. 1 is the numerator.

FRACTIONS EXERCISE 3

Write a mixed number that tells how many figures and parts of figures are shaded.

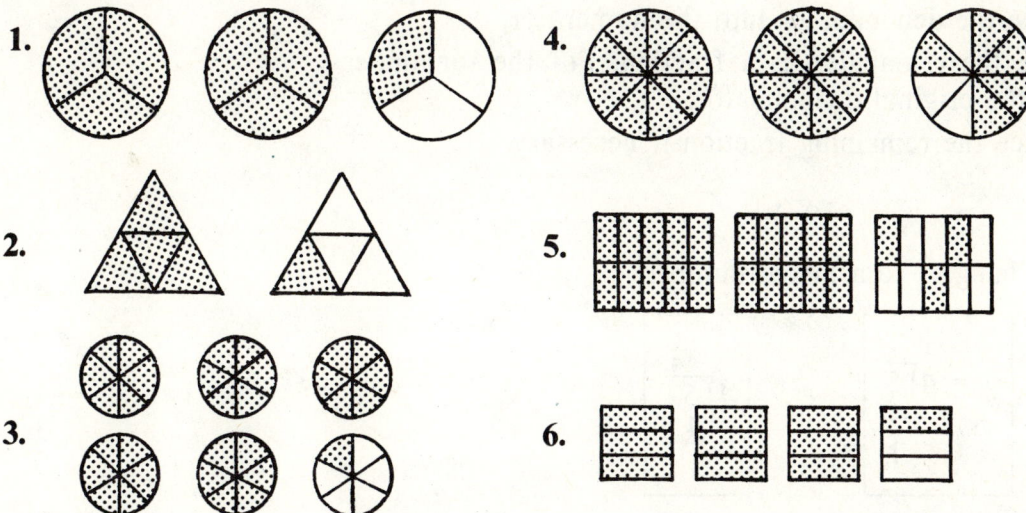

1.

2.

3.

4.

5.

6.

Answers and solutions start on page 126.

CHANGING IMPROPER FRACTIONS TO WHOLE OR MIXED NUMBERS

You have already seen $1.25 written as the mixed number $1\frac{1}{4}$. Think of $1.25 as five quarters. In fractional form, five quarters can be written as $\frac{5}{4}$. $\frac{5}{4}$ is called an **improper fraction.** In an improper fraction, the numerator is as big or bigger than the denominator. The value of an improper fraction is one whole or more.

In later sections, you will be working with a particular type of improper fraction—one in which the numerator and denominator are the same. For example, $\frac{4}{4}$ or $\frac{27}{27}$. Fractions whose numerator and denominator are the same are equal to 1.

The illustration to the right should make this clear.

$\frac{4}{4}$ of this box is shaded. The whole box is shaded. Therefore, $\frac{4}{4} = 1$. You can also see this by dividing the denominator into the numerator, $4 \div 4 = 1$.

In a **proper fraction,** the numerator is always <u>less</u> than the denominator. Until now, the fractions you worked with were all proper fractions. The value of a proper fraction is always less than one whole.

To change an improper fraction to a mixed number or a whole, divide. Remember that a fraction bar indicates that you can divide the denominator into the numerator.

**Rules for Changing Improper Fractions
to Whole or Mixed Numbers**

1. Divide the denominator into the numerator.
2. Write the remainder as a fraction. Put the remainder over the original denominator.
3. Reduce the remaining fraction if necessary.

EXAMPLE 1: Change $\frac{5}{4}$ to a mixed number.

$$\textcircled{1}\quad \frac{5}{4} = 4\overline{)5} \quad \begin{array}{r} 1 \\ \hline 5 \\ 4 \\ \hline 1 \end{array} \qquad \textcircled{2}\, \&\, \textcircled{3}\quad 4\overline{)5} \quad \begin{array}{r} 1\frac{1}{4} \\ \hline 5 \\ 4 \\ \hline 1 \end{array}$$

Step 1. Divide the denominator into the numerator.
Step 2. Write the remainder over the denominator.
Step 3. Try to reduce. $1\frac{1}{4}$ is already reduced to lowest terms.

EXAMPLE 2: Change $\frac{20}{6}$ to a mixed number.

$$\textcircled{1}\quad \frac{20}{6} = 6\overline{)20} \quad \begin{array}{r} 3 \\ \hline 20 \\ 18 \\ \hline 2 \end{array} \qquad \textcircled{2}\quad 6\overline{)20} \quad \begin{array}{r} 3\frac{2}{6} \\ \hline 20 \\ 18 \\ \hline 2 \end{array} \qquad \textcircled{3}\quad 3\frac{2}{6} = 3\frac{1}{3}$$

Step 1. Divide the denominator into the numerator.
Step 2. Write the remainder over the denominator.
Step 3. Try to reduce $3\frac{2}{6}$. $\frac{2}{6}$ can be reduced by 2.

If the solution to a fraction problem is an improper fraction, use this method to change your answer to a mixed number or a whole.

FRACTIONS EXERCISE 4 ————————————————————————

Change each of the following to whole or mixed numbers. Be sure to reduce any remaining fractions.

1. $\frac{15}{9} =$ 6. $\frac{28}{16} =$ 11. $\frac{21}{3} =$

2. $\frac{36}{8} =$ 7. $\frac{55}{30} =$ 12. $\frac{50}{12} =$

3. $\frac{13}{5} =$ 8. $\frac{15}{15} =$ 13. $\frac{82}{82} =$

4. $\frac{31}{6} =$ 9. $\frac{24}{18} =$ 14. $\frac{21}{6} =$

5. $\frac{6}{2} =$ 10. $\frac{7}{4} =$ 15. $\frac{44}{11} =$

Answers and solutions start on page 126.

CHANGING DECIMALS AND FRACTIONS

There is a relationship between decimals and fractions:

1. Both are used to represent parts.
2. They can be used to represent the same value:
   $\frac{1}{4}$ dollar = $.25 or .25 of a dollar.

Changing Decimals to Fractions

The decimal system represents parts based on tenths or multiples of tenths. .3 is three tenths and .017 is seventeen thousandths. You can write these as fractions with decimal place values as a denominator:

① $.3 = \frac{3}{10}$ ② $.017 = \frac{17}{1,000}$

EXAMPLE 1: Change .44 to a fraction.

① $.44 = \frac{44}{100}$ ② $\frac{44}{100} = \frac{11}{25}$

Step 1. Write a fraction using the original number (44) as the numerator and the decimal value (hundredths) as the denominator.

Step 2. Reduce, if possible.

EXAMPLE 2: Change .025 to a fraction.

① $.025 = \frac{25}{1,000}$ ② $\frac{25}{1,000} = \frac{1}{40}$

Step 1. Write the decimal as a fraction. If a zero is only used as a place holder, it can be dropped.

Step 2. Reduce, if necessary.

EXAMPLE 3: Change 4.06 to a mixed number.

$4.06 = 4\frac{6}{100} = 4\frac{3}{50}$

FRACTIONS EXERCISE 5

Change each of the following to fractions or mixed numbers and reduce to lowest terms.

1. .6 =

2. .5 =

3. .45 =

4. .80 =

5. .125 =

6. .065 =

7. 4.15 =

8. .96 =

9. .248 =

10. .0002 =

11. 15.004 =

12. 3.34 =

Answers and solutions start on page 127.

Changing Fractions to Decimals

How do you change $\frac{1}{4}$ to .25? By using the fraction bar as a division bar. $\frac{1}{4}$ also means $1 \div 4$ or $4\overline{)1}$.

Sometimes these work out evenly.

EXAMPLE 1: Change $\frac{9}{20}$ to a decimal.

$$\begin{array}{r} .45 = \mathbf{.45} \\ 20\overline{)9.00} \\ \underline{8\ 0} \\ 1\ 00 \\ \underline{1\ 00} \end{array}$$

Notice that you had to add two 0's to the 9 to carry out the division.

EXAMPLE 2: Change $\frac{3}{8}$ to a decimal.

$$8\overline{)3.00}^{.37\frac{4}{8}} = .37\frac{1}{2}$$

$$\frac{2\,4}{60}$$

$$\frac{56}{4}$$

There may be a remainder in division. Put this over the divisor to make a fraction.

EXAMPLE 3: Change $2\frac{3}{7}$ to a mixed decimal.

$$7\overline{)3.00}^{.42\frac{6}{7}} = .42\frac{6}{7} \qquad 2 + .42\frac{6}{7} = \textbf{2.42}\frac{6}{7}$$

$$\frac{2\,8}{20}$$

$$\frac{14}{6}$$

Notice that you added the decimal value of $\frac{3}{7}$ to the whole number 2.

FRACTIONS EXERCISE 6

Change each of the following to decimals or mixed decimals.

1.	$\frac{3}{4} =$	**5.**	$2\frac{1}{2} =$	**9.**	$\frac{1}{12} =$
2.	$\frac{1}{3} =$	**6.**	$\frac{1}{20} =$	**10.**	$5\frac{2}{3} =$
3.	$\frac{7}{10} =$	**7.**	$\frac{5}{6} =$	**11.**	$\frac{3}{25} =$
4.	$\frac{5}{8} =$	**8.**	$1\frac{4}{9} =$	**12.**	$10\frac{1}{6} =$

Answers and solutions start on page 127.

ADDING LIKE FRACTIONS AND MIXED NUMBERS

Like fractions are fractions that have the same denominators. $\frac{1}{5}$ and $\frac{2}{5}$ are like fractions. $\frac{2}{3}$ and $\frac{3}{4}$ are unlike fractions. When you add like fractions, the sum has the denominator of the fractions you added.

Rules for Adding Like Fractions

1. Add the numerator of each fraction and put the total over the denominator.
2. If the answer is an improper fraction, change that sum to a mixed number. Reduce, if necessary.

EXAMPLE 1: Add $\frac{1}{7} + \frac{2}{7}$.

①
$$\frac{1}{7}$$
$$+ \frac{2}{7}$$
$$\overline{7}$$

②
$$\frac{1}{7}$$
$$+ \frac{2}{7}$$
$$\frac{3}{7}$$

Step 1. Since the denominators are the same, bring 7 down as the denominator of the answer.

Step 2. Add the numerators: $1 + 2 = 3$. This is the numerator of your answer. $\frac{3}{7}$ is already reduced to lowest terms.

EXAMPLE 2: Find the sum of $\frac{7}{8}$ and $\frac{3}{8}$.

① $\frac{7}{8} + \frac{3}{8} = \frac{10}{8}$ ② $\frac{10}{8} = 1\frac{2}{8} = 1\frac{1}{4}$

Step 1. Add the numerators and put the total over the denominator.

Step 2. Change the improper fraction to a mixed number. Reduce the answer.

Note: The last two steps can be reversed. You can reduce $\frac{10}{8}$ by 2. $\frac{10 \div 2}{8 \div 2} = \frac{5}{4}$. Then change $\frac{5}{4}$ to a mixed number. $\frac{5}{4} = \mathbf{1\frac{1}{4}}$.

When adding mixed numbers, keep the fractions and the whole numbers separate until the end. First add the fractions and then add the whole numbers.

EXAMPLE 3: Add $7\frac{5}{12}$ and $8\frac{11}{12}$.

①
$$7\frac{5}{12}$$
$$+ 8\frac{11}{12}$$
$$15\frac{16}{12}$$

② $\frac{16}{12} = 1\frac{4}{12}$

③ $15 + 1\frac{4}{12} = 16\frac{4}{12} = \mathbf{16\frac{1}{3}}$

Step 1. Add the numerators and put the total over the denominator. Add the whole numbers.

Step 2. Change the improper fraction to a mixed number.

Step 3. Add the mixed number to the whole number and reduce the fraction.

FRACTIONS EXERCISE 7 ——————————————

Solve each problem.

1. $\begin{array}{r} \frac{2}{7} \\ +\frac{3}{7} \\ \hline \end{array}$

6. $\begin{array}{r} 9\frac{3}{10} \\ +4\frac{1}{10} \\ \hline \end{array}$

11. $\begin{array}{r} \frac{3}{4} \\ \frac{1}{4} \\ +\frac{3}{4} \\ \hline \end{array}$

2. $\begin{array}{r} \frac{2}{9} \\ +\frac{4}{9} \\ \hline \end{array}$

7. $\begin{array}{r} 10\frac{2}{3} \\ + 2\frac{1}{3} \\ \hline \end{array}$

12. $\begin{array}{r} 6\frac{3}{8} \\ 3\frac{1}{8} \\ +7\frac{5}{8} \\ \hline \end{array}$

3. $\begin{array}{r} \frac{8}{15} \\ +\frac{11}{15} \\ \hline \end{array}$

8. $\begin{array}{r} 6\frac{5}{9} \\ +7\frac{8}{9} \\ \hline \end{array}$

13. $2\frac{7}{12} + 4\frac{11}{12} + 9\frac{5}{12} =$

4. $\frac{5}{8} + \frac{5}{8} =$

9. $1\frac{7}{8} + 20\frac{3}{8} =$

14. $5\frac{4}{9} + 4\frac{5}{9} + 3\frac{8}{9} =$

5. $4\frac{1}{3} + 8\frac{1}{3} =$

10. $\frac{1}{16} + \frac{5}{16} + \frac{7}{16} =$

15. $13\frac{11}{15} + 16\frac{14}{15} + 9\frac{8}{15} =$

Answers and solutions start on page 128.

SUBTRACTING LIKE FRACTIONS WITHOUT BORROWING (RENAMING)

When like fractions are added, the sum has the same denominator as the fractions. Subtraction is similar. The difference between $\frac{4}{5}$ and $\frac{3}{5}$ is $\frac{1}{5}$.

With mixed numbers, subtract the fractions and the whole numbers separately.

Rules for Subtracting Like Fractions and Mixed Numbers, Without Borrowing

1. Subtract the numerators and put the difference over the denominator.
2. Subtract the whole numbers. Reduce, if necessary.

EXAMPLE: Subtract $3\frac{9}{16}$ from $10\frac{15}{16}$.   ① & ②

$$\begin{array}{r} 10\frac{15}{16} \\ -\ 3\frac{9}{16} \\ \hline 7\frac{6}{16} = 7\frac{3}{8} \end{array}$$

Step 1. Subtract the numerators. Put the difference over the denominator.

Step 2. Subtract the whole numbers. Reduce.

FRACTIONS EXERCISE 8

Solve each problem.

1. $\begin{array}{r} \frac{5}{9} \\ -\ \frac{2}{9} \\ \hline \end{array}$

5. $\begin{array}{r} \frac{23}{25} \\ -\ \frac{13}{25} \\ \hline \end{array}$

9. $\begin{array}{r} \frac{25}{36} \\ -\ \frac{17}{36} \\ \hline \end{array}$

2. $\begin{array}{r} 8\frac{5}{6} \\ -\ 2\frac{1}{6} \\ \hline \end{array}$

6. $\begin{array}{r} 10\frac{4}{5} \\ -\ 8\frac{1}{5} \\ \hline \end{array}$

10. $\begin{array}{r} 8\frac{5}{6} \\ -\ 1\frac{1}{6} \\ \hline \end{array}$

3. $\begin{array}{r} 12\frac{20}{27} \\ -\ 4\frac{11}{27} \\ \hline \end{array}$

7. $\begin{array}{r} \frac{19}{20} \\ -\ \frac{13}{20} \\ \hline \end{array}$

11. $\begin{array}{r} \frac{34}{35} \\ -\ \frac{13}{35} \\ \hline \end{array}$

4. $\begin{array}{r} 4\frac{7}{8} \\ -\ 2\frac{3}{8} \\ \hline \end{array}$

8. $\begin{array}{r} 6\frac{13}{15} \\ -\ 2\frac{4}{15} \\ \hline \end{array}$

12. $\begin{array}{r} 4\frac{17}{18} \\ -\ 4\frac{5}{18} \\ \hline \end{array}$

Answers and solutions start on page 128.

SUBTRACTING MIXED NUMBERS FROM WHOLE NUMBERS

Suppose you pay for something that costs $8.25 with a $10 bill. Your change should be $1.75. As a fraction problem this becomes:

$$\begin{array}{r} 10 \\ -\ 8\frac{1}{4} \\ \hline \end{array}$$

There is no fraction to subtract $\frac{1}{4}$ from. We must borrow 1 from the 10 and put it into fraction form. (Some of you may call this process "renaming.")

As in subtracting whole numbers, you can borrow with fractions. You borrow 1 from the whole number and rewrite it as a fraction with the same denominator as the fraction you are subtracting.

For instance, when you subtract $\frac{1}{4}$ from 10, borrow 1 from the 10 and rewrite 10 as $9\frac{4}{4}$. This has the same value as 10, but it is rewritten in a form that you can use to finish the subtraction.

Rules for Subtracting Mixed Numbers from Whole Numbers

1. Borrow 1 from the whole number at the top, and rewrite the number, making the 1 an improper fraction with the same denominator as the fraction you are subtracting.
2. Subtract the numerator and put the difference over the denominator. Subtract the whole numbers. Reduce if necessary.

EXAMPLE: Take $4\frac{5}{6}$ from 8.

①
$$8 = 7\frac{6}{6}$$
$$-4\frac{5}{6} = 4\frac{5}{6}$$

②
$$7\frac{6}{6}$$
$$-4\frac{5}{6}$$
$$\overline{3\frac{1}{6}}$$

Step 1. Borrow 1 from 8 and rewrite the 1 with the common denominator 6: $1 = \frac{6}{6}$. The top of the problem becomes: $8 = 7\frac{6}{6}$.

Step 2. Subtract the numerators: Put the difference over the denominator. Subtract the whole numbers.

FRACTIONS EXERCISE 9

Solve each problem.

1.
$$\begin{array}{r} 5 \\ -2\frac{8}{9} \\ \hline \end{array}$$

4.
$$\begin{array}{r} 8 \\ -5\frac{7}{12} \\ \hline \end{array}$$

7.
$$\begin{array}{r} 12 \\ -2\frac{5}{16} \\ \hline \end{array}$$

2.
$$\begin{array}{r} 9 \\ -1\frac{2}{3} \\ \hline \end{array}$$

5.
$$\begin{array}{r} 7 \\ -6\frac{1}{2} \\ \hline \end{array}$$

8.
$$\begin{array}{r} 43 \\ -31\frac{9}{10} \\ \hline \end{array}$$

3. $11 - 3\frac{5}{8} =$

6. $9 - 1\frac{1}{4} =$

9. $27 - 14\frac{9}{20} =$

Answers and solutions start on page 129.

BORROWING WITH LIKE DENOMINATORS

In mixed number subtraction problems, the fraction on the top is sometimes too small to subtract from.

Rules for Subtracting Mixed Numbers with Like Denominators: Borrowing

1. Borrow 1 from the top whole number. Rewrite the 1 with the same denominator as the fraction on the bottom, and add it to the top fraction.
2. Subtract the numerators and put the difference over the denominator. Subtract the whole numbers. Reduce.

Borrowing with fractions involves several steps. Be sure you understand the next example well before you try the exercise.

EXAMPLE: Take $6\frac{7}{8}$ from $12\frac{5}{8}$.

①
$$12\frac{5}{8} = 11\frac{8}{8} + \frac{5}{8} = 11\frac{13}{8}$$
$$- \ 6\frac{7}{8} = \qquad\qquad - \ 6\frac{7}{8}$$

②
$$11\frac{13}{8}$$
$$- \ 6\frac{7}{8}$$
$$\overline{\ 5\frac{6}{8} = \mathbf{5\frac{3}{4}}}$$

Step 1. Borrow 1 from 12. Rewrite the 1 as $\frac{8}{8}$ since 8 is the common denominator. Add $11\frac{8}{8} + \frac{5}{8}$.

Step 2. Subtract fractions and whole numbers. Reduce.

FRACTIONS EXERCISE 10

Solve each problem.

1. $\begin{array}{r} 5\frac{1}{8} \\ -2\frac{5}{8} \end{array}$

5. $\begin{array}{r} 8\frac{2}{5} \\ -1\frac{4}{5} \end{array}$

9. $\begin{array}{r} 13\frac{1}{4} \\ -12\frac{3}{4} \end{array}$

2. $\begin{array}{r} 28\frac{5}{9} \\ -13\frac{7}{9} \end{array}$

6. $\begin{array}{r} 112\frac{3}{10} \\ - \ 24\frac{9}{10} \end{array}$

10. $\begin{array}{r} 15\frac{7}{24} \\ -13\frac{19}{24} \end{array}$

3. $24\frac{11}{16} - 16\frac{15}{16} =$

7. $11\frac{7}{12} - 8\frac{11}{12} =$

11. $20\frac{5}{36} - 7\frac{25}{36} =$

4. $9\frac{1}{3} - 7\frac{2}{3} =$

8. $5\frac{2}{7} - 2\frac{6}{7} =$

12. $7\frac{5}{18} - 6\frac{17}{18} =$

Answers and solutions start on page 129.

FINDING COMMON DENOMINATORS

So far, you have added and subtracted like fractions. Often, the fractions in a problem are not like fractions. To add or subtract unlike fractions, you must first find a common denominator. A **common denominator** is a number each denominator in a problem will divide into evenly. The lowest number that each denominator will divide into evenly is called the **lowest common denominator.**

For the fractions $\frac{1}{2}$ and $\frac{2}{3}$, the lowest common denominator is 6. 6 is the lowest number both 2 <u>and</u> 3 will divide into evenly.

For the fractions $\frac{5}{6}$ and $\frac{3}{4}$, the lowest common denominator is 12. 12 is the lowest number both 6 <u>and</u> 4 divide into evenly.

Here are two tips for finding the lowest common denominator. With practice, you will find these tips easy to apply.

Tips on Finding a Lowest Common Denominator

1. First test the largest denominator as the common denominator.

2. If that doesn't work, go through the multiplication table of the largest denominator.

The examples below illustrate each method.

EXAMPLE 1: Find the lowest common denominator for $\frac{3}{4}$ and $\frac{5}{12}$.

   Solution: Look at the larger denominator, 12. 4 divides evenly into 12. Therefore, **12** is the lowest common denominator.

If this doesn't work or you have more than two fractions, use the second tip.

EXAMPLE 2: Find the lowest common denominator for $\frac{7}{9}$, $\frac{1}{12}$, and $\frac{5}{6}$.

   Solution: Although 6 divides evenly into 12, 9 does not. Go through the multiplication table of the largest denominator, 12. Look for a number that both 9 and 6 divide into evenly.

        $1 \times 12 = 12$, not evenly divisible by 9.
        $2 \times 12 = 24$, not evenly divisible by 9.
        $3 \times 12 = 36$, evenly divisible by both 9 and 6.

       36 is the lowest common denominator of 9, 12, and 6.

FRACTIONS EXERCISE 11 ━━━━━━━━━━

Find the lowest common denominator for each set of fractions. You do not need to add or subtract.

1. a) $\frac{2}{5}$ and $\frac{7}{10}$ b) $\frac{5}{6}$ and $\frac{3}{4}$ c) $\frac{5}{16}$ and $\frac{2}{3}$ d) $\frac{4}{9}$ and $\frac{5}{36}$

2. a) $\frac{3}{4}$ and $\frac{5}{9}$ b) $\frac{2}{5}$ and $\frac{5}{7}$ c) $\frac{1}{2}$ and $\frac{1}{3}$ d) $\frac{19}{100}$ and $\frac{4}{25}$

3. a) $\frac{1}{2}$ and $\frac{5}{9}$ b) $\frac{5}{8}$ and $\frac{3}{20}$ c) $\frac{2}{9}$ and $\frac{5}{6}$ d) $\frac{1}{8}$ and $\frac{4}{5}$

4. a) $\frac{1}{12}$ and $\frac{5}{16}$ b) $\frac{7}{12}$ and $\frac{5}{8}$ c) $\frac{1}{4}$ and $\frac{7}{10}$ d) $\frac{3}{10}$ and $\frac{9}{50}$

5. a) $\frac{1}{2}$, $\frac{3}{5}$, and $\frac{3}{4}$ b) $\frac{5}{6}$, $\frac{3}{4}$, and $\frac{2}{3}$ c) $\frac{3}{8}$, $\frac{5}{6}$, and $\frac{7}{12}$

6. a) $\frac{4}{9}$, $\frac{5}{6}$, and $\frac{1}{2}$ b) $\frac{7}{8}$, $\frac{11}{20}$, and $\frac{1}{2}$ c) $\frac{8}{9}$, $\frac{7}{12}$, and $\frac{2}{3}$

Answers and solutions start on page 130.

RAISING FRACTIONS TO HIGHER TERMS

When the fractions in an addition or subtraction problem are unlike, first find a common denominator. This process is called "raising a fraction to higher terms."

Suppose you want to change three quarters to an equivalent number of pennies. You know that three quarters equal 75¢. With fractions, you can solve this problem:

$$\frac{3}{4} = \frac{?}{100} \qquad \frac{3 \times 25}{4 \times 25} = \frac{75}{100}$$

The new denominator, 100, is how many times larger than the old denominator, 4? It is 25 times larger. Make the new numerator 25 times larger than the old numerator.

Earlier, you saw that to reduce a fraction, you must divide both the numerator and the denominator by the same number. To raise a fraction to higher terms, <u>multiply</u> both the numerator and the denominator by the same number.

Rules for Raising a Fraction to Higher Terms
1. Divide the old denominator into the new one to find the multiplier.
2. Multiply the result by the old numerator.

EXAMPLE: Raise $\frac{2}{3}$ to 15ths.

① $3\overline{)15}$ with quotient 5

② $\dfrac{2 \times 5}{3 \times 5} = \dfrac{10}{15}$

Step 1:   Divide the old denominator, 3, into the new denominator, 15. $15 \div 3 = 5$.

Step 2:   Multiply the numerator by that number.

FRACTIONS EXERCISE 12

Raise each fraction to higher terms. Use the new denominator shown in each problem.

1. $\frac{3}{5} = \frac{?}{20}$

2. $\frac{5}{12} = \frac{?}{24}$

3. $\frac{7}{8} = \frac{?}{40}$

4. $\frac{3}{4} = \frac{?}{16}$

5. $\frac{4}{9} = \frac{?}{27}$

6. $\frac{5}{11} = \frac{?}{44}$

7. $\frac{13}{20} = \frac{?}{40}$

8. $\frac{7}{10} = \frac{?}{60}$

9. $\frac{13}{20} = \frac{?}{100}$

10. $\frac{21}{25} = \frac{?}{75}$

11. $\frac{3}{4} = \frac{?}{100}$

12. $\frac{5}{9} = \frac{?}{36}$

13. $\frac{1}{8} = \frac{?}{24}$

14. $\frac{4}{5} = \frac{?}{45}$

15. $\frac{1}{3} = \frac{?}{36}$

Answers and solutions start on page 130.

ADDING UNLIKE FRACTIONS AND MIXED NUMBERS

The fractions you add must have the same denominators. When the fractions in an addition problem are unlike, find a common denominator before adding. If necessary, raise a fraction to higher terms. Follow the rules for adding like fractions and mixed numbers.

Rules for Adding Unlike Fractions and Mixed Numbers

1. Find the lowest common denominator. Raise the fractions to higher terms.
2. Add the numerators of the fractions and put the total over the common denominator.
3. If necessary, simplify the answer and reduce.

EXAMPLE 1: Add $\frac{2}{3} + \frac{7}{9}$.

①

$$\frac{2}{3} = \frac{6}{9}$$
$$+\frac{7}{9} = \frac{7}{9}$$

② & ③

$$\frac{2}{3} = \frac{6}{9}$$
$$+\frac{7}{9} = \frac{7}{9}$$
$$\frac{13}{9} = 1\frac{4}{9}$$

Step 1. Find the common denominator for 3 and 9. Since 3 divides evenly into 9, 9 is the common denominator. Change $\frac{2}{3}$ to $\frac{6}{9}$.

Step 2. Add the numerators and put the total over the common denominator.

Step 3. Simplify the improper fraction.

EXAMPLE 2: Find the sum of $7\frac{5}{8}$, $4\frac{2}{3}$, and $3\frac{5}{6}$.

① & ②

$$7\frac{5}{8} = 7\,\frac{5 \times 3}{8 \times 3} = 7\frac{15}{24}$$
$$4\frac{2}{3} = 4\,\frac{2 \times 8}{3 \times 8} = 4\frac{16}{24}$$
$$+3\frac{5}{6} = 3\,\frac{5 \times 4}{6 \times 4} = 3\frac{20}{24}$$
$$14\frac{51}{24}$$

③

$$14\frac{51}{24} = 14 + 2\frac{3}{24} =$$
$$16\frac{3}{24} = 16\frac{1}{8}$$

Step 1. Find the common denominator for 8, 3, and 6. Go through the multiplication tables of the largest denominator, 8:

$8 \times 1 = 8$, not divisible by 3 and 6
$8 \times 2 = 16$, not divisible by 3 and 6
$8 \times 3 = 24$, divisible also by 3 and 6

Raise each fraction to higher terms.

Step 2. Add the fractions and whole numbers.

Step 3. Change the improper fraction to a mixed number. Add the whole number and mixed number. Reduce the fraction.

FRACTIONS EXERCISE 13

Solve each problem.

1.
$$\frac{5}{6}$$
$$+\frac{7}{12}$$

2.
$$\frac{3}{8}$$
$$+\frac{5}{24}$$

3. $\frac{2}{3} + \frac{3}{4} =$

4. $9\frac{2}{3} + 4\frac{1}{2} =$

5.
$$10\frac{1}{8}$$
$$+ 3\frac{1}{3}$$

6.
$$2\frac{5}{6}$$
$$+ 4\frac{1}{2}$$

7. $\frac{2}{3} + \frac{7}{12} + \frac{5}{6} =$

9. $\begin{array}{r} \frac{5}{8} \\ + \frac{5}{6} \\ + \frac{3}{4} \\ \hline \end{array}$

11. $9\frac{1}{2} + 8\frac{3}{4} + 7\frac{3}{10} =$

8. $\frac{1}{2} + \frac{1}{4} + \frac{3}{5} =$

10. $4\frac{1}{3} + 1\frac{5}{6} + 6\frac{5}{8} =$

12. $4\frac{2}{25} + 3\frac{19}{100} + 2\frac{11}{50} =$

Answers and solutions start on page 130.

SUBTRACTING WITH UNLIKE DENOMINATORS

When subtracting fractions, you must also find a common denominator before finishing a problem.

Rules for Subtracting Fractions

1. Find the lowest common denominator and raise fractions to higher terms with the common denominator.
2. Borrow if necessary.
3. Subtract both the fractions and whole numbers. Reduce if necessary.

EXAMPLE: Subtract $9\frac{3}{4}$ from $15\frac{1}{3}$.

①
$$\begin{array}{r} 15\frac{1}{3} = 15\frac{4}{12} \\ - 9\frac{3}{4} = 9\frac{9}{12} \\ \hline \end{array}$$

② & ③
$$\begin{array}{r} 15\frac{4}{12} = 14\frac{16}{12} \\ - 9\frac{9}{12} = 9\frac{9}{12} \\ \hline 5\frac{7}{12} \end{array}$$

Step 1. Find the lowest common denominator, 12, and raise the fractions to higher terms.

Step 2. Borrow 1 from $15 = \frac{12}{12}$, and add: $14\frac{4}{12} + \frac{12}{12}$.

Step 3. Subtract whole numbers and fractions.

Work carefully through the next exercise. In some cases, you will need to borrow; in others you will not.

FRACTIONS EXERCISE 14

Solve each problem.

1. $\frac{7}{10}$
    $-\frac{1}{2}$

2. $2\frac{2}{3}$
    $-1\frac{3}{4}$

3. $\frac{5}{8} - \frac{3}{5} =$

4. $8\frac{1}{5} - 3\frac{3}{4} =$

5. $11\frac{3}{5}$
    $- 9\frac{1}{2}$

6. $9\frac{1}{2}$
    $-5\frac{2}{3}$

7. $16\frac{1}{4} - 7\frac{5}{8} =$

8. $4\frac{3}{10} - \frac{2}{3} =$

9. $7\frac{11}{12}$
    $-3\frac{5}{8}$

10. $8\frac{7}{20}$
    $-2\frac{3}{4}$

11. $9\frac{1}{6} - 3\frac{8}{9} =$

12. $4\frac{1}{2} - 2\frac{5}{16} =$

Answers and solutions start on page 131.

ADDING AND SUBTRACTING FRACTIONS: WORD PROBLEMS

The next problems will give you a chance to apply your skills to addition and subtraction of fractions. These are the kinds of problems you will find on the GED. First, read each problem carefully to decide whether to add or to subtract. Then, solve each problem and choose the correct answer from the five choices.

FRACTIONS EXERCISE 15

1. Carmen works part time at the Spring Street School. Last week, she worked $3\frac{1}{2}$ hours on Monday, $2\frac{3}{4}$ hours on Wednesday, and $4\frac{3}{4}$ hours on Friday. What was the total number of hours that Carmen worked at the school last week?

    (1) 9 hours
    (2) $10\frac{1}{4}$ hours
    (3) $10\frac{3}{4}$ hours
    (4) 11 hours
    (5) $11\frac{1}{2}$ hours

2. Ray had a piece of electric cable $4\frac{1}{2}$ yards long. He cut off a piece $3\frac{2}{3}$ yards long. How much cable was left?

    (1) $1\frac{1}{2}$ yards
    (2) $1\frac{1}{6}$ yards
    (3) $\frac{5}{6}$ yard
    (4) $\frac{3}{4}$ yard
    (5) $\frac{2}{3}$ yard

3. Mrs. Vega bought $2\frac{1}{2}$ pounds of chicken, $2\frac{5}{8}$ pounds of ground chuck, and $1\frac{3}{4}$ pounds of sausage. Find the pound weight of the food she bought.

(1) $6\frac{7}{8}$ pounds

(2) $6\frac{3}{4}$ pounds

(3) $7\frac{1}{8}$ pounds

(4) $5\frac{3}{4}$ pounds

(5) $5\frac{9}{16}$ pounds

4. On Monday, the cook at the Corner Kitchen Restaurant had 100 pounds of flour. He used $16\frac{3}{4}$ pounds of the flour on Tuesday and $20\frac{1}{2}$ pounds on Thursday. How many pounds of flour were left when the cook finished on Thursday?

(1) $65\frac{1}{4}$ pounds

(2) $62\frac{3}{4}$ pounds

(3) $64\frac{3}{4}$ pounds

(4) $64\frac{1}{2}$ pounds

(5) $63\frac{1}{4}$ pounds

5. The distance from Betty's house to her job is $12\frac{1}{2}$ miles. On her way to work, she stopped for gas $2\frac{9}{10}$ miles from her house. How far does Betty have to drive from the gas station to her job?

(1) $10\frac{4}{10}$ miles

(2) $9\frac{3}{5}$ miles

(3) $9\frac{3}{10}$ miles

(4) $8\frac{8}{10}$ miles

(5) $10\frac{6}{20}$ miles

6. By November, Paul's weight was down to $162\frac{1}{2}$ pounds. This was $43\frac{1}{2}$ pounds less than his weight in April. How much had Paul weighed in April?

(1) 119 pounds

(2) 120 pounds

(3) $200\frac{1}{2}$ pounds

(4) 205 pounds

(5) 206 pounds

Answers and solutions start on page 131.

MULTIPLYING FRACTIONS

Multiplying fractions is less complicated than adding or subtracting fractions. To multiply fractions you do not need a common denominator, and you never need to borrow.

Think about this problem. A dozen eggs cost $.80. How much do $1\frac{1}{2}$ dozen cost? To find the answer, you would automatically multiply:

$$\$.80 \times 1\frac{1}{2} = \$1.20.$$

To find a <u>fraction of</u> some quantity, you multiply.

When you multiply by a fraction, you find a part of something. When you multiply a whole number by a proper fraction, the result is <u>smaller</u> than the original number ($16 \times \frac{1}{4} = 4$). When you multiply two proper fractions, you are multiplying a part times a part, and the result is <u>smaller than either fraction</u> ($\frac{1}{2} \times \frac{1}{2} = \frac{1}{4}$).

Rules for Multiplying Fractions by Fractions

1. Multiply the numerators together.
2. Multiply the denominators together.
3. Reduce the answer if necessary.

EXAMPLE: Find the product of $\frac{3}{8}$ and $\frac{7}{10}$.

$$\frac{3}{8} \times \frac{7}{10} = \frac{21}{80}$$

Step 1 Multiply the numerators together.
Step 2. Multiply the denominators together.
Step 3. Try to reduce. $\frac{21}{80}$ is already reduced to lowest terms.

FRACTIONS EXERCISE 16

Solve each problem.

1. $\frac{3}{4} \times \frac{5}{7} =$ 4. $\frac{7}{8} \times \frac{1}{5} =$ 7. $\frac{2}{9} \times \frac{3}{5} =$

2. $\frac{2}{3} \times \frac{1}{2} =$ 5. $\frac{1}{4} \times \frac{5}{16} =$ 8. $\frac{5}{6} \times \frac{1}{8} =$

3. $\frac{1}{10} \times \frac{5}{8} =$ 6. $\frac{3}{10} \times \frac{3}{5} =$ 9. $\frac{3}{5} \times \frac{1}{6} =$

Answers and solutions start on page 132.

CANCELLATION

There is a shortcut for multiplying fractions, called **cancellation.** To cancel, find if a number can divide evenly into the numerator of one fraction and the denominator of the other.

EXAMPLE 1: Find the product of $\frac{3}{4} \times \frac{8}{15}$.

① $\frac{\cancel{3}^{1}}{4} \times \frac{8}{\cancel{15}_{5}} =$ ② & ③ $\frac{\cancel{3}^{1}}{\cancel{4}_{1}} \times \frac{\cancel{8}^{2}}{\cancel{15}_{5}} = \frac{2}{5}$

Step 1. See if a number divides evenly into both the numerator of one fraction and the denominator of the other. Cancel 3 and 15 by 3.

Step 2. Try the other numerator and denominator. Cancel 4 and 8 by 4.

Step 3. Multiply the new numerators and the new denominators. Reduce if necessary.

In some cases, you will be able to cancel only once or not at all.

Using cancellation makes it easy to multiply more than two fractions at a time. In such problems, you can cancel several times, but only one numerator or denominator at a time.

EXAMPLE 2: $\frac{5}{8} \times \frac{8}{9} \times \frac{7}{10} =$

$$\frac{\cancel{5}^{1}}{\cancel{8}_{1}} \times \frac{\cancel{8}^{1}}{9} \times \frac{7}{\cancel{10}_{2}} = \frac{7}{18}$$

Step 1. Cancel 5 and 10 by 5. Notice that you can "skip over" the middle fraction.

Step 2. Cancel 8 and 8 by 8.

Step 3. Multiply numerators and denominators. Reduce if necessary.

FRACTIONS EXERCISE 17 ――――――――――――――――

Solve each problem.

1. $\frac{3}{4} \times \frac{6}{7} =$ 5. $\frac{5}{8} \times \frac{2}{15} =$ 9. $\frac{2}{3} \times \frac{9}{20} =$

2. $\frac{14}{15} \times \frac{21}{25} =$ 6. $\frac{9}{10} \times \frac{2}{3} =$ 10. $\frac{6}{7} \times \frac{7}{8} \times \frac{4}{5} =$

3. $\frac{7}{8} \times \frac{14}{27} =$ 7. $\frac{3}{20} \times \frac{1}{3} =$ 11. $\frac{9}{10} \times \frac{1}{4} \times \frac{8}{9} =$

4. $\frac{4}{9} \times \frac{3}{8} =$ 8. $\frac{4}{5} \times \frac{5}{24} =$ 12. $\frac{3}{4} \times \frac{2}{9} \times \frac{15}{16} =$

Answers and solutions start on page 132.

MULTIPLYING FRACTIONS AND WHOLE NUMBERS

To multiply a fraction by a whole number, rewrite the whole number as a mixed number with a denominator of 1. The denominator 1 does not change the value of the whole number. (Remember, any number divided by 1 equals itself.) It only pushes the whole number into the numerator position.

For example, $\frac{6}{1}$ is equal to the whole number 6, because 6 divided by 1 equals 6.

EXAMPLE: $\frac{3}{4} \times 6 =$

①	②	③	④
$\frac{3}{4} \times \frac{6}{1} =$	$\frac{3}{\underset{2}{4}} \times \frac{\overset{3}{6}}{1} =$	$\frac{3}{\underset{2}{4}} \times \frac{\overset{3}{6}}{1} = \frac{9}{2}$	$\frac{9}{2} = 4\frac{1}{2}$

Step 1. Rewrite the whole number 6 as an improper fraction with a denominator of 1: $\frac{6}{1}$.

Step 2. Cancel 4 and 6 using 2.

Step 3. Multiply the new numerators, and multiply the new denominators.

Step 4. Change the improper fraction to a mixed number.

Remember that improper fraction answers should be changed to mixed numbers.

FRACTIONS EXERCISE 18 ━━━━━━━━━━━━━━

Solve each problem.

1. $\frac{1}{2} \times 16 =$

2. $10 \times \frac{2}{3} =$

3. $\frac{3}{8} \times 12 =$

4. $8 \times \frac{7}{10} =$

5. $\frac{2}{5} \times 10 =$

6. $9 \times \frac{11}{20} =$

7. $\frac{5}{6} \times 9 =$

8. $15 \times \frac{7}{100} =$

9. $18 \times \frac{2}{3} =$

10. $20 \times \frac{4}{5} =$

11. $\frac{1}{15} \times 36 =$

12. $24 \times \frac{3}{10} =$

Answers and solutions start on page 132.

CHANGING MIXED NUMBERS TO IMPROPER FRACTIONS

To multiply with mixed numbers, first change the mixed numbers into improper fractions.

Rules for Changing Mixed Numbers to Improper Fractions

1. Find the product of the denominator of the fraction and the whole number. Put the product over the denominator.

2. Add the two fractions.

EXAMPLE: Change $4\frac{2}{3}$ into an improper fraction.

$$①\,\&\,② \quad \boxed{4\frac{2}{3} = \frac{12}{3} + \frac{2}{3} = \frac{14}{3}}$$

Step 1. Multiply the denominator (3) by the whole number (4). Put the product (12) over the denominator (3).

Step 2. Add the two fractions.

FRACTIONS EXERCISE 19

Change each mixed number to an improper fraction.

1. $2\frac{2}{3} =$ 6. $5\frac{4}{7} =$ 11. $12\frac{3}{5} =$

2. $1\frac{5}{8} =$ 7. $3\frac{1}{2} =$ 12. $9\frac{1}{4} =$

3. $8\frac{2}{5} =$ 8. $7\frac{1}{3} =$ 13. $13\frac{2}{3} =$

4. $3\frac{1}{4} =$ 9. $6\frac{2}{9} =$ 14. $15\frac{1}{3} =$

5. $3\frac{5}{6} =$ 10. $10\frac{1}{3} =$ 15. $4\frac{3}{8} =$

Answers and solutions start on page 133.

MULTIPLYING MIXED NUMBERS

Remember to change every mixed number to an improper fraction before multiplying.

Rules for Multiplying Fractions, Whole Numbers, and Mixed Numbers

1. Change every mixed number or whole number into an improper fraction.
2. Cancel.
3. Multiply the numerators together, and multiply the denominators together.
4. If the answer is an improper fraction, change it to a mixed number and reduce, if possible.

EXAMPLE 1: $\frac{1}{3} \times 2\frac{1}{4} =$

① $\boxed{\frac{1}{3} \times \frac{9}{4} =}$ ② & ③ $\boxed{\frac{1}{3} \times \frac{\overset{3}{\cancel{9}}}{4} = \frac{3}{4}}$

Step 1. Change $2\frac{1}{4}$ to an improper fraction.

Step 2. Cancel.

Step 3. Multiply numerators and denominators. $\frac{3}{4}$ is already reduced to lowest terms.

EXAMPLE 2: $3\frac{7}{10} \times 4\frac{1}{6} =$

① $\boxed{\frac{37}{10} \times \frac{25}{6} =}$ ② & ③ $\boxed{\frac{37}{\underset{2}{\cancel{10}}} \times \frac{\overset{5}{\cancel{25}}}{6} = \frac{185}{12}}$ ④ $\boxed{\frac{185}{12} = \mathbf{15\frac{5}{12}}}$

Step 1. Change the mixed numbers to improper fractions.

Step 2. Cancel.

Step 3. Multiply numerators and denominators.

Step 4. Change the improper fraction to a mixed number.

EXAMPLE 3: Find the product of $\frac{1}{3}$, 16, and $1\frac{1}{2}$.

① $\boxed{\frac{1}{3} \times \frac{16}{1} \times \frac{3}{2} =}$ ② $\boxed{\frac{1}{\underset{1}{\cancel{3}}} \times \frac{\overset{8}{\cancel{16}}}{1} \times \frac{\overset{1}{\cancel{3}}}{\underset{1}{\cancel{2}}} =}$ ③ $\boxed{\frac{1}{\underset{1}{\cancel{3}}} \times \frac{\overset{8}{\cancel{16}}}{1} \times \frac{\overset{1}{\cancel{3}}}{\underset{1}{\cancel{2}}} = \frac{8}{1} = \mathbf{8}}$

Step 1. Change the mixed and whole numbers to improper fractions.

Step 2. Cancel.

Step 3. Multiply numerators and denominators.

FRACTIONS EXERCISE 20 ━━━━━━━━━━━━━━━

Solve each problem.

1. $1\frac{2}{3} \times 3\frac{1}{2} =$

2. $1\frac{1}{2} \times \frac{2}{3} =$

3. $6\frac{1}{2} \times 3\frac{1}{4} =$

4. $\frac{5}{8} \times 2\frac{2}{9} =$

5. $\frac{2}{3} \times 3\frac{3}{4} =$

6. $2\frac{5}{8} \times 1\frac{1}{3} =$

7. $1\frac{1}{2} \times \frac{4}{5} \times 2\frac{1}{4} =$

8. $\frac{5}{8} \times 4 \times 1\frac{2}{5} =$

9. $2\frac{1}{2} \times 1\frac{3}{5} \times 6 =$

10. $2\frac{1}{2} \times 1\frac{3}{5} \times \frac{7}{11} =$

11. $1\frac{7}{20} \times 5\frac{1}{3} \times 1\frac{1}{4} =$

12. $1\frac{7}{8} \times \frac{3}{4} \times 2\frac{2}{9} =$

13. $\frac{3}{8} \times 2\frac{5}{6} \times 1\frac{1}{3} =$

14. $1\frac{5}{16} \times 2\frac{6}{7} \times 2\frac{2}{5} =$

Answers and solutions start on page 133.

DIVIDING FRACTIONS

Suppose you had to divide five pies into quarters. Mathematically this would be written as $5 \div \frac{1}{4}$. If you draw a picture, you can see what happens.

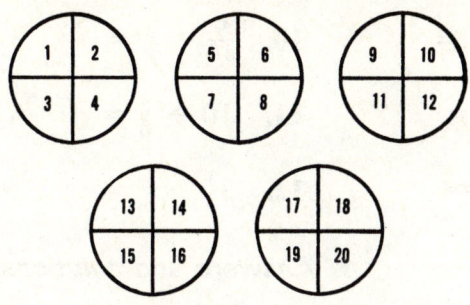

Five pies divided into quarters

This division results in a total of 20 pieces. You can see that you get the same results with $5 \div \frac{1}{4}$ as with 5×4.

When dividing by a fraction, take the second number and invert it (turn it upside down). The "inverted" number is called the **reciprocal**. Then multiply.

Unlike work with whole numbers, division by a proper fraction results in an answer that is <u>larger than either number</u>. This is true whether you are dividing a fraction into a whole number or into another proper fraction.

Rules for Dividing by a Fraction

1. Change any mixed number or whole number into an improper fraction.
2. Invert the divisor (the number at the <u>right</u>) and change the $\div$ sign to a $\times$ sign.
3. Cancel if possible.
4. Multiply the numerators and denominators.
5. Change an improper fraction answer to a mixed number and reduce if possible.

EXAMPLE: $3\frac{1}{2} \div \frac{3}{8} =$

① $\boxed{\frac{7}{2} \div \frac{3}{8} =}$ ② $\boxed{\frac{7}{2} \times \frac{8}{3} =}$ ③ & ④ $\boxed{\frac{7}{2} \times \frac{\overset{4}{\cancel{8}}}{\cancel{3}} = \frac{28}{3}}$ ⑤ $\boxed{\frac{28}{3} = 9\frac{1}{3}}$

Step 1. Change mixed numbers to improper fractions.
Step 2. Invert the divisor and change the $\div$ sign to $\times$.
Step 3. Cancel.
Step 4. Multiply the numerators and denominators.
Step 5. Change the improper fraction to a mixed number.

FRACTIONS EXERCISE 21

Solve each problem.

1. $\frac{1}{3} \div \frac{1}{6} =$ 5. $\frac{1}{3} \div \frac{2}{3} =$ 9. $5\frac{5}{6} \div \frac{7}{8} =$

2. $5 \div \frac{5}{6} =$ 6. $\frac{5}{7} \div \frac{5}{14} =$ 10. $\frac{9}{10} \div \frac{3}{5} =$

3. $4\frac{1}{2} \div \frac{3}{4} =$ 7. $4 \div \frac{3}{8} =$ 11. $10 \div \frac{5}{6} =$

4. $2\frac{2}{3} \div \frac{2}{15} =$ 8. $\frac{5}{9} \div \frac{3}{4} =$ 12. $3\frac{1}{3} \div \frac{1}{3} =$

Answers and solutions start on page 134.

DIVIDING A FRACTION BY WHOLE NUMBERS OR MIXED NUMBERS

You can divide a fraction or a mixed number by a whole number. Simply put the number over a denominator of 1, then invert the fraction and multiply.

EXAMPLE 1: $4\frac{2}{3} \div 6$

① $\boxed{\dfrac{14}{3} \div \dfrac{6}{1}}$ ② $\boxed{\dfrac{\overset{7}{\cancel{14}}}{3} \times \dfrac{1}{\underset{3}{\cancel{6}}} = \dfrac{7}{9}}$

Step 1. Change the mixed number to an improper fraction. Put the whole number over 1.

Step 2. Invert the divisor and change the ÷ to ×. Cancel and multiply.

When you divide by a mixed number, change the mixed number to an improper fraction before you invert and multiply.

EXAMPLE 2: $4\frac{1}{2} \div 1\frac{1}{2} =$

① $\boxed{\dfrac{9}{2} \div \dfrac{3}{2}}$ ② $\boxed{\dfrac{9}{2} \times \dfrac{2}{3}}$ ③ & ④ $\boxed{\dfrac{\overset{3}{\cancel{9}}}{\underset{1}{\cancel{2}}} \times \dfrac{\overset{1}{\cancel{2}}}{\underset{1}{\cancel{3}}} = \dfrac{3}{1}}$ ⑤ $\boxed{\dfrac{3}{1} = 3}$

Step 1. Change mixed numbers to improper fractions.

Step 2. Invert the divisor and change the ÷ sign to ×.

Step 3. Cancel.

Step 4. Multiply the numerators and denominators.
Step 5. Change the improper fraction to a whole number.

FRACTIONS EXERCISE 22

Solve each problem.

1. $10 \div 1\frac{1}{2} =$ 5. $21 \div 4\frac{1}{5} =$ 9. $5\frac{5}{6} \div 7 =$

2. $1\frac{1}{3} \div 3\frac{1}{5} =$ 6. $2\frac{2}{9} \div 2 =$ 10. $\frac{3}{4} \div 3\frac{1}{5} =$

3. $6 \div 1\frac{1}{3} =$ 7. $\frac{9}{10} \div 3 =$ 11. $13\frac{3}{4} \div 4 =$

4. $2\frac{1}{2} \div 3\frac{1}{4} =$ 8. $1\frac{3}{4} \div 7 =$ 12. $10 \div 1\frac{2}{3} =$

Answers and solutions start on page 134.

MULTIPLICATION AND DIVISION OF FRACTIONS: WORD PROBLEMS

These problems will give you a chance to apply your multiplication and division of fractions skills. Read each problem carefully to decide whether to multiply or to divide. Remember in division problems, whatever is being divided must come first when you set up problems.

FRACTIONS EXERCISE 23

Solve each problem. Then choose the correct answer from the five choices.

1. One year, the town of Middleville received $1\frac{1}{2}$ million for community programs. The money was shared equally among six programs. How much did each program receive?

(1) $9 million
(2) $\frac{1}{6}$ million
(3) $\frac{3}{4}$ million
(4) $\frac{1}{2}$ million
(5) $\frac{1}{4}$ million

2. How many $\frac{1}{8}''$ strips of metal can be cut from a piece of tubing that is $\frac{3}{4}''$ long?

(1) 3
(2) 4
(3) 6
(4) 8
(5) 10

3. Hiro and Fumiko together earned $22,500 last year. They spent $\frac{1}{5}$ of their salaries on rent. How much did they spend on rent last year?

    (1) $4,500
    (2) $3,750
    (3) $1,800
    (4) $375
    (5) $450

4. In the November election, only $\frac{2}{5}$ of the registered voters in Middletown went to the polls. There are 85,000 registered voters in Middletown. How many of them stayed home in November?

    (1) 17,000
    (2) 45,000
    (3) 51,000
    (4) 68,000
    (5) 34,000

5. Donna worked for $8\frac{1}{2}$ hours at $8.20 an hour. How much did she earn?

    (1) $65.60
    (2) $70.60
    (3) $66.70
    (4) $68.60
    (5) $69.70

6. Manny bought a pair of trousers on sale for $\frac{1}{4}$ off the regular price. The regular price was $32. Find the sale price of the trousers.

    (1) $8
    (2) $16
    (3) $24
    (4) $28
    (5) $40

Answers and solutions start on page 135.

FRACTION WORD PROBLEMS

FRACTION EXERCISE 24

Solve each problem. Then choose the correct answer from the five choices.

1. A share of Consolidated Utilities stock sold for $32\frac{1}{4}$ on Monday. A share went up $1\frac{1}{8}$ on Tuesday, but then went down $2\frac{1}{2}$ by the end of Tuesday's work day. Find the closing price on Tuesday.

    (1) $35\frac{7}{8}$
    (2) $33\frac{5}{8}$
    (3) $31\frac{3}{8}$
    (4) $30\frac{7}{8}$
    (5) $28\frac{5}{8}$

2. The Murphys want to divide part of their farm into $1\frac{1}{2}$ acre lots. How many lots can they get from a 30-acre field?

    (1) 10
    (2) 20
    (3) 30
    (4) 45
    (5) 15

3. Chie had 9 yards of material. She used $2\frac{1}{2}$ yards to make curtains and $3\frac{3}{4}$ yards to make skirts for her daughters. How much material did she have left?

(1) $2\frac{1}{4}$ yards
(2) $2\frac{3}{4}$ yards
(3) $3\frac{1}{4}$ yards
(4) $3\frac{3}{4}$ yards
(5) $3\frac{1}{2}$ yards

4. Jeff paid $38 for $9\frac{1}{2}$ yards of lumber. Find the price of one yard.

(1) $3.50
(2) $4
(3) $4.50
(4) $5
(5) $3.75

5. Lucas takes home $256 a week. His family spends $\frac{3}{8}$ of his pay on food. He also spends $\frac{1}{4}$ of his income on rent. How much money does he spend on food and rent?

(1) $88
(2) $96
(3) $123
(4) $146
(5) $160

6. Dorothy usually works 3 hours a day, 5 days a week in a department store. The week before Christmas she worked an additional $1\frac{1}{4}$ hours on Monday, $2\frac{3}{4}$ hours on Wednesday, and $3\frac{1}{2}$ on Thursday. How many hours did she work altogether that week?

(1) $22\frac{1}{2}$ hours
(2) $18\frac{1}{2}$ hours
(3) $15\frac{1}{2}$ hours
(4) $10\frac{1}{2}$ hours
(5) $7\frac{1}{2}$ hours

7. One cubic foot of water weighs $62\frac{1}{2}$ pounds. Find the weight of 8 cubic feet of water.

(1) 600 pounds
(2) 500 pounds
(3) 625 pounds
(4) 550 pounds
(5) 700 pounds

8. The Casa family spends $\frac{1}{3}$ of their income for food, $\frac{1}{4}$ for mortgage payments, and $\frac{1}{8}$ for utilities. What fraction of their income goes for everything besides these three items?

(1) $\frac{3}{4}$
(2) $\frac{17}{24}$
(3) $\frac{1}{4}$
(4) $\frac{7}{24}$
(5) $\frac{5}{12}$

9. Mary bought $5\frac{1}{4}$ yards of material for $6.40 a yard. How much did she pay for the material?

(1) $32
(2) $33.60
(3) $35.20
(4) $30.25
(5) $32.80

10. Nick cut a board 82 inches long into 4 equal pieces. How long was each piece?

(1) $20\frac{1}{2}$ inches
(2) 24 inches
(3) 14 inches
(4) $18\frac{1}{2}$ inches
(5) 22 inches

Answers and solutions start on page 135.

FRACTIONS TEST

For each problem choose the best answer.

1. A yard has 36 inches. 13 inches are what fraction of a yard?

(1) $\frac{13}{25}$
(2) $\frac{25}{36}$
(3) $\frac{1}{2}$
(4) $\frac{1}{3}$
(5) $\frac{13}{36}$

1 ① ② ③ ④ ⑤

2. On a test with 30 problems, Don got 3 problems wrong. What fraction of the problems did he get right?

(1) $\frac{1}{2}$
(2) $\frac{1}{3}$
(3) $\frac{1}{10}$
(4) $\frac{9}{10}$
(5) $\frac{29}{30}$

2 ① ② ③ ④ ⑤

3. Change $\frac{3}{5}$ to an equivalent fraction with a denominator of 60.

(1) $\frac{36}{60}$
(2) $\frac{30}{60}$
(3) $\frac{5}{60}$
(4) $\frac{3}{60}$
(5) $\frac{60}{100}$

3 ① ② ③ ④ ⑤

4. Which of the following is equal to $\frac{48}{9}$?

(1) $5\frac{2}{3}$
(2) $5\frac{1}{3}$
(3) $5\frac{1}{9}$
(4) $\frac{9}{48}$
(5) $\frac{1}{5}$

4 ① ② ③ ④ ⑤

5. Which of the following is equal to $4\frac{3}{8}$?

   (1) $\frac{3}{8}$
   (2) $\frac{15}{8}$
   (3) $\frac{32}{8}$
   (4) $\frac{35}{8}$
   (5) $\frac{7}{8}$

6. 4.075 is equal to

   (1) $\frac{3}{4}$
   (2) $\frac{3}{40}$
   (3) 4
   (4) $4\frac{3}{4}$
   (5) $4\frac{3}{40}$

7. $\frac{5}{6}$ is equal to

   (1) $.83\frac{1}{3}$
   (2) $.83\frac{2}{3}$
   (3) $.16\frac{1}{3}$
   (4) $.16$
   (5) $.16\frac{2}{3}$

8. Find the sum of $8\frac{1}{4}$, $2\frac{5}{6}$, and $4\frac{5}{8}$.

   (1) $14\frac{17}{24}$
   (2) $14\frac{11}{18}$
   (3) $15\frac{11}{24}$
   (4) $15\frac{1}{6}$
   (5) $15\frac{17}{24}$

9. Find the difference between $10\frac{1}{4}$ and $15\frac{1}{3}$.

   (1) $4\frac{1}{3}$
   (2) $4\frac{11}{12}$
   (3) $5\frac{1}{12}$
   (4) $5\frac{11}{12}$
   (5) $4\frac{1}{12}$

10. For Thanksgiving, Mrs. Bennett roasted a turkey weighing $24\frac{3}{4}$ pounds. Her daughter-in-law Faye roasted another turkey weighing $18\frac{1}{2}$ pounds. Find the combined weight of the turkeys.

    (1) $5\frac{1}{4}$ pounds

    (2) $42\frac{1}{4}$ pounds

    (3) $42\frac{3}{4}$ pounds

    (4) $43\frac{1}{4}$ pounds

    (5) 44 pounds

    1 ① ② ③ ④ ⑤

11. From a 10-pound sack of flour, Sylvia used $4\frac{1}{3}$ pounds of flour to bake bread. How much flour did Sylvia have left?

    (1) $6\frac{1}{3}$ pounds

    (2) $5\frac{2}{3}$ pounds

    (3) $4\frac{2}{3}$ pounds

    (4) $6\frac{2}{3}$ pounds

    (5) $14\frac{1}{3}$ pounds

    2 ① ② ③ ④ ⑤

12. Find the product of $3\frac{3}{4}$ and $3\frac{1}{3}$.

    (1) 5

    (2) $9\frac{1}{4}$

    (3) $7\frac{1}{12}$

    (4) $12\frac{1}{2}$

    (5) $6\frac{1}{4}$

    3 ① ② ③ ④ ⑤

13. What is $24\frac{1}{2}$ divided by $3\frac{1}{2}$?

    (1) $\frac{1}{7}$

    (2) 8

    (3) $\frac{1}{8}$

    (4) 7

    (5) $3\frac{1}{2}$

    4 ① ② ③ ④ ⑤

14. The Andersons spend $\frac{1}{3}$ of their income on food. In a week, the Andersons bring home $267. How much do they spend each week on food?

    (1) $123

    (2) $112

    (3) $109

    (4) $92

    (5) $89

    5 ① ② ③ ④ ⑤

15. Pete paid $41.40 for $4\frac{1}{2}$ yards of lumber. What was the price of one yard?

    (1) $2.30

    (2) $8.60

    (3) $10.20

    (4) $9.20

    (5) $4.60

6 ① ② ③ ④ ⑤

Answers and solutions start on page 136.

FRACTIONS TEST EVALUATION

Problem	Section	Starting Page
1	Writing Fractions	87
2	Reducing Fractions	90
3	Raising Fractions to Higher Terms	104
4	Changing Improper Fractions to Whole or Mixed Numbers	93
5	Changing Mixed Numbers to Improper Fractions	112
6	Changing Decimals to Fractions	95
7	Changing Fractions to Decimals	96
8	Adding Mixed Numbers	97
9	Subtracting with Unlike Denominators	107
10, 11	Addition and Subtraction of Fractions Word Problems	108
12	Multiplying Mixed Numbers	113
13	Dividing by Mixed Numbers	116
14, 15	Multiplication and Division of Fractions Word Problems	117

<u>Passing score:</u> <u>11</u> right out of 15 problems.
<u>Your score:</u>   __ right out of 15 problems.

If you had less than a passing score, review the sections for the problems you got wrong. Then repeat this test before you go on to the mixed review.

If you had a passing score, correct any problem you got wrong. Then go to the mixed review.

MIXED REVIEW ─────────────────────────────────────

These problems give you a chance to practice the skills you have learned so far in this book. For each problem, choose the best answer.

1. What is the difference between 693 and 4,000?   **1** ① ② ③ ④ ⑤
   (1) 3,407
   (2) 2,407
   (3) 3,307
   (4) 3,393
   (5) 4,307

2. Round off 485,296 to the nearest ten-thousand.   **2** ① ② ③ ④ ⑤
   (1) 500,000
   (2) 485,000
   (3) 480,000
   (4) 486,000
   (5) 490,000

3. The attendance at the Uptown Theatre was 483 people on Friday   **3** ① ② ③ ④ ⑤
   night, 527 people on Saturday night, and 364 people on Sunday
   afternoon. What was the average attendance for these three times?
   (1) 483
   (2) 375
   (3) 450
   (4) 458
   (5) 511

4. Find the largest decimal in this group.   **4** ① ② ③ ④ ⑤
   (1) .045
   (2) .04
   (3) .05
   (4) .0455
   (5) .054

5. Find the difference between 13 and .68.   **5** ① ② ③ ④ ⑤
   (1) 12.32
   (2) .55
   (3) 6.2
   (4) 12.68
   (5) 12.48

6. In January, Sam traveled 500.7 miles. The next month, he traveled   **6** ① ② ③ ④ ⑤
   200.25 miles, and in March he traveled 300.3 miles. What was his
   total mileage for three months?
   (1) 1,000.35 miles
   (2) 10.0125 miles
   (3) 100,125 miles
   (4) 1,001.25 miles
   (5) 1,001.15 miles

7. Allen bought 2.2 pounds of wood screws at $.68 a pound. How much did he pay for the screws? Round off your answer to the nearest cent.

(1) $1.36
(2) $1.42
(3) $1.50
(4) $1.56
(5) $1.62

8. Mark usually works $7\frac{1}{2}$ hours a day. Thursday, he worked $2\frac{3}{4}$ hours overtime. How many hours did Mark work altogether on Thursday?

(1) $9\frac{1}{4}$ hours
(2) $9\frac{3}{4}$ hours
(3) 10 hours
(4) $10\frac{1}{4}$ hours
(5) $10\frac{3}{4}$ hours

9. Mark worked for $7\frac{1}{2}$ hours at $4.80 an hour and for $2\frac{3}{4}$ hours at $7.20 an hour. How much did he make altogether?

(1) $48.00
(2) $52.60
(3) $55.80
(4) $62.40
(5) $73.80

10. A 12-foot board was split into four equal pieces. Later, each piece was cut in half. How long was each piece?

(1) $1\frac{1}{2}$ feet
(2) 4 feet
(3) 8 feet
(4) 3 feet
(5) 96 feet

Answers and solutions start on page 137.

ANSWERS AND SOLUTIONS

Fractions Exercise 1

1. a) $\frac{1}{3}$ b) $\frac{5}{6}$ c) $\frac{1}{2}$ d) $\frac{3}{8}$ e) $\frac{2}{5}$

2. $\frac{5}{12}$

3. $\frac{2}{7}$

4. $\frac{27}{100}$

5. $\frac{209}{250}$

6. $\frac{41}{250}$

$$\begin{array}{r} 250 \text{ gross salary} \\ -209 \text{ take-home} \\ \hline 41 \text{ amount withheld} \end{array}$$

7. $\frac{333}{1000}$

8. $\frac{13}{24}$

$$\begin{array}{r} 24 \text{ total miles in marathon} \\ -11 \text{ miles run so far} \\ \hline 13 \text{ miles to go} \end{array}$$

9. $\frac{164}{745}$

10. $\frac{83}{600}$

$$\begin{array}{r} \$600 \text{ total down payment} \\ -\ 517 \text{ amount saved} \\ \hline \$\ 83 \text{ amount left to save} \end{array}$$

Fractions Exercise 2

1. $\frac{5}{7}$

$$\frac{25 \div 5}{35 \div 5} = \frac{5}{7}$$

2. $\frac{1}{3}$

$$\frac{8 \div 8}{24 \div 8} = \frac{1}{3}$$

3. $\frac{4}{9}$

$$\frac{12 \div 3}{27 \div 3} = \frac{4}{9}$$

4. $\frac{5}{8}$

$$\frac{50 \div 10}{80 \div 10} = \frac{5}{8}$$

5. $\frac{3}{20}$

$$\frac{30 \div 10}{200 \div 10} = \frac{3}{20}$$

6. $\frac{15}{16}$

$$\frac{90 \div 2}{96 \div 2} =$$
$$\frac{45 \div 3}{48 \div 3} = \frac{15}{16}$$

7. $\frac{9}{11}$

$$\frac{36 \div 4}{44 \div 4} = \frac{9}{11}$$

8. $\frac{4}{5}$

$$\frac{48 \div 6}{60 \div 6} =$$
$$\frac{8 \div 2}{10 \div 2} = \frac{4}{5}$$

9. $\frac{1}{5}$

$$\frac{50 \div 10}{250 \div 10} =$$
$$\frac{5 \div 5}{25 \div 5} = \frac{1}{5}$$

10. $\frac{5}{7}$

$$\frac{55 \div 11}{77 \div 11} = \frac{5}{7}$$

11. $\frac{1}{2}$

$$\frac{80 \div 80}{160 \div 80} = \frac{1}{2}$$

12. $\frac{3}{10}$

$$\frac{15 \div 5}{50 \div 5} = \frac{3}{10}$$

13. $\frac{1}{2}$

$$\frac{18 \div 18}{36 \div 18} = \frac{1}{2}$$

14. $\frac{9}{10}$

$$\frac{81 \div 9}{90 \div 9} = \frac{9}{10}$$

15. $\frac{9}{14}$

$$\frac{45 \div 5}{70 \div 5} = \frac{9}{14}$$

16. $\frac{9}{10}$

$$\frac{45 \div 5}{50 \div 5} = \frac{9}{10}$$

17. $\frac{2}{3}$

$$\frac{8 \div 4}{12 \div 4} = \frac{2}{3}$$

18. $\frac{2}{7}$

The number of games lost is $21-15 = 6$.

$$\frac{6 \div 3}{21 \div 3} = \frac{2}{7}$$

19. $\frac{13}{20}$

$$\frac{650 \div 10}{1,000 \div 10} = \frac{65}{100}$$
$$\frac{65 \div 5}{100 \div 5} = \frac{13}{20}$$

20. $\frac{4}{7}$

The total number of students is $8 + 6 = 14$.

$$\frac{\text{women}}{\text{total}} = \frac{8 \div 2}{14 \div 2} = \frac{4}{7}$$

Fractions Exercise 3

1. $2\frac{1}{3}$ 2. $1\frac{1}{4}$ 3. $5\frac{1}{6}$ 4. $2\frac{5}{8}$ 5. $2\frac{3}{10}$ 6. $3\frac{1}{3}$

Fractions Exercise 4

1. $1\frac{2}{3}$

$$\begin{array}{r} 1\frac{6}{9} = 1\frac{2}{3} \\ 9{\overline{)15}} \\ \underline{9} \\ 6 \end{array}$$

2. $4\frac{1}{2}$

$$\begin{array}{r} 4\frac{4}{8} = 4\frac{1}{2} \\ 8{\overline{)36}} \\ \underline{32} \\ 4 \end{array}$$

3. $2\frac{3}{5}$

$$\begin{array}{r} 2\frac{3}{5} \\ 5{\overline{)13}} \\ \underline{10} \\ 3 \end{array}$$

4. $5\frac{1}{6}$

$$\begin{array}{r} 5\frac{1}{6} \\ 6{\overline{)31}} \\ \underline{30} \\ 1 \end{array}$$

5. 3

$$2\overline{)6}^{\;3}$$
$$\underline{6}$$

6. $1\frac{3}{4}$

$$16\overline{)28}^{\;1\frac{12}{16}} = 1\frac{3}{4}$$
$$\underline{16}$$
$$12$$

7. $1\frac{5}{6}$

$$30\overline{)55}^{\;1\frac{25}{30}} = 1\frac{5}{6}$$
$$\underline{30}$$
$$25$$

8. 1

$$15\overline{)15}^{\;1}$$
$$\underline{15}$$

9. $1\frac{1}{3}$

$$18\overline{)24}^{\;1\frac{6}{18}} = 1\frac{1}{3}$$
$$\underline{18}$$
$$6$$

10. $1\frac{3}{4}$

$$4\overline{)7}^{\;1\frac{3}{4}}$$
$$\underline{4}$$
$$3$$

11. 7

$$3\overline{)21}^{\;7}$$
$$\underline{21}$$

12. $4\frac{1}{6}$

$$12\overline{)50}^{\;4\frac{2}{12}} = 4\frac{1}{6}$$
$$\underline{48}$$
$$2$$

13. 1

$$82\overline{)82}^{\;1}$$
$$\underline{82}$$

14. $3\frac{1}{2}$

$$6\overline{)21}^{\;3\frac{3}{6}} = 3\frac{1}{2}$$
$$\underline{18}$$
$$3$$

15. 4

$$11\overline{)44}^{\;4}$$
$$\underline{44}$$

Fractions Exercise 5

1. $\frac{3}{5}$

$.6 = \frac{6}{10} = \frac{3}{5}$

2. $\frac{1}{2}$

$.5 = \frac{5}{10} = \frac{1}{2}$

3. $\frac{9}{20}$

$.45 = \frac{45}{100} = \frac{9}{20}$

4. $\frac{4}{5}$

$.80 = \frac{80}{100} = \frac{4}{5}$

5. $\frac{1}{8}$

$.125 = \frac{125}{1,000} = \frac{1}{8}$

6. $\frac{13}{200}$

$.065 = \frac{65}{1,000} =$
$\frac{13}{200}$

7. $4\frac{3}{20}$

$4.15 = 4\frac{15}{100} =$
$4\frac{3}{20}$

8. $\frac{24}{25}$

$.96 = \frac{96}{100} =$
$\frac{24}{25}$

9. $\frac{31}{125}$

$.248 = \frac{248}{1,000} =$
$\frac{31}{125}$

10. $\frac{1}{5,000}$

$.0002 = \frac{2}{10,000} =$
$\frac{1}{5,000}$

11. $15\frac{1}{250}$

$15.004 = 15\frac{4}{1,000} =$
$15\frac{1}{250}$

12. $3\frac{17}{50}$

$3.34 = 3\frac{34}{100} =$
$3\frac{17}{50}$

Fractions Exercise 6

1. .75

$$4\overline{)3.00}^{\;.75}$$
$$\underline{2\,8}$$
$$20$$
$$\underline{20}$$

2. $.33\frac{1}{3}$

$$3\overline{)1.00}^{\;.33\frac{1}{3}}$$
$$\underline{9}$$
$$10$$
$$\underline{9}$$
$$1$$

3. .7 or .70

$$10\overline{)7.0}^{\;.7}$$
$$\underline{7\,0}$$

4. $.62\frac{1}{2}$ or .625

$$8\overline{)5.00}^{\;.62\frac{1}{2}}$$
$$\underline{4\,8}$$
$$20$$
$$\underline{16}$$
$$\frac{4}{4} = \frac{4}{8} = \frac{1}{2}$$

5. 2.5

$$2\overline{)1.0}^{\;.5} \qquad 2 + .5 =$$
$$\underline{1\,0} \qquad\qquad 2.5$$

6. .05

$$20\overline{)1.00}^{\;.05}$$
$$\underline{1\,00}$$

7. $.83\frac{1}{3}$

$$6\overline{)5.00}^{\;.83\frac{1}{3}}$$
$$\underline{4\,8}$$
$$20$$
$$\underline{18}$$
$$\frac{2}{2} = \frac{2}{6} = \frac{1}{3}$$

8. $1.44\frac{4}{9}$

$$9\overline{)4.00}^{\;.44\frac{4}{9}}$$
$$\underline{3\,6}$$
$$40$$
$$\underline{36}$$
$$4$$

$1 + .44\frac{4}{9} =$
$1.44\frac{4}{9}$

9. $.08\frac{1}{3}$

$$12\overline{)1.00}^{\;.08\frac{1}{3}}$$
$$\underline{96}$$
$$\frac{4}{4} = \frac{4}{12} = \frac{1}{3}$$

10. $5.66\frac{2}{3}$

$$3\overline{)2.00}^{\;.66\frac{2}{3}}$$
$$\underline{1\,8}$$
$$20$$
$$\underline{18}$$
$$2$$

$5 + .66\frac{2}{3} =$
$5.66\frac{2}{3}$

Exercise 6 cont'd.

11. .12

$$25\overline{)\,3.00}^{.12}$$
$$\underline{2\,5}$$
$$50$$
$$\underline{50}$$

12. $10.16\frac{2}{3}$

$$6\overline{)\,1.00}^{.16\frac{2}{3}}$$
$$\underline{6}$$
$$40$$
$$\underline{36}$$
$$4 = \frac{4}{6} = \frac{2}{3}$$

$10 + .16\frac{2}{3} =$

$10.16\frac{2}{3}$

12. $17\frac{1}{8}$

$$6\frac{3}{8}$$
$$3\frac{1}{8}$$
$$+\,7\frac{5}{8}$$
$$\overline{16\frac{9}{8} = 17\frac{1}{8}}$$

13. $16\frac{11}{12}$

$$2\frac{7}{12}$$
$$4\frac{11}{12}$$
$$+\,9\frac{5}{12}$$
$$\overline{15\frac{23}{12} = 16\frac{11}{12}}$$

Fractions Exercise 7

1. $\frac{5}{7}$

$$\frac{2}{7}$$
$$+\frac{3}{7}$$
$$\overline{\frac{5}{7}}$$

2. $\frac{2}{3}$

$$\frac{2}{9}$$
$$+\frac{4}{9}$$
$$\overline{\frac{6}{9} = \frac{2}{3}}$$

3. $1\frac{4}{15}$

$$\frac{8}{15}$$
$$+\frac{11}{15}$$
$$\overline{\frac{19}{15} = 1\frac{4}{15}}$$

4. $1\frac{1}{4}$

$$\frac{5}{8}$$
$$+\frac{5}{8}$$
$$\overline{\frac{10}{8} = 1\frac{2}{8} = 1\frac{1}{4}}$$

5. $12\frac{2}{3}$

$$4\frac{1}{3}$$
$$+8\frac{1}{3}$$
$$\overline{12\frac{2}{3}}$$

6. $13\frac{2}{5}$

$$9\frac{3}{10}$$
$$+4\frac{1}{10}$$
$$\overline{13\frac{4}{10} = 13\frac{2}{5}}$$

7. 13

$$10\frac{2}{3}$$
$$+2\frac{1}{3}$$
$$\overline{12\frac{3}{3} = 13}$$

8. $14\frac{4}{9}$

$$6\frac{5}{9}$$
$$+\,7\frac{8}{9}$$
$$\overline{13\frac{13}{9} = 14\frac{4}{9}}$$

9. $22\frac{1}{4}$

$$1\frac{7}{8}$$
$$+20\frac{3}{8}$$
$$\overline{21\frac{10}{8} = 22\frac{2}{8}}$$
$$= 22\frac{1}{4}$$

10. $\frac{13}{16}$

$$\frac{1}{16}$$
$$\frac{5}{16}$$
$$+\frac{7}{16}$$
$$\overline{\frac{13}{16}}$$

11. $1\frac{3}{4}$

$$\frac{3}{4}$$
$$\frac{1}{4}$$
$$+\frac{3}{4}$$
$$\overline{\frac{7}{4} = 1\frac{3}{4}}$$

14. $13\frac{8}{9}$

$$5\frac{4}{9}$$
$$4\frac{5}{9}$$
$$+\,3\frac{8}{9}$$
$$\overline{12\frac{17}{9} = 13\frac{8}{9}}$$

15. $40\frac{1}{5}$

$$13\frac{11}{15}$$
$$16\frac{14}{15}$$
$$+\,9\frac{8}{15}$$
$$\overline{38\frac{33}{15} = 40\frac{3}{15}}$$
$$= 40\frac{1}{5}$$

Fractions Exercise 8

1. $\frac{1}{3}$

$$\frac{5}{9}$$
$$-\frac{2}{9}$$
$$\overline{\frac{3}{9} = \frac{1}{3}}$$

2. $6\frac{2}{3}$

$$8\frac{5}{6}$$
$$-2\frac{1}{6}$$
$$\overline{6\frac{4}{6} = 6\frac{2}{3}}$$

3. $8\frac{1}{3}$

$$12\frac{20}{27}$$
$$-\,4\frac{11}{27}$$
$$\overline{8\frac{9}{27} = 8\frac{1}{3}}$$

4. $2\frac{1}{2}$

$$4\frac{7}{8}$$
$$-2\frac{3}{8}$$
$$\overline{2\frac{4}{8} = 2\frac{1}{2}}$$

5. $\frac{2}{5}$

$$\frac{23}{25}$$
$$-\frac{13}{25}$$
$$\overline{\frac{10}{25} = \frac{2}{5}}$$

6. $2\frac{3}{5}$

$$10\frac{4}{5}$$
$$-\,8\frac{1}{5}$$
$$\overline{2\frac{3}{5}}$$

7. $\frac{3}{10}$

$$\frac{19}{20}$$
$$-\frac{13}{20}$$
$$\overline{\frac{6}{20} = \frac{3}{10}}$$

8. $4\frac{3}{5}$

$$6\frac{13}{15}$$
$$-2\frac{4}{15}$$
$$\overline{4\frac{9}{15} = 4\frac{3}{5}}$$

9. $\frac{2}{9}$

$$\frac{25}{36}$$
$$-\frac{17}{36}$$
$$\frac{8}{36} = \frac{2}{9}$$

10. $7\frac{2}{3}$

$$8\frac{5}{6}$$
$$-1\frac{1}{6}$$
$$7\frac{4}{6} = 7\frac{2}{3}$$

11. $\frac{3}{5}$

$$\frac{34}{35}$$
$$-\frac{13}{35}$$
$$\frac{21}{35} = \frac{3}{5}$$

12. $\frac{2}{3}$

$$4\frac{17}{18}$$
$$-4\frac{5}{18}$$
$$\frac{12}{18} = \frac{2}{3}$$

Fractions Exercise 9

1. $2\frac{1}{9}$

$$5 = 4\frac{9}{9}$$
$$-2\frac{8}{9} = 2\frac{8}{9}$$
$$2\frac{1}{9}$$

2. $7\frac{1}{3}$

$$9 = 8\frac{3}{3}$$
$$-1\frac{2}{3} = 1\frac{2}{3}$$
$$7\frac{1}{3}$$

3. $7\frac{3}{8}$

$$11 = 10\frac{8}{8}$$
$$- 3\frac{5}{8} = 3\frac{5}{8}$$
$$7\frac{3}{8}$$

4. $2\frac{5}{12}$

$$8 = 7\frac{12}{12}$$
$$-5\frac{7}{12} = 5\frac{7}{12}$$
$$2\frac{5}{12}$$

5. $\frac{1}{2}$

$$7 = 6\frac{2}{2}$$
$$-6\frac{1}{2} = 6\frac{1}{2}$$
$$\frac{1}{2}$$

6. $7\frac{3}{4}$

$$9 = 8\frac{4}{4}$$
$$-1\frac{1}{4} = 1\frac{1}{4}$$
$$7\frac{3}{4}$$

7. $9\frac{11}{16}$

$$12 = 11\frac{16}{16}$$
$$- 2\frac{5}{16} = 2\frac{5}{16}$$
$$9\frac{11}{16}$$

8. $11\frac{1}{10}$

$$43 = 42\frac{10}{10}$$
$$-31\frac{9}{10} = 31\frac{9}{10}$$
$$11\frac{1}{10}$$

9. $12\frac{11}{20}$

$$27 = 26\frac{20}{20}$$
$$-14\frac{9}{20} = 14\frac{9}{20}$$
$$12\frac{11}{20}$$

Fractions Exercise 10

1. $2\frac{1}{2}$

$$5\frac{1}{8} = 4\frac{1}{8} + \frac{8}{8} = 4\frac{9}{8}$$
$$-2\frac{5}{8} = \qquad\qquad 2\frac{5}{8}$$
$$2\frac{4}{8} = 2\frac{1}{2}$$

2. $14\frac{7}{9}$

$$28\frac{5}{9} = 27\frac{5}{9} + \frac{9}{9} = 27\frac{14}{9}$$
$$-13\frac{7}{9} = \qquad\qquad 13\frac{7}{9}$$
$$14\frac{7}{9}$$

3. $7\frac{3}{4}$

$$24\frac{11}{16} = 23\frac{11}{16} + \frac{16}{16} = 23\frac{27}{16}$$
$$-16\frac{15}{16} = \qquad\qquad 16\frac{15}{16}$$
$$7\frac{12}{16} = 7\frac{3}{4}$$

4. $1\frac{2}{3}$

$$9\frac{1}{3} = 8\frac{1}{3} + \frac{3}{3} = 8\frac{4}{3}$$
$$-7\frac{2}{3} = \qquad\qquad 7\frac{2}{3}$$
$$1\frac{2}{3}$$

5. $6\frac{3}{5}$

$$8\frac{2}{5} = 7\frac{2}{5} + \frac{5}{5} = 7\frac{7}{5}$$
$$-1\frac{4}{5} = \qquad\qquad 1\frac{4}{5}$$
$$6\frac{3}{5}$$

6. $87\frac{2}{5}$

$$112\frac{3}{10} = 111\frac{3}{10} + \frac{10}{10} = 111\frac{13}{10}$$
$$-24\frac{9}{10} = \qquad\qquad 24\frac{9}{10}$$
$$87\frac{4}{10} = 87\frac{2}{5}$$

7. $2\frac{2}{3}$

$$11\frac{7}{12} = 10\frac{7}{12} + \frac{12}{12} = 10\frac{19}{12}$$
$$-8\frac{11}{12} = \qquad\qquad 8\frac{11}{12}$$
$$2\frac{8}{12} = 2\frac{2}{3}$$

8. $2\frac{3}{7}$

$$5\frac{2}{7} = 4\frac{2}{7} + \frac{7}{7} = 4\frac{9}{7}$$
$$-2\frac{6}{7} = \qquad\qquad 2\frac{6}{7}$$
$$2\frac{3}{7}$$

9. $\frac{1}{2}$

$$13\frac{1}{4} = 12\frac{1}{4} + \frac{4}{4} = 12\frac{5}{4}$$
$$-12\frac{3}{4} = \qquad\qquad 12\frac{3}{4}$$
$$\frac{2}{4} = \frac{1}{2}$$

10. $1\frac{1}{2}$

$$15\frac{7}{24} = 14\frac{7}{24} + \frac{24}{24} = 14\frac{31}{24}$$
$$-13\frac{19}{24} = \qquad\qquad 13\frac{19}{24}$$
$$1\frac{12}{24} = 1\frac{1}{2}$$

Exercise 10 cont'd.

11. $12\frac{4}{9}$

$$20\frac{5}{36} = 19\frac{5}{36} + \frac{36}{36} = 19\frac{41}{36}$$
$$-7\frac{25}{36} = \qquad\qquad 7\frac{25}{36}$$
$$\qquad\qquad 12\frac{16}{36} = 12\frac{4}{9}$$

12. $\frac{1}{3}$

$$7\frac{5}{18} = 6\frac{5}{18} + \frac{18}{18} = 6\frac{23}{18}$$
$$-6\frac{17}{18} = \qquad\qquad 6\frac{17}{18}$$
$$\qquad\qquad \frac{6}{18} = \frac{1}{3}$$

Fractions Exercise 11

1. a) 10	b) 12	c) 48	d) 36	**4.** a) 48	b) 24	c) 20	d) 50
2. a) 36	b) 35	c) 6	d) 100	**5.** a) 20	b) 12	c) 24	
3. a) 18	b) 40	c) 18	d) 40	**6.** a) 18	b) 40	c) 36	

Fractions Exercise 12

1. $\frac{12}{20}$

$$\frac{3 \times 4}{5 \times 4} = \frac{12}{20}$$

2. $\frac{10}{24}$

$$\frac{5 \times 2}{12 \times 2} = \frac{10}{24}$$

3. $\frac{35}{40}$

$$\frac{7 \times 5}{8 \times 5} = \frac{35}{40}$$

4. $\frac{12}{16}$

$$\frac{3 \times 4}{4 \times 4} = \frac{12}{16}$$

5. $\frac{12}{27}$

$$\frac{4 \times 3}{9 \times 3} = \frac{12}{27}$$

6. $\frac{20}{44}$

$$\frac{5 \times 4}{11 \times 4} = \frac{20}{44}$$

7. $\frac{26}{40}$

$$\frac{13 \times 2}{20 \times 2} = \frac{26}{40}$$

8. $\frac{42}{60}$

$$\frac{7 \times 6}{10 \times 6} = \frac{42}{60}$$

9. $\frac{65}{100}$

$$\frac{13 \times 5}{20 \times 5} = \frac{65}{100}$$

10. $\frac{63}{75}$

$$\frac{21 \times 3}{25 \times 3} = \frac{63}{75}$$

11. $\frac{75}{100}$

$$\frac{3 \times 25}{4 \times 25} = \frac{75}{100}$$

12. $\frac{20}{36}$

$$\frac{5 \times 4}{9 \times 4} = \frac{20}{36}$$

13. $\frac{3}{24}$

$$\frac{1 \times 3}{8 \times 3} = \frac{3}{24}$$

14. $\frac{36}{45}$

$$\frac{4 \times 9}{5 \times 9} = \frac{36}{45}$$

15. $\frac{12}{36}$

$$\frac{1 \times 12}{3 \times 12} = \frac{12}{36}$$

Fractions Exercise 13

1. $1\frac{5}{12}$

$$\frac{5}{6} = \frac{10}{12}$$
$$+ \frac{7}{12} = \frac{7}{12}$$
$$\frac{17}{12} = 1\frac{5}{12}$$

2. $\frac{7}{12}$

$$\frac{3}{8} = \frac{9}{24}$$
$$+ \frac{5}{24} = \frac{5}{24}$$
$$\frac{14}{24} = \frac{7}{12}$$

3. $1\frac{5}{12}$

$$\frac{2}{3} = \frac{8}{12}$$
$$+ \frac{3}{4} = \frac{9}{12}$$
$$\frac{17}{12} = 1\frac{5}{12}$$

4. $14\frac{1}{6}$

$$9\frac{2}{3} = 9\frac{4}{6}$$
$$+ 4\frac{1}{2} = 4\frac{3}{6}$$
$$13\frac{7}{6} = 14\frac{1}{6}$$

5. $13\frac{11}{24}$

$$10\frac{1}{8} = 10\frac{3}{24}$$
$$+ 3\frac{1}{3} = 3\frac{8}{24}$$
$$13\frac{11}{24}$$

6. $7\frac{1}{3}$

$$2\frac{5}{6} = 2\frac{5}{6}$$
$$+ 4\frac{1}{2} = 4\frac{3}{6}$$
$$6\frac{8}{6} = 7\frac{2}{6} = 7\frac{1}{3}$$

7. $2\frac{1}{12}$

$$\frac{2}{3} = \frac{8}{12}$$
$$\frac{7}{12} = \frac{7}{12}$$
$$+ \frac{5}{6} = \frac{10}{12}$$
$$\frac{25}{12} = 2\frac{1}{12}$$

9. $2\frac{5}{24}$

$$\frac{5}{8} = \frac{15}{24}$$
$$\frac{5}{6} = \frac{20}{24}$$
$$+ \frac{3}{4} = \frac{18}{24}$$
$$\frac{53}{24} = 2\frac{5}{24}$$

11. $25\frac{11}{20}$

$$9\frac{1}{2} = 9\frac{10}{20}$$
$$8\frac{3}{4} = 8\frac{15}{20}$$
$$+ 7\frac{3}{10} = 7\frac{6}{20}$$
$$24\frac{31}{20} = 25\frac{11}{20}$$

8. $1\frac{7}{20}$

$$\frac{1}{2} = \frac{10}{20}$$
$$\frac{1}{4} = \frac{5}{20}$$
$$+ \frac{3}{5} = \frac{12}{20}$$
$$\frac{27}{20} = 1\frac{7}{20}$$

10. $12\frac{19}{24}$

$$4\frac{1}{3} = 4\frac{8}{24}$$
$$1\frac{5}{6} = 1\frac{20}{24}$$
$$+ 6\frac{5}{8} = 6\frac{15}{24}$$
$$11\frac{43}{24} = 12\frac{19}{24}$$

12. $9\frac{49}{100}$

$$4\frac{2}{25} = 4\frac{8}{100}$$
$$3\frac{19}{100} = 3\frac{19}{100}$$
$$+ 2\frac{11}{50} = 2\frac{22}{100}$$
$$9\frac{49}{100}$$

Fractions Exercise 14

1. $\frac{1}{5}$

$$\frac{7}{10} = \frac{7}{10}$$
$$- \frac{1}{2} = \frac{5}{10}$$
$$\frac{2}{10} = \frac{1}{5}$$

7. $8\frac{5}{8}$

$$16\frac{1}{4} = 15\frac{2}{8} + \frac{8}{8} = 15\frac{10}{8}$$
$$- 7\frac{5}{8} = \qquad\qquad 7\frac{5}{8}$$
$$8\frac{5}{8}$$

2. $\frac{11}{12}$

$$2\frac{2}{3} = 1\frac{8}{12} + \frac{12}{12} = 1\frac{20}{12}$$
$$- 1\frac{3}{4} = 1\frac{9}{12} \qquad\quad 1\frac{9}{12}$$
$$\frac{11}{12}$$

8. $3\frac{19}{30}$

$$4\frac{3}{10} = 3\frac{9}{30} + \frac{30}{30} = 3\frac{39}{30}$$
$$- \frac{2}{3} = \qquad\qquad \frac{20}{30}$$
$$3\frac{19}{30}$$

3. $\frac{1}{40}$

$$\frac{5}{8} = \frac{25}{40}$$
$$- \frac{3}{5} = \frac{24}{40}$$
$$\frac{1}{40}$$

9. $4\frac{7}{24}$

$$7\frac{11}{12} = 7\frac{22}{24}$$
$$- 3\frac{5}{8} = 3\frac{15}{24}$$
$$4\frac{7}{24}$$

4. $4\frac{9}{20}$

$$8\frac{1}{5} = 7\frac{4}{20} + \frac{20}{20} = 7\frac{24}{20}$$
$$- 3\frac{3}{4} = \qquad\qquad 3\frac{15}{20}$$
$$4\frac{9}{20}$$

10. $5\frac{3}{5}$

$$8\frac{7}{20} = 7\frac{7}{20} + \frac{20}{20} = 7\frac{27}{20}$$
$$- 2\frac{3}{4} = \qquad\qquad 2\frac{15}{20}$$
$$5\frac{12}{20} = 5\frac{3}{5}$$

5. $2\frac{1}{10}$

$$11\frac{3}{5} = 11\frac{6}{10}$$
$$- 9\frac{1}{2} = 9\frac{5}{10}$$
$$2\frac{1}{10}$$

11. $5\frac{5}{18}$

$$9\frac{1}{6} = 8\frac{3}{18} + \frac{18}{18} = 8\frac{21}{18}$$
$$- 3\frac{8}{9} = \qquad\qquad 3\frac{16}{18}$$
$$5\frac{5}{18}$$

6. $3\frac{5}{6}$

$$9\frac{1}{2} = 8\frac{3}{6} + \frac{6}{6} = 8\frac{9}{6}$$
$$- 5\frac{2}{3} = \qquad\qquad 5\frac{4}{6}$$
$$3\frac{5}{6}$$

12. $2\frac{3}{16}$

$$4\frac{1}{2} = 4\frac{8}{16}$$
$$- 2\frac{5}{16} = 2\frac{5}{16}$$
$$2\frac{3}{16}$$

Fractions Exercise 15

1. (4) 11 hours

$$3\frac{1}{2} = 3\frac{2}{4}$$
$$2\frac{3}{4} = 2\frac{3}{4}$$
$$+ 4\frac{3}{4} = 4\frac{3}{4}$$
$$9\frac{8}{4} = 11$$

2. (3) $\frac{5}{6}$ yard

$$4\frac{1}{2} = 3\frac{3}{6} + \frac{6}{6} = 3\frac{9}{6}$$
$$- 3\frac{2}{3} = \qquad\qquad 3\frac{4}{6}$$
$$\frac{5}{6}$$

Exercise 15 cont'd.

3. (1) $6\frac{7}{8}$ pounds

$$2\frac{1}{2} = 2\frac{4}{8}$$
$$2\frac{5}{8} = 2\frac{5}{8}$$
$$+ 1\frac{3}{4} = 1\frac{6}{8}$$
$$5\frac{15}{8} = 6\frac{7}{8}$$

5. (2) $9\frac{3}{5}$ miles

$$12\frac{1}{2} = 11\frac{5}{10} + \frac{10}{10} = 11\frac{15}{10}$$
$$- 2\frac{9}{10} = \qquad\qquad\qquad 2\frac{9}{10}$$
$$9\frac{6}{10} = 9\frac{3}{5}$$

4. (2) $62\frac{3}{4}$ pounds

$$100 = 99\frac{4}{4}$$
$$- 16\frac{3}{4} = 16\frac{3}{4}$$
$$83\frac{1}{4}$$

$$83\frac{1}{4} = 82\frac{1}{4} + \frac{4}{4} = 82\frac{5}{4}$$
$$- 20\frac{1}{2} = \qquad\qquad\quad 20\frac{2}{4}$$
$$62\frac{3}{4}$$

6. (5) 206 pounds

$$162\frac{1}{2}$$
$$+ 43\frac{1}{2}$$
$$205\frac{2}{2} = 206$$

Fractions Exercise 16

1. $\frac{15}{28}$

$$\frac{3}{4} \times \frac{5}{7} = \frac{15}{28}$$

2. $\frac{1}{3}$

$$\frac{2}{3} \times \frac{1}{2} = \frac{2}{6} = \frac{1}{3}$$

3. $\frac{1}{16}$

$$\frac{1}{10} \times \frac{5}{8} = \frac{5}{80} = \frac{1}{16}$$

4. $\frac{7}{40}$

$$\frac{7}{8} \times \frac{1}{5} = \frac{7}{40}$$

5. $\frac{5}{64}$

$$\frac{1}{4} \times \frac{5}{16} = \frac{5}{64}$$

6. $\frac{9}{50}$

$$\frac{3}{10} \times \frac{3}{5} = \frac{9}{50}$$

7. $\frac{2}{15}$

$$\frac{2}{9} \times \frac{3}{5} = \frac{6}{45} = \frac{2}{15}$$

8. $\frac{5}{48}$

$$\frac{5}{6} \times \frac{1}{8} = \frac{5}{48}$$

9. $\frac{1}{10}$

$$\frac{3}{5} \times \frac{1}{6} = \frac{3}{30} = \frac{1}{10}$$

Fractions Exercise 17

1. $\frac{9}{14}$

$$\frac{3}{4} \times \frac{6}{7} = \frac{9}{14}$$

2. $\frac{98}{125}$

$$\frac{14}{15} \times \frac{21}{25} = \frac{98}{125}$$

3. $\frac{49}{108}$

$$\frac{7}{8} \times \frac{14}{27} = \frac{49}{108}$$

4. $\frac{1}{6}$

$$\frac{4}{9} \times \frac{3}{8} = \frac{1}{6}$$

5. $\frac{1}{12}$

$$\frac{8}{8} \times \frac{2}{13} = \frac{1}{12}$$

6. $\frac{3}{5}$

$$\frac{9}{10} \times \frac{2}{3} = \frac{3}{5}$$

7. $\frac{1}{20}$

$$\frac{5}{20} \times \frac{1}{2} = \frac{1}{20}$$

8. $\frac{1}{6}$

$$\frac{4}{5} \times \frac{5}{24} = \frac{1}{6}$$

9. $\frac{3}{10}$

$$\frac{2}{3} \times \frac{9}{20} = \frac{3}{10}$$

10. $\frac{3}{5}$

$$\frac{6}{7} \times \frac{7}{8} \times \frac{4}{5} = \frac{3}{5}$$

11. $\frac{1}{5}$

$$\frac{9}{10} \times \frac{1}{4} \times \frac{8}{9} = \frac{1}{5}$$

12. $\frac{5}{32}$

$$\frac{3}{4} \times \frac{2}{9} \times \frac{15}{16} = \frac{5}{32}$$

Fractions Exercise 18

1. 8

$$\frac{1}{2} \times \frac{16}{1} = \frac{8}{1} = 8$$

2. $6\frac{2}{3}$

$$\frac{10}{1} \times \frac{2}{3} = \frac{20}{3} = 6\frac{2}{3}$$

3. $4\frac{1}{2}$

$$\frac{3}{8} \times \frac{12}{1} = \frac{9}{2} = 4\frac{1}{2}$$

4. $5\frac{3}{5}$

$$\frac{8}{1} \times \frac{7}{10} = \frac{28}{5} = 5\frac{3}{5}$$

5. 4

$$\frac{2}{5} \times \frac{10}{1} = \frac{4}{1} = 4$$

6. $4\frac{19}{20}$

$$\frac{9}{1} \times \frac{11}{20} = \frac{99}{20} = 4\frac{19}{20}$$

7. $7\frac{1}{2}$

$$\frac{\cancel{5}}{\cancel{6}_2} \times \frac{\cancel{9}^3}{1} = \frac{15}{2} = 7\frac{1}{2}$$

9. 12

$$\frac{\cancel{18}^6}{1} \times \frac{2}{\cancel{3}} = \frac{12}{1} = 12$$

11. $9\frac{3}{5}$

$$\frac{4}{\cancel{15}_5} \times \frac{\cancel{36}^{12}}{1} = \frac{48}{5} = 9\frac{3}{5}$$

8. $1\frac{1}{20}$

$$\frac{\cancel{15}^3}{1} \times \frac{7}{\cancel{100}_{20}} = \frac{21}{20} = 1\frac{1}{20}$$

10. 16

$$\frac{\cancel{20}^4}{1} \times \frac{4}{\cancel{5}} = \frac{16}{1} = 16$$

12. $7\frac{1}{5}$

$$\frac{\cancel{24}^{12}}{1} \times \frac{3}{\cancel{10}_5} = \frac{36}{5} = 7\frac{1}{5}$$

Fractions Exercise 19

1. $\frac{8}{3}$

$$2\frac{2}{3} = \frac{6}{3} + \frac{2}{3} = \frac{8}{3}$$

2. $\frac{13}{8}$

$$1\frac{5}{8} = \frac{8}{8} + \frac{5}{8} = \frac{13}{8}$$

3. $\frac{42}{5}$

$$8\frac{2}{5} = \frac{40}{5} + \frac{2}{5} = \frac{42}{5}$$

4. $\frac{13}{4}$

$$3\frac{1}{4} = \frac{12}{4} + \frac{1}{4} = \frac{13}{4}$$

5. $\frac{23}{6}$

$$3\frac{5}{6} = \frac{18}{6} + \frac{5}{6} = \frac{23}{6}$$

6. $\frac{39}{7}$

$$5\frac{4}{7} = \frac{35}{7} + \frac{4}{7} = \frac{39}{7}$$

7. $\frac{7}{2}$

$$3\frac{1}{2} = \frac{6}{2} + \frac{1}{2} = \frac{7}{2}$$

8. $\frac{22}{3}$

$$7\frac{1}{3} = \frac{21}{3} + \frac{1}{3} = \frac{22}{3}$$

9. $\frac{56}{9}$

$$6\frac{2}{9} = \frac{54}{9} + \frac{2}{9} = \frac{56}{9}$$

10. $\frac{31}{3}$

$$10\frac{1}{3} = \frac{30}{3} + \frac{1}{3} = \frac{31}{3}$$

11. $\frac{63}{5}$

$$12\frac{3}{5} = \frac{60}{5} + \frac{3}{5} = \frac{63}{5}$$

12. $\frac{37}{4}$

$$9\frac{1}{4} = \frac{36}{4} + \frac{1}{4} = \frac{37}{4}$$

13. $\frac{41}{3}$

$$13\frac{2}{3} = \frac{39}{3} + \frac{2}{3} = \frac{41}{3}$$

14. $\frac{46}{3}$

$$15\frac{1}{3} = \frac{45}{3} + \frac{1}{3} = \frac{46}{3}$$

15. $\frac{35}{8}$

$$4\frac{3}{8} = \frac{32}{8} + \frac{3}{8} = \frac{35}{8}$$

Fractions Exercise 20

1. $5\frac{5}{6}$ $1\frac{2}{3} \times 3\frac{1}{2} =$

$$\frac{5}{3} \times \frac{7}{2} = \frac{35}{6} = 5\frac{5}{6}$$

2. 1 $1\frac{1}{2} \times \frac{2}{3} =$

$$\frac{\cancel{3}}{\cancel{2}_1} \times \frac{\cancel{2}}{\cancel{3}_1} = \frac{1}{1} = 1$$

3. $21\frac{1}{8}$ $6\frac{1}{2} \times 3\frac{1}{4} =$

$$\frac{13}{2} \times \frac{13}{4} = \frac{169}{8} = 21\frac{1}{8}$$

4. $1\frac{7}{18}$ $\frac{5}{8} \times 2\frac{2}{9} =$

$$\frac{5}{\cancel{8}_2} \times \frac{\cancel{20}^5}{9} = \frac{25}{18} = 1\frac{7}{18}$$

5. $2\frac{1}{2}$ $\frac{2}{3} \times 3\frac{3}{4} =$

$$\frac{\cancel{2}^1}{\cancel{3}_1} \times \frac{\cancel{15}^5}{\cancel{4}_2} = \frac{5}{2} = 2\frac{1}{2}$$

6. $3\frac{1}{2}$ $2\frac{5}{8} \times 1\frac{1}{3} =$

$$\frac{\cancel{21}^7}{\cancel{8}_2} \times \frac{\cancel{4}^1}{\cancel{3}_1} = \frac{7}{2} = 3\frac{1}{2}$$

7. $2\frac{7}{10}$ $1\frac{1}{2} \times \frac{4}{5} \times 2\frac{1}{4} =$

$$\frac{3}{2} \times \frac{\cancel{4}^1}{5} \times \frac{9}{\cancel{4}_1} = \frac{27}{10} = 2\frac{7}{10}$$

8. $3\frac{1}{2}$ $\frac{5}{8} \times 4 \times 1\frac{2}{5} =$

$$\frac{\cancel{5}^1}{\cancel{8}_2} \times \frac{\cancel{4}^1}{1} \times \frac{7}{\cancel{5}_1} = \frac{7}{2} = 3\frac{1}{2}$$

9. 24 $2\frac{1}{2} \times 1\frac{3}{5} \times 6 =$

$$\frac{\cancel{5}}{\cancel{2}_1} \times \frac{\cancel{8}^4}{\cancel{5}_1} \times \frac{6}{1} = \frac{24}{1} = 24$$

10. $2\frac{6}{11}$ $2\frac{1}{2} \times 1\frac{3}{5} \times \frac{7}{11} =$

$$\frac{\cancel{5}}{\cancel{2}_1} \times \frac{\cancel{8}^4}{\cancel{5}_1} \times \frac{7}{11} = \frac{28}{11} = 2\frac{6}{11}$$

11. 9 $1\frac{7}{20} \times 5\frac{1}{3} \times 1\frac{1}{4} =$

$$\frac{\cancel{27}}{\cancel{20}_4} \times \frac{\cancel{16}^1}{\cancel{3}_1} \times \frac{\cancel{5}^1}{\cancel{4}_1} = \frac{9}{1} = 9$$

Exercise 20 cont'd.

12. $3\frac{1}{8}$ $1\frac{7}{8} \times \frac{3}{4} \times 2\frac{2}{9} =$

$\frac{15}{8} \times \frac{3}{4} \times \frac{20}{9} = \frac{25}{8} = 3\frac{1}{8}$

13. $1\frac{5}{12}$ $\frac{3}{8} \times 2\frac{5}{6} \times 1\frac{1}{3} =$

$\frac{3}{8} \times \frac{17}{6} \times \frac{4}{3} = \frac{17}{12} = 1\frac{5}{12}$

14. 9 $1\frac{5}{16} \times 2\frac{6}{7} \times 2\frac{2}{5} =$

$\frac{21}{16} \times \frac{20}{7} \times \frac{12}{5} = \frac{9}{1} = 9$

Fractions Exercise 21

1. 2

$\frac{1}{3} \div \frac{1}{6} =$

$\frac{1}{3} \times \frac{6}{1} = \frac{2}{1} = 2$

2. 6

$5 \div \frac{5}{6} =$

$\frac{5}{1} \times \frac{6}{5} = \frac{6}{1} = 6$

3. 6

$4\frac{1}{2} \div \frac{3}{4} =$

$\frac{9}{2} \times \frac{4}{3} = \frac{6}{1} = 6$

4. 20

$2\frac{2}{3} \div \frac{2}{15} =$

$\frac{8}{3} \times \frac{15}{2} = \frac{20}{1} = 20$

5. $\frac{1}{2}$

$\frac{1}{3} \div \frac{2}{3} =$

$\frac{1}{3} \times \frac{3}{2} = \frac{1}{2}$

6. 2

$\frac{5}{7} \div \frac{5}{14} =$

$\frac{5}{7} \times \frac{14}{5} = \frac{2}{1} = 2$

7. $10\frac{2}{3}$

$4 \div \frac{3}{8} =$

$\frac{4}{1} \times \frac{8}{3} = \frac{32}{3} = 10\frac{2}{3}$

8. $\frac{20}{27}$

$\frac{5}{9} \div \frac{3}{4} =$

$\frac{5}{9} \times \frac{4}{3} = \frac{20}{27}$

9. $6\frac{2}{3}$

$5\frac{5}{6} \div \frac{7}{8} =$

$\frac{35}{6} \times \frac{8}{7} = \frac{20}{3} = 6\frac{2}{3}$

10. $1\frac{1}{2}$

$\frac{9}{10} \div \frac{3}{5} =$

$\frac{9}{10} \times \frac{5}{3} = \frac{3}{2} = 1\frac{1}{2}$

11. 12

$10 \div \frac{5}{6} =$

$\frac{10}{1} \times \frac{6}{5} = \frac{12}{1} = 12$

12. 10

$3\frac{1}{3} \div \frac{1}{3} =$

$\frac{10}{3} \times \frac{3}{1} = \frac{10}{1} = 10$

Fractions Exercise 22

1. $6\frac{2}{3}$

$10 \div 1\frac{1}{2} =$

$\frac{10}{1} \div \frac{3}{2} =$

$\frac{10}{1} \times \frac{2}{3} = \frac{20}{3} = 6\frac{2}{3}$

2. $\frac{5}{12}$

$1\frac{1}{3} \div 3\frac{1}{5} =$

$\frac{4}{3} \div \frac{16}{5} =$

$\frac{4}{3} \times \frac{5}{16} = \frac{5}{12}$

3. $4\frac{1}{2}$

$6 \div 1\frac{1}{3} =$

$\frac{6}{1} \div \frac{4}{3} =$

$\frac{6}{1} \times \frac{3}{4} = \frac{9}{2} = 4\frac{1}{2}$

4. $\frac{10}{13}$

$2\frac{1}{2} \div 3\frac{1}{4} =$

$\frac{5}{2} \div \frac{13}{4} =$

$\frac{5}{2} \times \frac{4}{13} = \frac{10}{13}$

5. 5

$21 \div 4\frac{1}{5} =$

$\frac{21}{1} \div \frac{21}{5} =$

$\frac{21}{1} \times \frac{5}{21} = \frac{5}{1} = 5$

6. $1\frac{1}{9}$

$2\frac{2}{9} \div 2 =$

$\frac{20}{9} \div \frac{2}{1} =$

$\frac{20}{9} \times \frac{1}{2} = \frac{10}{9} = 1\frac{1}{9}$

7. $\frac{3}{10}$

$\frac{9}{10} \div 3 =$

$\frac{9}{10} \div \frac{3}{1} =$

$\frac{9}{10} \times \frac{1}{3} = \frac{3}{10}$

8. $\frac{1}{4}$

$1\frac{3}{4} \div 7 =$

$\frac{7}{4} \div \frac{7}{1} =$

$\frac{7}{4} \times \frac{1}{7} = \frac{1}{4}$

9. $\frac{5}{6}$

$5\frac{5}{6} \div 7 =$

$\frac{35}{6} \div \frac{7}{1} =$

$\frac{35}{6} \times \frac{1}{7} = \frac{5}{6}$

10. $\frac{15}{64}$

$\frac{3}{4} \div 3\frac{1}{5} =$

$\frac{3}{4} \div \frac{16}{5} =$

$\frac{3}{4} \times \frac{5}{16} = \frac{15}{64}$

11. $3\frac{7}{16}$

$13\frac{3}{4} \div 4 =$

$\frac{55}{4} \div \frac{4}{1} =$

$\frac{55}{4} \times \frac{1}{4} = \frac{55}{16} = 3\frac{7}{16}$

12. 6

$10 \div 1\frac{2}{3} =$

$\frac{10}{1} \div \frac{5}{3} =$

$\frac{10}{1} \times \frac{3}{5} = \frac{6}{1} = 6$

Fractions Exercise 23

1. (5) $\$\frac{1}{4}$ million

$$1\frac{1}{2} \div 6 =$$

$$\frac{3}{2} \div \frac{6}{1} =$$

$$\frac{\overset{1}{\cancel{3}}}{2} \times \frac{1}{\underset{2}{\cancel{6}}} = \$\frac{1}{4} \text{ million}$$

2. (3) 6

$$\frac{3}{4} \div \frac{1}{8} =$$

$$\frac{3}{\cancel{4}} \times \frac{\overset{2}{\cancel{8}}}{1} = \frac{6}{1} = 6$$

3. (1) $\$4,500$

$$\frac{1}{\cancel{8}} \times \frac{\overset{4,500}{\cancel{22,500}}}{1} = \$4,500$$

4. (3) 51,000

$$\frac{2}{\cancel{5}} \times \frac{\overset{17,000}{\cancel{85,000}}}{1} = 34,000$$

$$\begin{array}{r} 85,000 \\ -34,000 \\ \hline 51,000 \end{array}$$

5. (5) $\$69.70$

$$8\frac{1}{2} \times 8.20 =$$

$$\frac{17}{\cancel{2}} \times \frac{\overset{4.10}{\cancel{8.20}}}{1} = \$69.70$$

6. (3) $\$24$

$$\frac{1}{\cancel{4}} \times \frac{\overset{8}{\cancel{32}}}{1} = 8$$

$$\begin{array}{r} \$32 \\ - \quad 8 \\ \hline \$24 \end{array}$$

Fractions Exercise 24

1. (4) $30\frac{7}{8}$

$$\begin{array}{r} 32\frac{1}{4} = 32\frac{2}{8} \\ + 1\frac{1}{8} = 1\frac{1}{8} \\ \hline 33\frac{3}{8} \end{array}$$

$$\begin{array}{r} 33\frac{3}{8} = 32\frac{3}{8} + \frac{8}{8} = 32\frac{11}{8} \\ - 2\frac{1}{2} = \qquad 2\frac{4}{8} \\ \hline 30\frac{7}{8} \end{array}$$

2. (2) 20

$$30 \div 1\frac{1}{2} = \frac{30}{1} \div \frac{3}{2} =$$

$$\frac{\overset{10}{\cancel{30}}}{1} \times \frac{2}{\cancel{3}} = 20$$

Notice the 30-acre field is being divided. 30 comes first in the solution.

3. (2) $2\frac{3}{4}$ yds.

$$\begin{array}{r} 9 = 8\frac{2}{2} \\ - 2\frac{1}{2} = 2\frac{1}{2} \\ \hline 6\frac{1}{2} \end{array}$$

$$\begin{array}{r} 6\frac{1}{2} = 5\frac{2}{4} + \frac{4}{4} = 5\frac{6}{4} \\ - 3\frac{3}{4} = \qquad 3\frac{3}{4} \\ \hline 2\frac{3}{4} \end{array}$$

4. (2) $\$4$

$$38 \div 9\frac{1}{2} = \frac{38}{1} \div \frac{19}{2} =$$

$$\frac{\overset{2}{\cancel{38}}}{1} \times \frac{2}{\cancel{19}} = \$4$$

5. (5) $\$160$

$$\frac{3}{\cancel{8}} \times \frac{\overset{32}{\cancel{256}}}{1} = 96$$

$$\frac{1}{\cancel{4}} \times \frac{\overset{64}{\cancel{256}}}{1} = 64$$

$$\begin{array}{r} 96 \\ + 64 \\ \hline 160 \end{array}$$

6. (1) $22\frac{1}{2}$ hours. $\quad 3 \times 5 = 15$ *hours per week*

$$\begin{array}{r} 15 = 15 \\ 1\frac{1}{4} = 1\frac{1}{4} \\ 2\frac{3}{4} = 2\frac{3}{4} \\ + 3\frac{1}{2} = 3\frac{2}{4} \\ \hline 21\frac{6}{4} = 22\frac{2}{4} = 22\frac{1}{2} \end{array}$$

7. (2) 500 pounds $\quad 8 \times 62\frac{1}{2}$

$$\frac{\overset{4}{\cancel{8}}}{1} \times \frac{125}{\cancel{4}} = 500$$

8. (4) $\frac{7}{24}$

$$\begin{array}{r} \frac{1}{3} = \frac{8}{24} \\ \frac{1}{4} = \frac{6}{24} \\ + \frac{1}{8} = \frac{3}{24} \\ \hline \frac{17}{24} \end{array}$$

$$\begin{array}{r} 1 = \frac{24}{24} \\ - \frac{17}{24} = \frac{17}{24} \\ \hline \frac{7}{24} \end{array}$$

Exercise 24 cont'd.

9. (2) $33.60 $\quad 5\frac{1}{4} \times 6.40 =$

$$\frac{21}{4} \times \frac{\overset{1.60}{6.40}}{1} = \$33.60$$

10. (1) $20\frac{1}{2}$ inches $\quad 82 \div 4 =$

$$\frac{\overset{41}{82}}{1} \times \frac{1}{\underset{2}{4}} = \frac{41}{2} = 20\frac{1}{2}$$

Fractions Test

1. (5) $\frac{13}{36}$

2. (4) $\frac{9}{10}$ $\quad 30 - 3 = 27$

$\qquad \frac{27}{30} = \frac{9}{10}$

3. (1) $\frac{36}{60}$ $\quad \frac{3 \times 12}{5 \times 12} = \frac{36}{60}$

4. (2) $5\frac{1}{3}$ $\quad \begin{array}{c} 5\frac{3}{9} = 5\frac{1}{3} \\ 9\overline{)48} \end{array}$

5. (4) $\frac{35}{8}$ $\quad 8 \times 4 = 32$

$\qquad 32 + 3 = 35$

$\qquad 4\frac{3}{8} = \frac{35}{8}$

6. (5) $4\frac{3}{40}$ $\quad 4.075 = 4\frac{75}{1,000} = 4\frac{3}{40}$

7. (1) $.83\frac{1}{3}$ $\quad \begin{array}{r} .83\frac{2}{6} = .83\frac{1}{3} \\ 6\overline{)5.00} \\ \underline{4\,8} \\ 20 \\ \underline{18} \\ 2 \end{array}$

8. (5) $15\frac{17}{24}$ $\quad \begin{array}{r} 8\frac{1}{4} = 8\frac{6}{24} \\ 2\frac{5}{6} = 2\frac{20}{24} \\ +4\frac{5}{8} = 4\frac{15}{24} \\ \hline 14\frac{41}{24} = 15\frac{17}{24} \end{array}$

9. (3) $5\frac{1}{12}$ $\quad \begin{array}{r} 15\frac{1}{3} = 15\frac{4}{12} \\ -10\frac{1}{4} = 10\frac{3}{12} \\ \hline 5\frac{1}{12} \end{array}$

10. (4) $43\frac{1}{4}$ pounds $\quad \begin{array}{r} 24\frac{3}{4} = 24\frac{3}{4} \\ +18\frac{1}{2} = 18\frac{2}{4} \\ \hline 42\frac{5}{4} = 43\frac{1}{4} \end{array}$

11. (2) $5\frac{2}{3}$ pounds $\quad \begin{array}{r} 10 = 9\frac{3}{3} \\ -4\frac{1}{3} = 4\frac{1}{3} \\ \hline 5\frac{2}{3} \end{array}$

12. (4) $12\frac{1}{2}$ $\quad 3\frac{3}{4} \times 3\frac{1}{3} =$

$$\frac{\overset{5}{15}}{\underset{2}{4}} \times \frac{\overset{5}{10}}{\underset{1}{3}} = \frac{25}{2} = 12\frac{1}{2}$$

13. (4) 7 $\quad 24\frac{1}{2} \div 3\frac{1}{2} = \frac{49}{2} \div \frac{7}{2}$

$$\frac{\overset{7}{49}}{\underset{1}{2}} \times \frac{\overset{1}{2}}{\underset{1}{7}} = \frac{7}{1} = 7$$

14. (5) $89 $\quad \frac{1}{3} \times \frac{\overset{89}{267}}{1} = \frac{89}{1} = 89$

15. (4) $9.20 $\quad 41.40 \div 4\frac{1}{2} = \frac{41.40}{1} \div \frac{9}{2} =$

$$\frac{\overset{4.60}{41.40}}{1} \times \frac{2}{\underset{1}{9}} = \frac{9.20}{1} = 9.20$$

Mixed Review

1. (3) 3,307

$$\begin{array}{r} 4,000 \\ -\ \ 693 \\ \hline 3,307 \end{array}$$

2. (5) 490,000

3. (4) 458

$$\begin{array}{r} 483 \\ 527 \\ +364 \\ \hline 1,374 \end{array} \qquad \begin{array}{r} 458 \\ 3\overline{)1,374} \end{array}$$

4. (5) .054

5. (1) 12.32

$$\begin{array}{r} 13.00 \\ -\ \ .68 \\ \hline 12.32 \end{array}$$

6. (4) 1,001.25 miles

$$\begin{array}{r} 500.7 \\ 200.25 \\ 300.3 \\ \hline 1,001.25 \end{array}$$

7. (3) $1.50

$$\begin{array}{r} \$.6\,8 \\ \times 2.2 \\ \hline 136 \\ 136 \\ \hline \end{array}$$

$1.496 to the
nearest cent =
$1.50

8. (4) $10\frac{1}{4}$ hours

$$\begin{array}{r} 7\frac{1}{2} = 7\frac{2}{4} \\ +2\frac{3}{4} = 2\frac{3}{4} \\ \hline 9\frac{5}{4} = 10\frac{1}{4} \end{array}$$

9. (3) $55.80

$$7\frac{1}{2} \times 4.80 = \frac{15}{\cancel{2}_{1}} \times \frac{\overset{2.40}{\cancel{4.80}}}{1} =$$

$$\frac{36.00}{1} = 36.00$$

$$2\frac{3}{4} \times 7.20 = \frac{11}{\cancel{4}_{1}} \times \frac{\overset{1.80}{\cancel{7.20}}}{1} =$$

$$\frac{19.80}{1} = 19.80$$

$$\begin{array}{r} \$36.00 \\ +\ 19.80 \\ \hline \$55.80 \end{array}$$

10. (1) $1\frac{1}{2}$ feet

$$4\overline{)12}\ \ ^{3}$$

$$\frac{3}{1} \times \frac{1}{2} = \frac{3}{2} = 1\frac{1}{2}$$

Percents

THE USES OF PERCENTS

Percents are commonly used in the business world and by consumers. Percents are used to measure commissions, taxes, interest, markups, and discounts.

In the previous sections of this book, you have used fractions and decimals to indicate parts of a whole. You can also use percents to indicate parts.

Percent is based on the idea of 100 equal parts. The **percent sign**, %, means "per 100" or "out of 100."

For example, if you worked 87 out of 100 days, you could write this fractionally as $\frac{87}{100}$ or as the decimal .87—both meaning eighty-seven hundredths. You could also say that you worked 87% of the days.

THE MEANING OF 100%

The preceding explanation shows that 87% represents 87 out of 100 parts. In addition to representing 100 equal parts, 100% also represents a total amount or a whole.

For instance, if a club has 57 members and all 57 are at a picnic, this could be written fractionally as $\frac{57}{57}$. You know that $\frac{57}{57} = 1 = 100\%$. You could say that 100%, or all of the members are there.

The concept of 100% as a total will be useful throughout your work with percents.

PERCENTS LESS THAN 1% AND PERCENTS MORE THAN 100%

Percents Less Than 1%

You have seen that 1% represents 1 out of 100 equal parts. Can there be a number smaller than 1%? Certainly. The box below is divided into 100 smaller boxes. One-half of one of the smaller boxes is shaded.

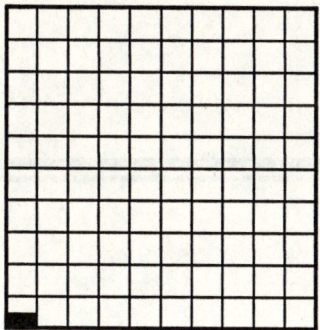

Each small box represents 1% of the entire box. The shaded portion is $\frac{1}{2}$ of 1% and can be written as $\frac{1}{2}$%. The decimal .5% means the same, since $.5\% = \frac{5}{10}\% = \frac{1}{2}\%$.

Percents More Than 100%

If 100% is a total, can there be a number greater than 100%? Think of this situation. If a class grows in size from 15 to 30 students, it grew to 200% of the original class size.

At the beginning, the class was $\frac{15}{15}$, or a whole — 100%. The new class size became $\frac{30}{15} = 2$, or 2 times the total of the original class size.

$$2 \times 100\% = 200\%.$$

These types of numbers may not be as familiar to you as the percents from 1% to 100%. Be sure to read every percent carefully to understand its value.

PERCENT EXERCISE 1

Match the correct percent with the value written in the column on the left side.

1. $\frac{12}{12}$ _____(a) $700\frac{1}{3}\%$

2. $\frac{24}{12}$ _____(b) $\frac{1}{3}\%$

3. one-third of one percent   _____(c) $7\frac{1}{3}\%$

4. seven and one-third percent   _____(d)   100%

5. seventy and one-third percent   _____(e) $70\frac{1}{3}\%$

6. seven hundred and one-third percent _____(f)   200%

Answers and solutions start on page 168.

DECIMALS AND PERCENTS

Changing A Percent To A Decimal

In some problems, you may need to change a percent to a decimal or a decimal to a percent. The following examples will show you how to do this.

Rule for Changing a Percent to a Decimal

Drop the percent sign (%) and move the decimal point two places to the left. Remember from your work with decimals that a whole number written without a decimal point is understood to have a decimal point to the right of the units' digit.

> EXAMPLES: 25% = .25. = **.25**
>   165% = 1.65. = **1.65**

Sometimes you can drop zeros to the right of the digits after you move the decimal point.

Note: Changing a percent to a decimal is like dividing by 100.

> EXAMPLES: 20% = .20. = **.2**
>   250% = 2.50. = **2.5**

Sometimes, you may have to write extra zeros to the left of the digits in order to move the point two places.

> EXAMPLES: 3.2% = .03.2 = **.032**
>   .25% = .00.25 = **.0025**

If a percent contains a fraction, act as if there were a decimal point to the right of the last whole number digit.

EXAMPLE: $33\frac{1}{3}\% = .33.\frac{1}{3} = .33\frac{1}{3}$

PERCENT EXERCISE 2

Change each percent to a decimal or a whole number.

1. $9\% =$ **5.** $87\frac{1}{2}\% =$ **9.** $2.7\% =$

2. $24\% =$ **6.** $8\frac{1}{3}\% =$ **10.** $3.95\% =$

3. $100\% =$ **7.** $.15\% =$ **11.** $57\% =$

4. $.3\% =$ **8.** $275\% =$ **12.** $1,000\% =$

Answers and solutions start on page 168.

Changing A Decimal To A Percent

Rule for Changing a Decimal to a Percent

Move the decimal point two places to the <u>right</u>, and write the percent sign after the last digit.

EXAMPLES:
$.25 = .25. = \mathbf{25\%}$
$.19 = .19. = \mathbf{19\%}$
$3.65 = 3.65. = \mathbf{365\%}$
$.625 = .62.5 = \mathbf{62.5\%}$

Notice that you do not have to write the decimal point if it moves to the end. The point is not written to the right of the units place in whole numbers.

Sometimes you will have to write extra zeros in order to move two places.

EXAMPLES:
$.6 = .60. = \mathbf{60\%}$
$2.7 = 2.70. = \mathbf{270\%}$
$36 = 36.00. = \mathbf{3,600\%}$

When the point moves between a digit and a fraction, you do not write the point.

$$\text{EXAMPLES:} \quad .37\tfrac{1}{2} = .37.\tfrac{1}{2} = 37\tfrac{1}{2}\%$$
$$.14\tfrac{1}{4} = .14.\tfrac{1}{4} = 14\tfrac{1}{4}\%$$
$$.05\tfrac{1}{3} = .05.\tfrac{1}{3} = 5\tfrac{1}{3}\%$$

You can drop unnecessary zeros after you move the point.

$$\text{EXAMPLES:} \quad .045 = .04.5 = 4.5\%$$
$$.003 = .00.3 = .3\%$$
$$.0008 = .00.08 = .08\%$$

If you are unsure about which way to move the decimal point in a problem, think of this diagram.

$$\boxed{\begin{array}{c} D \longleftrightarrow P \\ .25 = 25\% \end{array}}$$

Written in alphabetical order, the D is to the left of the P. Therefore, to change a percent to a decimal, move the point to the left, and to change a decimal to a percent, move the point to the right.

PERCENT EXERCISE 3

Change each decimal to a percent.

1.	.81 =	**5.**	.0009 =	**9.**	2.1 =
2.	.37½ =	**6.**	.217 =	**10.**	4.85 =
3.	.5 =	**7.**	.03 =	**11.**	3.924 =
4.	.004 =	**8.**	.33⅓ =	**12.**	.015 =

Answers and solutions start on page 168.

PERCENTS AND FRACTIONS

Changing a Fraction to a Percent

Suppose Alfonso usually works five days a week. If he is out sick one day,

Percents **143**

you could write this as a fraction: $\frac{1}{5}$. This means that he was sick $\frac{1}{5}$ of the total (100%) work week.

In the section on fractions, you learned that you multiply to find a "fraction of" some quantity. When you change a fraction to a percent, you are finding a part of 100%. This is the basis for understanding the following rule.

Rule for Changing a Fraction to a Percent
Multiply the fraction by 100%.

EXAMPLE 1: Change $\frac{3}{4}$ to a percent.

$$\frac{3}{4} = \frac{3}{\overset{}{\underset{1}{\cancel{4}}}} \times \frac{\overset{25}{\cancel{100\%}}}{1} = 75\%$$

Sometimes, the denominator of the fraction does not divide evenly into 100 and the percent will contain a fraction.

EXAMPLE 2: Change $\frac{5}{7}$ to a percent.

$$\frac{5}{7} = \frac{5}{7} \times \frac{100}{1} = \frac{500}{7} = 71\frac{3}{10}\%$$

When working with an improper fraction, it is worthwhile to reduce the fraction first, if possible, before multiplying.

EXAMPLE 3: Change $\frac{30}{25}$ to a percent.

① $\boxed{\dfrac{30}{25} = \dfrac{6}{5}}$ ② $\boxed{\dfrac{6}{\underset{1}{\cancel{5}}} \times \dfrac{\overset{20}{\cancel{100\%}}}{1} = 120\%}$

Step 1. Reduce the fraction.
Step 2. Multiply the reduced fraction by 100%.

PERCENT EXERCISE 4 ——————————————

Change each fraction to a percent.

1. $\frac{1}{5} =$ 5. $\frac{14}{8} =$ 9. $\frac{1}{6} =$

2. $\frac{5}{6} =$ 6. $\frac{9}{10} =$ 10. $\frac{10}{5} =$

3. $\frac{3}{8} =$ 7. $\frac{5}{12} =$ 11. $\frac{1}{12} =$

4. $\frac{2}{3} =$ 8. $\frac{6}{7} =$ 12. $\frac{2}{11} =$

Answers and solutions start on page 168.

Changing a Percent to a Fraction

A percent is similar to a fraction with a denominator of 100. For example, 20% is the same as $\frac{20}{100}$.

> ### Rule for Changing a Percent to a Fraction or Mixed Number
> Replace the % sign with a denominator of 100 and reduce.

EXAMPLE 4: Change 75% to a fraction.

$$75\% = \frac{75}{100} = \frac{3}{4}$$

EXAMPLE 5: Change 125% to a mixed number.

$$125\% = \frac{125}{100} = 1\frac{25}{100} = 1\frac{1}{4}$$

Some percents, like $16\frac{2}{3}\%$, are hard to reduce when you replace the percent sign with 100. To simplify these percents, <u>divide</u> by 100 instead of trying to reduce a fraction. It is possible to divide by 100 because a fraction bar also means "divided by." $\frac{16\frac{2}{3}}{100}$ means $16\frac{2}{3}$ divided by 100.

EXAMPLE 6: Change $16\frac{2}{3}\%$ to a fraction.

① $\boxed{16\frac{2}{3}\% = \frac{16\frac{2}{3}}{100}}$ ② $\boxed{\frac{16\frac{2}{3}}{100} = 16\frac{2}{3} \div 100}$ ③ $\boxed{\frac{50}{3} \div \frac{100}{1} = \frac{\overset{1}{\cancel{50}}}{3} \times \frac{1}{\underset{2}{\cancel{100}}} = \frac{1}{6}}$

Step 1. Write $16\frac{2}{3}\%$ over a denominator of 100.

Step 2. Rewrite this as a division problem, since the fraction bar means "divided by."

Step 3. Change $16\frac{2}{3}$ to an improper fraction, invert the 100 to $\frac{1}{100}$, and multiply.

Percents containing decimals are also difficult to change to fractions. First, change these percents to decimals. Then, change the decimals to fractions and reduce.

EXAMPLE 7: Change 2.5% to a fraction.

① $\boxed{2.5\% = .02.5 = .025}$ ② $\boxed{.025 = \frac{25}{1,000} = \frac{1}{40}}$

Step 1. Change 2.5% to a decimal by dropping the percent sign and moving the decimal point two places to the left.

Step 2. Change .025 to a fraction. Three places are thousandths. 25 is the numerator and 1,000 is the denominator. Reduce the fraction to lowest terms.

PERCENT EXERCISE 5

Change each percent to a fraction or a mixed number.

1. $45\% =$

2. $37\frac{1}{2}\% =$

3. $6\frac{2}{3}\% =$

4. $8\% =$

5. $2\% =$

6. $83\frac{1}{3}\% =$

7. $24\% =$

8. $33\frac{1}{3}\% =$

9. $28\frac{4}{7}\% =$

10. $80\% =$

11. $150\% =$

12. $12\frac{1}{2}\% =$

13. $1.5\% =$

14. $.09\% =$

15. $.6\% =$

16. $1\% =$

Answers and solutions start on page 168.

COMMON FRACTIONS, DECIMALS, AND PERCENTS

The next exercise is a chart that you should fill in. When you finish, check your answers. Then, take the time to memorize this list. The values on this chart are some of the most commonly used percents, decimals, and fractions. You will save time on many problems if you know these equivalencies. For example, you may find it easier in some problems to use the fraction $\frac{1}{4}$ than to use .25 or 25%.

Notice that the first line has already been filled in.

PERCENT EXERCISE 6 ────────────────────

fraction	decimal	percent
$\frac{1}{4}$	.25	25%
$\frac{1}{2}$		
		75%
	.125 or .12$\frac{1}{2}$	
		37$\frac{1}{2}$% or 37.5%
$\frac{5}{8}$		
	.875 or .87$\frac{1}{2}$	
		20%
	.4	
$\frac{3}{5}$		
		80%
	.1	
$\frac{3}{10}$		
		70%
	.9	
	.33$\frac{1}{3}$	
$\frac{2}{3}$		
		16$\frac{2}{3}$%
$\frac{5}{6}$		

Answers start on page 169.

══════════════════════════════════════

THE PERCENT BOX

3 cans are 50% of a six-pack. The statement "3 is 50% of 6" has three numbers. 3 is the **part.** 6 is the **whole.** And 50 is the **percent.** The parts of a percent problem fit into a diagram called "the percent box."

The Percent Box	**part**	**%**
	whole	**100**

The percent box is composed of two equivalent fractions:

$$\frac{\text{Part (3 cans)}}{\text{(6 cans) Whole}} = \frac{\text{Percent (50\%)}}{\text{(100\%) Total}}$$

In the example above, $\frac{3}{6} = \frac{50}{100}$. These are equivalent because both can be reduced to $\frac{1}{2}$. $\frac{50}{100}$ is a fractional representation of 50%.

The percent box is a tool for organizing the information in a percent problem. The numbers in the statement "3 is 50% of 6" fit into the percent box as follows:

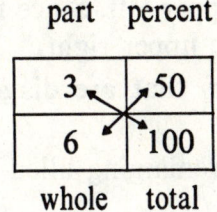

You can use the percent box to solve almost any percent problem.

Rules for Using the Percent Box

part	%
whole	100

1. Fill in the information given in a percent problem as shown in the diagram above.

2. Multiply the numbers in the two filled-in boxes that are diagonal (criss-cross) to each other.

3. Divide by the number in the remaining filled-in box.

100 is the easiest part of the percent box to fill in. 100 <u>always</u> goes in the lower right corner because 100% is a total. The percent is also easy to fill in because in a problem, the percent usually has the % sign. The part and the whole are harder to recognize. Usually, the whole is preceded by the word *of*.

USING THE PERCENT BOX TO FIND A PERCENT, A PART, OR A WHOLE

One of the advantages of the percent box is that it can be used to find either a percent, a part, or a whole. The key to successful use of the percent box is putting the numbers in the right boxes.

Finding the Percent

EXAMPLE 1:   27 is what % of 45?

①
27	%
45	100

②
$$
\begin{array}{r}
27 \\
\times 100 \\
\hline
2,700
\end{array}
$$

③
$$
\begin{array}{r}
60\% \\
45\overline{)2,700} \\
2\,70 \\
\hline
00
\end{array}
$$

Step 1.   Fill in the percent box. 100 always goes in the lower right portion of the box. 45 is the whole. Put 45 in the lower left. 27 is the part. Put 27 in the upper left. Since the percent is asked for, put a % sign in the upper right.

Step 2.   Multiply the filled-in numbers that are diagonal to each other, 27 and 100.

Step 3.   Divide by the number in the remaining filled-in portion of the box, 45. 27 is **60%** of 45.

Finding the Part

You can also use the percent box to find the part when the percent and the whole are given.

EXAMPLE 2:   What is 15% of 60?

①
part	15
60	100

②
$$
\begin{array}{r}
15 \\
\times 60 \\
\hline
900
\end{array}
$$

③
$$
\begin{array}{r}
9 \\
100\overline{)900}
\end{array}
$$

Step 1.   Fill in the percent box. Put 100 in the lower right. Put 15 in the upper right. 15 stands for the percent. 60 follows the word *of*. 60 is the whole. Put 60 in the lower left. Find the part.

Step 2.   Multiply the numbers that are diagonal to each other, 60 and 15.

Step 3.   Divide by the number in the remaining filled-in portion of the box, 100. 900 ÷ 100 = 9. 15% of 60 is **9**.

EXAMPLE 3:   Find $62\frac{1}{2}$% of 56.

①
part	$62\frac{1}{2}$
56	100

②
$$
56 \times 62\tfrac{1}{2} =
$$
$$
\frac{\overset{28}{\cancel{56}}}{1} \times \frac{125}{\underset{1}{\cancel{2}}} = 3,500
$$

③
$$
\begin{array}{r}
35 \\
100\overline{)3,500}
\end{array}
$$

Step 1. Fill in the percent box. Put 100 in the lower right. $62\frac{1}{2}$ is the percent. Put $62\frac{1}{2}$ in the upper right. 56 follows the word *of*. 56 is the whole. Put 56 in the lower left. You must find the part.

Step 2. Multiply the numbers that are diagonal to each other. $56 \times 62\frac{1}{2}$. Multiply these in the same way that you have learned to multiply mixed numbers.

Step 3. Divide by the number in the remaining filled-in portion of the box, 100. $62\frac{1}{2}\%$ of 56 is **35**.

Note: The percent box is a useful tool, but it is not always the easiest way to find a percent of a number. When you filled in the chart on page 146, you found that $62\frac{1}{2}\%$ is equal to $\frac{5}{8}$. In percent problems where you find the part, it may be easier to use a fraction ($62\frac{1}{2}\% = \frac{5}{8}$) or change the percent to a decimal ($25\% = .25$) than to use the percent box.

For example, you could have used $\frac{5}{8}$ to find $62\frac{1}{2}\%$ of 56. Remember, the word *of* following a fraction means to multiply.

$$\frac{5}{\underset{1}{\cancel{8}}} \times \frac{\overset{7}{\cancel{56}}}{1} = 35$$

In the solutions to the next exercises, you will see only the percent box. However, if you can save time by using the equivalent fractions or decimals to find a percent of a number, by all means do so.

You may find instances when the number that we have been calling the part is larger than the whole. This happens in cases where you are asked to find more than 100%.

EXAMPLE 4: What is 180% of 250.

①
part	180
250	100

②
$$\begin{array}{r} 250 \\ \times 180 \\ \hline 45{,}000 \end{array}$$

③
$$\begin{array}{r} 450 \\ 100\overline{)45{,}000} \end{array}$$

Step 1. Fill in the percent box. Put 100 in the lower right. Put 180, the percent, in the upper right. 250 follows the word *of*. Put 250, the whole, in the lower left. Find the part.

Step 2. Multiply the numbers that are diagonal to each other, 250 and 180.

Step 3. Divide by the number in the remaining filled-in portion of the box, 100. 180% of 250 is 450. Notice that the "part" is bigger than the whole because you were asked to find more than 100% of 250.

Finding the Whole

You can also use the percent box to find the whole when the percent and the part are given.

EXAMPLE 5: 8 is 50% of what number?

①
8	50
whole	100

②
$$\begin{array}{r} 100 \\ \times\ 8 \\ \hline 800 \end{array}$$

③
$$50{\overline{\smash{\big)}\,800}}\ \overset{16}{}$$

Step 1. Fill in the percent box. Put 100 in the lower right. Put 50, the percent, in the upper right. The words *what number* follow the word *of*. The whole is missing. 8 is the part. Put 8 in the upper left.

Step 2. Multiply the numbers that are diagonal to each other. 8 × 100 = 800.

Step 3. Divide by the number in the remaining filled-in portion of the box, 50. 8 is 50% of **16**.

Types of Percent Problems

You have learned to solve three kinds of percent problems:

1. finding the **part**,
2. finding the **percent**, and
3. finding the **whole.**

Remember, the whole usually follows the word *of*, and when the percent is given, it has the percent sign (%). Before you work with percent word problems, take the time to see that you can tell what you are being asked to solve for.

EXAMPLE 1: Tell what is missing in the following problem: the part, the percent, or the whole—

Find 60% of 45.

Answer: 60 is the percent because it has a percent sign. 45 is the whole because it follows the word *of*. **The part is missing.**

EXAMPLE 2: Tell what is missing in the following problem: the part, the percent, or the whole—

12 is 50% of what number?

Answer: The words *what number* follow the word *of*. **The whole is missing.** 50 is the percent, and 12 is the part.

EXAMPLE 3: Tell what is missing in the following problem: the part, the percent, or the whole—

16 is what percent of 64?

Answer: No number has a percent sign. **The percent is missing.** 16 is the part, and 64 is the whole.

PERCENT EXERCISE 7 ————————————————————

Read each problem carefully. Write down what is being asked for: the part, the percent, or the whole. Then go back and solve each problem.

1. 16 is what percent of 32?

2. Find 80% of 90.

3. 30 is 60% of what number?

4. What is $4\frac{1}{2}$% of 800?

5. What percent of 50 is 14?

6. $12\frac{1}{2}$% of what number is 15?

7. 15 is what percent of 45?

8. What is 3.6% of 900?

9. 120 is what percent of 80?

10. $33\frac{1}{3}$% of what number is 45?

11. What is 8.6% of 200?

12. 100% of 38.7 equals what?

Answers and solutions start on page 169.

PERCENT WORD PROBLEMS

Finding the Percent, Part, or Whole

When you solve percent word problems, first decide whether you are looking for the part, the percent, or the whole. Carefully study the next examples that illustrate some common uses of percents.

EXAMPLE 1: Mr. Gomez pays $80 for a suit. He puts a 30% markup on each suit that he sells in his store. Find the amount of the markup.

① & ②	③	④
part / 30 80 / 100	80 ×30 2,400	$ 24 100) 2,400

Step 1. Decide whether you are looking for the part, the percent, or the whole. 30 is the percent, and $80, the cost of the suit, is the whole. You are looking for the part, the amount of the markup.

Step 2. Fill in the percent box.

Step 3. Multiply the diagonal numbers, 80 and 30.

Step 4. Divide by the number in the remaining filled-in portion of the box, 100

The amount of the markup is **$24.**

EXAMPLE 2: Mrs. Jackson makes $200 a week. She spends $60 a week for food. Food represents what percent of Mrs. Jackson's weekly income?

① & ②	③	④
60 \ % 200 \ 100	60 ×100 6,000	30% 200) 6,000

Step 1. Decide whether you are looking for the part, the percent, or the whole. $60 is the part that she spends on food. $200 is her whole income. You are asked to find what percent of her income she pays for food.

Step 2. Fill in the percent box.

Step 3. Multiply the diagonal numbers, 60 and 100.

Step 4. Divide by the number in the remaining filled-in portion of the box, 200.

Mrs. Jackson spends **30%** of her weekly income for food.

EXAMPLE 3: Lois got a 6% commission for selling a house. Her commission was $3,000. Find the selling price of the house.

① & ②	③	④
3,000 | 6 whole | 100	3,000 ×100 300,000	$ 50,000 6) 300,000

Step 1. Decide whether you are looking for the part, the percent, or the whole. 6 is given as the percent. $3,000 is the part Lois received as a commission. You are looking for the whole, the selling price of the house.

Step 2. Fill in the percent box.

Step 3. Multiply the diagonal numbers, 3,000 and 100.

Step 4. Divide by the number in the remaining filled-in portion of the box, 6.

The selling price of the house was **$50,000.**

PERCENT EXERCISE 8

Read each problem carefully. First write down what is being asked for: the part, the percent, or the whole. Then, solve each problem.

1. A jacket originally selling for $40 was on sale for 15% off the original price. How much was saved by buying the jacket on sale?

2. Mr. and Mrs. Shin need $8,000 for a down payment on a house. So far they have saved $6,000. What percent of the total amount have they saved?

3. Alfredo earns $250 a week. His employer deducts 12% of his earnings for taxes and social security. How much is deducted from Alfredo's weekly pay?

4. John now weighs 172 pounds. This is 80% of what John weighed a year ago. How much did John weigh a year ago?

5. Fiona makes $600 a month and pays $150 a month for rent. Rent is what percent of her income?

6. The sales tax in Muhammed's state is 6%. How much tax does Muhammed pay for a television that costs $240?

7. Eighteen people showed up for David's evening math class. This represents 75% of the number registered for the class. How many people are registered for the class?

8. Mr. Kee pays $20 for a pair of shoes. He puts a $6 markup on every pair of shoes in his store. The markup is what percent of the price Mr. Kee pays?

Answers and solutions start on page 170.

More Complicated Percent Problems

Some percent problems require more than simply solving for the percent, part, or whole. Read through these next examples carefully.

EXAMPLE 4: A chair originally sold for $130. It was on sale at a 20% discount. What was the sale price of the chair?

①
part	20
130	100

②
130
×20
2,600

③
$$100\overline{)2,600} = 26$$

④
$130
−26
$104

Step 1. Decide that to find the sale price of the chair, you just have to find the discount, the part. Fill in the percent box: 130 is the whole, 20 is the percent.

Step 2. Multiply the diagonal numbers, 130 and 20.

Step 3. Divide by 100, the number in the remaining filled-in portion of the box.

Step 4. The part, or discount, is $26. Subtract $26 from the original price.

The sale price of the chair was **$104.**

EXAMPLE 5: The owner of the Victoria Clothing Store pays $75 each for men's suits. He marks up 40% on each suit. How much does each suit sell for?

①
part	40
75	100

②
75
× 40
3,000

③
$$100\overline{)3,000} = 30$$

④
$75
+30
$105

Step 1. Decide whether you are looking for the part, the percent, or the whole. Eventually, you want to find the selling price of the suit. First, though, you need to find the part, the amount of the markup. Fill in the percent box. 40 is the percent. 75 is the whole.

Step 2. Multiply the diagonal numbers, 75 and 40.

Step 3. Divide by 100, the number in the remaining filled-in portion of the box.

Step 4. The part, or markup, is $30. Add $30 to the original price of $75.

The selling price of a suit is **$105.**

EXAMPLE 6: Martin bought a new car last year for $5,500. This year, the car is worth $4,400. By what percent did the value of the car depreciate?

①
1,100	%
5,500	100

②
$$1,100$$
$$\times \quad 100$$
$$\overline{110,000}$$

③
$$5,500 \overline{)\begin{array}{r} 20 \\ 110,000 \end{array}}$$
$$\underline{110\ 00}$$
$$00$$

Step 1. You need to find the percent of depreciation. Before you can do that, find the part, the difference between the two car values. $5,500 − $4,400 = $1,100.

Fill in the percent box. You have found the part, $1,100. $5,500, the original price, is the whole.

Step 2. Multiply the diagonal numbers, 1,100 and 100.

Step 3. Divide by 5,500, the number in the remaining filled-in portion of the box.

The value of the car depreciated **20%**.

Percent problems may be difficult. Read every percent problem carefully.

Hints: When a problem asks for a **rate,** such as a markup rate or a discount rate, you must find a percent. Remember that the whole in these problems is often the original amount.

When a problem asks for the amount of a markup or a discount, you must find the part.

Problems in which you must find the whole are sometimes harder to recognize. In these problems, you are often looking for an amount that is larger than the amount given in the problem.

PERCENT EXERCISE 9 ━━━━━━━━━━━━━━━━━

Choose the correct answer for each problem. Be careful. Some problems just ask for the percent, part, or whole, while others require more steps. Before you start, decide whether you are looking for the percent, the part, or the whole.

1. Eva sells Mountain Dew Cosmetics for a 9% commission. In November she sold cosmetics worth $3,840. How much did she make in commissions for November?

    (1) $223.36
    (2) $336.70
    (3) $345.60
    (4) $353.28
    (5) $418.56

2. A tent sold originally for $150. It was on sale for $90 off the original price. What was the discount rate?

(1) 30%
(2) 60%
(3) $33\frac{1}{3}$%
(4) $66\frac{2}{3}$%
(5) 50%

3. Floria makes $15,600 a year. She spends 25% of her salary on rent. How much rent does she pay each month?

(1) $130
(2) $260
(3) $305
(4) $325
(5) $390

4. A real estate agent made a $2,100 commission for selling a house. His commission rate is 5%. What was the selling price of the house?

(1) $42,000
(2) $50,000
(3) $36,700
(4) $21,000
(5) $19,950

5. The selling price of a portable typewriter is $90. The store owner originally paid the dealer $75 for the typewriter. Find the markup rate on the typewriter.

(1) 90%
(2) 80%
(3) 25%
(4) 20%
(5) 15%

6. A farm with a market value of $120,000 was assessed for 60% of its market value. Farms are taxed at 2% of their assessed value. How much does the owner have to pay in taxes?

(1) $14,400
(2) $7,200
(3) $5,600
(4) $4,100
(5) $1,440

7. Alejo worked for $250 a week. He received a $30 a week raise. Find the percent that his salary increased.

(1) 10%
(2) 12%
(3) 15%
(4) 20%
(5) 30%

8. During the last election in Middletown, 4,800 voters went to the polls. This represents 60% of the registered voters. How many registered voters in Middletown did not go to the polls?

(1) 3,200
(2) 6,400
(3) 2,880
(4) 9,600
(5) 28,800

9. The George Street School bought 100 new desks originally listed at $65 each. The desks were on sale at 10% off the list price. The school received an additional 5% discount off the list price for buying in large quantity and another 3% discount off the list price for placing the order early in the spring. How much did the school pay for the 100 desks?

(1) $5,850
(2) $5,525
(3) $5,655
(4) $5,330
(5) $5,225

10. The Vernon Tool and Die Company bought a new lathe for $3,600. One year later, the lathe was worth $3,312. What was the yearly depreciation rate on the value of the lathe?

(1) 8%
(2) 9.2%
(3) 10%
(4) 12.5%
(5) 15%

11. Mark's take-home pay is 80% of his gross salary. He takes home $230 each week. Find his gross salary for a week.

(1) $184
(2) $310
(3) $252.70
(4) $262.80
(5) $287.50

12. The owners of Waleks' Shoe Store pay $12 for a pair of children's shoes. They put a 65% markup on their shoes. Find the selling price of a pair of children's shoes at Waleks.

(1) $20.40
(2) $19.80
(3) $16.20
(4) $17.80
(5) $14.80

13. Alberto borrowed $1,250. A year later, he paid back $1,400 including interest. What rate of interest had he paid on the loan?

(1) 15%
(2) 14%
(3) 12%
(4) 10%
(5) 8%

14. Arthur's gross salary is $1,040 a month. His employer withholds 9% for federal tax, 5% for social security, and 4% for state tax. Find Arthur's net salary for a month.

(1) $821.60
(2) $852.80
(3) $894.40
(4) $946.40
(5) $1,081.60

15. A day care center has 60 boys. Boys make up 40% of the total. Find the total number of children at the day care center.

(1) 240
(2) 210
(3) 180
(4) 150
(5) 120

16. Fran bought new furniture selling for $1,800. She paid 15% down and $50 a month for 36 months. What total amount did she pay for the furniture?

(1) $2,070
(2) $2,700
(3) $1,800
(4) $2,100
(5) $1,860

Answers and solutions start on page 170.

RATE OF INCREASE AND RATE OF DECREASE

A special type of percent word problem requires that you find the **rate of increase** or the **rate of decrease** over a period of time. The key to solving these problems is to compare the difference between the two amounts to the <u>original quantity</u>. In the percent box, the original quantity is represented as the whole.

EXAMPLE 1: Martin bought a new car last year for $3,500. This year, it is worth $2,800. What is the rate of decrease of the value of the car?

①	②	③	④
3500 −2800 — 700	700 | % ————— 3,500 | 100	700 × 100 ————— 70,000	20 $3{,}500 \overline{)70{,}000}$

Step 1. Subtract to find the amount of the decrease.

Step 2. Fill in the percent box. $3,500, the original amount, is the whole, and $700, the difference, is the part. You are looking for the percent.

Step 3. Multiply the numbers that are diagonal to each other.

Step 4. Divide by the remaining number.

The rate of decrease is **20%**.

EXAMPLE 2: Over the past ten years, the population of Little Lake, Minnesota, has increased from 1,200 to 1,500 people. What was the rate of increase in the population?

①	②	③	④
1,500 −1,200 — 300	300 | % ————— 1,200 | 100	300 × 100 ————— 30,000	25 $1{,}200 \overline{)30{,}000}$

Step 1. Find the amount of the increase.

Step 2. Fill in the percent box to find the percent of increase. The original amount, 1,200, is the whole. The difference, 300, is the part. You are looking for the percent.

Step 3. Multiply 300 and 100.

Step 4. Divide by 1,200.

The rate of increase is **25%**.

PERCENT EXERCISE 10 ────────────────

Solve each problem.

1. The price of a dozen eggs increased from 84¢ to 90¢. What was the percent of increase in the price of the eggs?

2. Sal wanted to know the rate of decrease of the value of his motorcycle. He had bought it for $1,200 and can only get $900 if he resells it. What is the rate of decrease?

3. The unemployment rate rose from 10% to 12%. What was the percent increase?

4. A company's workers have been asked to take a pay cut. While discussing this, they want to know by what percent their wages will decrease. An assembly line worker's weekly wages would decrease from $300 to $275 per week. What is the percent decrease?

Answers and solutions start on page 172.

SIMPLE INTEREST

Interest for Whole Years

Interest is money that money makes. You earn interest when your money is in a savings account. You pay interest when you borrow money. Interest is one of the most common applications of percents.

To find interest you need three things:

1. the **principal**—the amount of money borrowed or saved,
2. the **rate**—the percent used to find the interest, and
3. the **time**—the number of years (or part of a year) that money is borrowed or saved.

Rule for Finding Interest

Multiply the principal by the rate by the time:

Interest = principal × rate × time, often written as I = *prt*.

The easiest way to find interest is to set up a fraction multiplication problem.

EXAMPLE 1: Find the interest on $800 at 6% annual interest for one year.

① $$800 \times \frac{6}{100} \times 1 =$$

② $$\frac{\overset{8}{\cancel{800}}}{1} \times \frac{6}{\cancel{100}} \times \frac{1}{1} = \textbf{\$48}$$

Step 1. Set the problem up like a fraction problem. $800 is the principal. 6% is the rate. Use 100 as the denominator for the rate since percents are fractions with 100 in the denominator. $6\% = \frac{6}{100}$. One year is the time.

Step 2. Cancel and multiply across.

The interest is **$48**.

Note: If the time in an interest problem is exactly one year, you do not need to put 1 into the problem. The answer will be exactly the same.

EXAMPLE 2: Find the interest on $500 at $8\frac{1}{2}$% annual interest for one year.

① $$500 \times \frac{8\frac{1}{2}}{100} =$$

② $$\frac{\overset{5}{\cancel{500}}}{1} \times \frac{8\frac{1}{2}}{\cancel{100}} =$$

③ $$5 \times \frac{17}{2} = \textbf{\$42.50}$$

Step 1. Set this up as a fraction problem. Write the interest rate, $8\frac{1}{2}$%, as a fraction with a denominator of 100.

Step 2. Cancel the 100 and 500. You now have the simplified problem of $5 \times 8\frac{1}{2}$.

Step 3. Change $8\frac{1}{2}$ to an improper fraction. Multiply the numerators and divide by 2.

The interest is **$42.50**.

Note: If it is not possible to cancel in such a problem, proceed by multiplying the numerators and dividing by the denominator, 100.

In some cases, you will be asked for a new amount that is the interest plus the original principal.

EXAMPLE 3: Roger borrowed $14,000 for one year for home improvement at the rate of $13\frac{1}{2}$%. How much did he have to repay at the end of the year?

① $$14{,}000 \times \frac{13\frac{1}{2}}{100}$$

② $$\frac{\overset{140}{\cancel{14{,}000}}}{1} \times \frac{13\frac{1}{2}}{\cancel{100}} = \frac{\overset{70}{\cancel{140}}}{1} \times \frac{27}{\cancel{2}} = \textit{\$1,890}$$

③ $$\begin{array}{r} 14{,}000 \\ +\ 1{,}890 \\ \hline \textbf{\$15,890} \end{array}$$

Step 1. Set this up as a fraction problem with 100 as the denominator for the interest rate $13\frac{1}{2}\%$.

Step 2. Once the 100 has been cancelled, turn the mixed number into an improper fraction. Multiply and solve. The amount of interest is *$1,890*.

Step 3. Add the interest to the principal to get the total that Roger must repay. He must repay **$15,890.**

Interest for Parts of a Year

The time in a percent problem is measured in years. Six months is $\frac{6}{12} = \frac{1}{2}$ year. Two years and four months is written as $2\frac{4}{12} = 2\frac{1}{3}$ years.

EXAMPLE 4: Find the interest on $500 at 9% annual interest for eight months.

① $\boxed{\text{8 months} = \frac{8}{12} = \frac{2}{3} \text{ year}}$ ② $\boxed{\dfrac{\overset{5}{\cancel{\$500}}}{1} \times \dfrac{\overset{3}{\cancel{9}}}{\underset{1}{\cancel{100}}} \times \dfrac{2}{\underset{1}{\cancel{3}}} = \$30}$

Step 1. Change 8 months to a fraction of a year.

Step 2. Set up the problem as a fraction problem. $500 is the principal. 9% is the rate. $\frac{2}{3}$ year is the time. Cancel and multiply across.

PERCENT EXERCISE 11

Solve each problem.

1. Find the interest on $3,000 at 12.5% annual interest for one year.

2. How much money would Sara have at the end of one year on $800 deposited in a savings account earning $5\frac{1}{4}\%$ annual interest?

3. What is the interest on $5,000 at 9% annual interest for two years?

4. Find the interest on $800 at 6% annual interest for nine months.

5. How much interest did José pay on $900 at 11.5% annual interest for six months?

6. The Millers paid interest on $500 borrowed at 14% annual interest for one year and six months. How much interest did they pay?

7. Sally had $2,000 deposited in her savings account for 2 years and six months. If she had been earning 6% interest, how much did she have at the end of that time?

8. The Lewis family borrowed $900 at 13% annual interest over one year and 8 months. How much did they repay at the end of this time?

9. To the nearest penny, find the interest on $4,000 in a savings account at 10% annual interest for two years and 4 months.

10. If you had borrowed $1,200 at 14.75% annual interest for nine months, how much would you repay at the end of that period?

Answers and solutions start on page 172.

PERCENT TEST ——————————————————

Fill in the circle that corresponds to the correct answer.

1. Which of the following has the same value as .95?
   (1) .095%
   (2) 0.95%
   (3) 9.5%
   (4) 95%
   (5) .0095%

   1 ① ② ③ ④ ⑤

2. Which of the following is equal to 3.2%?
   (1) .32
   (2) 3.2
   (3) 3.20
   (4) .032
   (5) .0032

   2 ① ② ③ ④ ⑤

3. Which of the following equals $\frac{5}{8}$?
   (1) $62\frac{1}{2}\%$
   (2) 58%
   (3) 625%
   (4) 85%
   (5) 6.25%

   3 ① ② ③ ④ ⑤

4. Which of the following is equal to $37\frac{1}{2}\%$?
   (1) $\frac{1}{4}$
   (2) $\frac{3}{8}$
   (3) $\frac{5}{16}$
   (4) $\frac{7}{16}$
   (5) $\frac{9}{16}$

   4 ① ② ③ ④ ⑤

5. Find 2.7% of 360.   5 ① ② ③ ④ ⑤
   (1) .972
   (2) 9.72
   (3) 97.20
   (4) 972
   (5) .0972

6. What percent of 35 is 21?   6 ① ② ③ ④ ⑤
   (1) 60%
   (2) 16%
   (3) 167%
   (4) 40%
   (5) 30%

7. 75 is 20% of what number?   7 ① ② ③ ④ ⑤
   (1) 150
   (2) 15
   (3) 1,500
   (4) 375
   (5) 37.5

8. 150% of what number is 48?   8 ① ② ③ ④ ⑤
   (1) 144
   (2) 72
   (3) 64
   (4) 56
   (5) 32

9. The price of a gallon of gasoline went from $1.20 to $1.05. By   9 ① ② ③ ④ ⑤
   what percent did the price of a gallon drop?
   (1) $12\frac{1}{2}$%
   (2) 15%
   (3) 20%
   (4) .85%
   (5) $87\frac{1}{2}$%

10. Mr. Seltzer sells shoes for a 5% commission. He earned $115 in   10 ① ② ③ ④ ⑤
    commissions one week. Find the total value of the shoes he sold
    that week.
    (1) $5.75
    (2) $57.50
    (3) $575
    (4) $580.75
    (5) $2,300

11. A stereo listed for $350 is on sale for 20% off the list price. The sales tax is 6%. What final price does a customer pay for the stereo if he buys it on sale?

11 ① ② ③ ④ ⑤

(1) $324.00
(2) $259.00
(3) $262.20
(4) $296.80
(5) $301.00

12. A store bought a tie at $10.00 and marked it up 20%. At a one-day-only sale, the tie was reduced by $\frac{1}{3}$. What did the tie cost the day of the sale?

12 ① ② ③ ④ ⑤

(1) $2.00
(2) $8.00
(3) $10.00
(4) $11.00
(5) $12.00

13. Adrian wants to buy a used car that sells for $1,200. He has already saved $1,000. What percent of the total price has he already saved?

13 ① ② ③ ④ ⑤

(1) $16\frac{2}{3}$%
(2) 20%
(3) 80%
(4) $83\frac{1}{3}$%
(5) $16\frac{2}{3}$%

14. Ann put $850 in her savings account. She received a total of $935 one year later. What rate of interest did the bank pay?

14 ① ② ③ ④ ⑤

(1) 12%
(2) 10%
(3) 9%
(4) 8%
(5) 7%

15. Find the interest on $450 at 8% annual interest rate for nine months.

15 ① ② ③ ④ ⑤

(1) $18
(2) $27
(3) $36
(4) $477
(5) $486

Answers and solutions start on page 173.

PERCENT TEST EVALUATION

Problem	Section	Starting Page
1, 2	Decimals and Percents	140
3, 4	Fractions and Percents	142
5	Percent Problems: Finding the Part	148
6	Percent Problems: Finding the Percent	148
7, 8	Percent Problems: Finding the Whole	150
9-13	Percent Word Problems	151
14, 15	Interest	159

<u>Passing score:</u> __12__ right out of 15 problems.
<u>Your score:</u> ____ right out of 15 problems.

If you had less than a passing score, review the sections for the problems you got wrong.

If you had a passing score, correct any problem you got wrong. Then, go on to the mixed review.

MIXED REVIEW

These problems give you a chance to practice the skills you have learned so far in this book. For each problem, choose the best answer.

1. Which of the following correctly expresses "four hundred three and nineteen thousandths"?
   (1) 43,019
   (2) 403.019
   (3) 403.19
   (4) .4319
   (5) 4,319

   1 ① ② ③ ④ ⑤

2. What is 385.68 rounded off to the nearest tenth?
   (1) 390
   (2) 390.68
   (3) 385.78
   (4) 0.7
   (5) 385.7

   2 ① ② ③ ④ ⑤

3. Beverley bought 1.4 pounds of cheese at $2.69 a pound. How much did she pay for the cheese?
   (1) $4.09
   (2) $3.80
   (3) $3.77
   (4) $3.60
   (5) $3.59

   3 ① ② ③ ④ ⑤

4. Find the combined weight of $2\frac{1}{4}$ pounds of fish, $3\frac{1}{2}$ pounds of chicken, and $3\frac{3}{4}$ pounds of pork.
   (1) $9\frac{3}{4}$ lb.
   (2) $9\frac{1}{4}$ lb.
   (3) $8\frac{3}{4}$ lb.
   (4) $8\frac{1}{2}$ lb.
   (5) $9\frac{1}{2}$ lb.

   4 ① ② ③ ④ ⑤

5. The Tibenskys spend $\frac{3}{8}$ of their income for food. Together the Tibenskys bring home $280 a week. How much do they have left in their weekly budget after they pay for food?
   (1) $205
   (2) $185
   (3) $175
   (4) $145
   (5) $105

   5 ① ② ③ ④ ⑤

6. The price of a shirt at Sammy's is $14. Sammy pays the manufacturer $8 for a shirt. Find the markup rate for a shirt at Sammy's.

    6 ① ② ③ ④ ⑤

    (1) 60%
    (2) 75%
    (3) 80%
    (4) 40%
    (5) 25%

7. Find the interest on $8,000 at 12% annual interest for five years and six months.

    7 ① ② ③ ④ ⑤

    (1) $9,600
    (2) $5,280
    (3) $4,800
    (4) $960
    (5) $528

8. When she sent a package, Marion paid $4.60 in shipping costs. If the package weighed 11.5 pounds, what was the cost per pound?

    8 ① ② ③ ④ ⑤

    (1) $16.10
    (2) $4.00
    (3) $1.61
    (4) $.40
    (5) $.04

9. Francine bought 10 feet of wood. She used $6\frac{1}{2}$ feet in one project, $2\frac{3}{4}$ feet in another, and wasted $\frac{1}{8}$ foot of wood. How much wood did she have left?

    9 ① ② ③ ④ ⑤

    (1) $\frac{5}{8}$ ft.
    (2) $3\frac{5}{8}$ ft.
    (3) $9\frac{1}{2}$ ft.
    (4) $9\frac{3}{8}$ ft.
    (5) $9\frac{5}{8}$ ft.

10. A cable TV company makes the following offer: $3.00 for the basic package, $9.00 for an additional "family package", and $6.50 for each additional channel. Which statement shows how to find the total for both the basic and family packages plus a sports station and a movie channel?

    10 ① ② ③ ④ ⑤

    (1) 3 + 9 + 6.50
    (2) 3 + 9 + 2(6.50)
    (3) 3 × 9 × 6.50
    (4) 3 + 9 − 6.50
    (5) 3 × 9 + 6.50

Answers and solutions start on page 174.

ANSWERS AND SOLUTIONS

Percent Exercise 1

1. d	**2.** f	**3.** b	**4.** c	**5.** e	**6.** a

Percent Exercise 2

1. .09	**3.** 1	**5.** $.87\frac{1}{2}$	**7.** .0015	**9.** .027	**11.** .57
2. .24	**4.** .003	**6.** $.08\frac{1}{3}$	**8.** 2.75	**10.** .0395	**12.** 10

Percent Exercise 3

1. 81%	**3.** 50%	**5.** .09%	**7.** 3%	**9.** 210%	**11.** 392.4%
2. $37\frac{1}{2}\%$	**4.** .4%	**6.** 21.7%	**8.** $33\frac{1}{3}\%$	**10.** 485%	**12.** 1.5%

Percent Exercise 4

1. 20% $\qquad \dfrac{1}{\cancel{5}} \times \dfrac{\overset{20}{\cancel{100\%}}}{1} = 20\%$

2. $83\frac{1}{3}\%$ $\qquad \dfrac{5}{\underset{3}{\cancel{6}}} \times \dfrac{\overset{50}{\cancel{100\%}}}{1} = \dfrac{250}{3} = 83\frac{1}{3}\%$

3. $37\frac{1}{2}\%$ $\qquad \dfrac{3}{\underset{2}{\cancel{8}}} \times \dfrac{\overset{25}{\cancel{100\%}}}{1} = \dfrac{75}{2} = 37\frac{1}{2}\%$

4. $66\frac{2}{3}\%$ $\qquad \dfrac{2}{3} \times \dfrac{100\%}{1} = \dfrac{200}{3} = 66\frac{2}{3}\%$

5. 175% $\qquad \dfrac{14}{8} = \dfrac{7}{4}$

$\qquad\qquad \dfrac{7}{\cancel{4}} \times \dfrac{\overset{25}{\cancel{100\%}}}{1} = 175\%$

6. 90% $\qquad \dfrac{9}{\underset{1}{\cancel{10}}} \times \dfrac{\overset{10}{\cancel{100\%}}}{1} = 90\%$

7. $41\frac{2}{3}\%$ $\qquad \dfrac{5}{\underset{3}{\cancel{12}}} \times \dfrac{\overset{25}{\cancel{100\%}}}{1} = \dfrac{125}{3} = 41\frac{2}{3}\%$

8. $85\frac{5}{7}\%$ $\qquad \dfrac{6}{7} \times \dfrac{100\%}{1} = \dfrac{600}{7} = 85\frac{5}{7}\%$

9. $16\frac{2}{3}\%$ $\qquad \dfrac{1}{\underset{3}{\cancel{6}}} \times \dfrac{\overset{50}{\cancel{100\%}}}{1} = \dfrac{50}{3} = 16\frac{2}{3}\%$

10. 200% $\qquad \dfrac{10}{5} = \dfrac{2}{1}$

$\qquad\qquad \dfrac{2}{1} \times \dfrac{100\%}{1} = 200\%$

11. $8\frac{1}{3}\%$ $\qquad \dfrac{1}{\underset{3}{\cancel{12}}} \times \dfrac{\overset{25}{\cancel{100\%}}}{1} = \dfrac{25}{3} = 8\frac{1}{3}\%$

12. $18\frac{2}{11}\%$ $\qquad \dfrac{2}{11} \times \dfrac{100\%}{1} = \dfrac{200}{11} = 18\frac{2}{11}\%$

Percent Exercise 5

1. $\dfrac{9}{20}$

$\qquad \dfrac{45}{100} = \dfrac{9}{20}$

2. $\dfrac{3}{8}$

$\qquad 37\frac{1}{2} \div 100 =$

$\qquad \dfrac{\overset{3}{\cancel{75}}}{2} \times \dfrac{1}{\underset{4}{\cancel{100}}} = \dfrac{3}{8}$

3. $\dfrac{1}{15}$

$\qquad 6\frac{2}{3} \div 100 =$

$\qquad \dfrac{\overset{1}{\cancel{20}}}{3} \times \dfrac{1}{\underset{5}{\cancel{100}}} = \dfrac{1}{15}$

4. $\dfrac{2}{25}$

$\qquad \dfrac{8}{100} = \dfrac{2}{25}$

5. $\dfrac{1}{50}$

$\qquad \dfrac{2}{100} = \dfrac{1}{50}$

6. $\dfrac{5}{6}$

$\qquad 83\frac{1}{3} \div 100 =$

$\qquad \dfrac{\overset{5}{\cancel{250}}}{3} \times \dfrac{1}{\underset{2}{\cancel{100}}} = \dfrac{5}{6}$

7. $\dfrac{6}{25}$

$\qquad \dfrac{24}{100} = \dfrac{6}{25}$

8. $\dfrac{1}{3}$

$\qquad 33\frac{1}{3} \div 100 =$

$\qquad \dfrac{\overset{1}{\cancel{100}}}{3} \times \dfrac{1}{\cancel{100}} = \dfrac{1}{3}$

9. $\dfrac{2}{7}$

$\qquad 28\frac{4}{7} \div 100 =$

$\qquad \dfrac{\overset{2}{\cancel{200}}}{7} \times \dfrac{1}{\underset{1}{\cancel{100}}} = \dfrac{2}{7}$

10. $\dfrac{4}{5}$

$\qquad \dfrac{80}{100} = \dfrac{4}{5}$

11. $1\frac{1}{2}$

$\qquad \dfrac{150}{100} = 1\frac{50}{100} = 1\frac{1}{2}$

12. $\dfrac{1}{8}$

$\qquad 12\frac{1}{2} \div 100$

$\qquad \dfrac{\overset{1}{\cancel{25}}}{2} \times \dfrac{1}{\underset{4}{\cancel{100}}} = \dfrac{1}{8}$

13. $\dfrac{3}{200}$

$\qquad 1.5\% = .015 =$

$\qquad \dfrac{15}{1,000} = \dfrac{3}{200}$

14. $\dfrac{9}{10,000}$

$\qquad .09\% = .0009 =$

$\qquad \dfrac{9}{10,000}$

15. $\dfrac{3}{500}$

$\qquad .6\% = .006 =$

$\qquad \dfrac{6}{1,000} = \dfrac{3}{500}$

16. $\dfrac{1}{100}$

Percent Exercise 6

fraction	decimal	percent
$\frac{1}{4}$	.25	25%
$\frac{1}{2}$	.5	50%
$\frac{3}{4}$	.75	75%
$\frac{1}{8}$	.125 or .12$\frac{1}{2}$	12$\frac{1}{2}$% or 12.5%
$\frac{3}{8}$	.375 or .37$\frac{1}{2}$	37$\frac{1}{2}$% or 37.5%
$\frac{5}{8}$	.625 or .62$\frac{1}{2}$	62$\frac{1}{2}$% or 62.5%
$\frac{7}{8}$	.875 or .87$\frac{1}{2}$	87$\frac{1}{2}$% or 87.5%
$\frac{1}{5}$	.2	20%
$\frac{2}{5}$	.4	40%
$\frac{3}{5}$	.6	60%
$\frac{4}{5}$	.8	80%
$\frac{1}{10}$	.1	10%
$\frac{3}{10}$	.3	30%
$\frac{7}{10}$	.7	70%
$\frac{9}{10}$	.9	90%
$\frac{1}{3}$	.33$\frac{1}{3}$	33$\frac{1}{3}$%
$\frac{2}{3}$	.66$\frac{2}{3}$	66$\frac{2}{3}$%
$\frac{1}{6}$	.16$\frac{2}{3}$	16$\frac{2}{3}$%
$\frac{5}{6}$	.83$\frac{1}{3}$	83$\frac{1}{3}$%

Percent Exercise 7

1. percent: 50%

16	%
32	100

$$\begin{array}{r} 16 \\ \times\ 100 \\ \hline 1,600 \end{array} \qquad \begin{array}{r} 50 \\ 32\overline{)1,600} \end{array}$$

2. part: 72

part	80
90	100

$$\begin{array}{r} 80 \\ \times\ 90 \\ \hline 7,200 \end{array} \qquad \begin{array}{r} 72 \\ 100\overline{)7,200} \end{array}$$

3. whole: 50

30	60
whole	100

$$\begin{array}{r} 30 \\ \times\ 100 \\ \hline 3,000 \end{array} \qquad \begin{array}{r} 50 \\ 60\overline{)3,000} \end{array}$$

4. part: 36

part	$4\frac{1}{2}$
800	100

$$4\frac{1}{2} \times 800 =$$
$$\frac{9}{\overset{2}{\cancel{2}}} \times \frac{\overset{400}{\cancel{800}}}{1} = 3,600 \qquad \begin{array}{r} 36 \\ 100\overline{)3,600} \end{array}$$

5. percent: 28

14	%
50	100

$$\begin{array}{r} 14 \\ \times\ 100 \\ \hline 1,400 \end{array} \qquad \begin{array}{r} 28 \\ 50\overline{)1,400} \end{array}$$

6. whole: 120

15	$12\frac{1}{2}$
whole	100

$$\begin{array}{r} 15 \\ \times\ 100 \\ \hline 1,500 \end{array} \qquad \begin{array}{l} 1,500 \div 12\frac{1}{2} = \\ \frac{1,500}{1} \div \frac{25}{2} = \end{array}$$
$$\frac{\overset{60}{\cancel{1,500}}}{1} \times \frac{2}{\cancel{25}} = 120$$

7. percent: $33\frac{1}{3}$%

15	%
45	100

$$\begin{array}{r} 15 \\ \times\ 100 \\ \hline 1,500 \end{array} \qquad 33\frac{15}{45} = 33\frac{1}{3}\%$$
$$\begin{array}{r} 45\overline{)1,500} \\ \underline{135} \\ 150 \\ \underline{135} \\ 15 \end{array}$$

8. part: 32.4

part	3.6
900	100

$$\begin{array}{r} 3.6 \\ \times\ 900 \\ \hline 3,240.0 \end{array} \qquad \begin{array}{r} 32.4 \\ 100\overline{)3,240.0} \end{array}$$

9. percent: 150%

120	%
80	100

$$\begin{array}{r} 120 \\ \times\ 100 \\ \hline 12,000 \end{array} \qquad \begin{array}{r} 150 \\ 80\overline{)12,000} \end{array}$$

10. whole: 135

45	$33\frac{1}{3}$
whole	100

$$\begin{array}{r} 45 \\ \times\ 100 \\ \hline 4,500 \end{array} \qquad \begin{array}{l} 4,500 \div 33\frac{1}{3} = \\ \frac{4,500}{1} \div \frac{100}{3} = \end{array}$$
$$\frac{\overset{45}{\cancel{4,500}}}{1} \times \frac{3}{\cancel{100}} = 135$$

11. part: 17.2

part	8.6
200	100

$$\begin{array}{r} 8.6 \\ \times\ 200 \\ \hline 1,720.0 \end{array} \qquad \begin{array}{r} 17.2 \\ 100\overline{)1,720.0} \end{array}$$

12. part: 38.7

part	100
38.7	100

$$\begin{array}{r} 38.7 \\ \times\ 100 \\ \hline 3,870.0 \end{array} \qquad \begin{array}{r} 38.7 \\ 100\overline{)3,870.0} \end{array}$$

Percent Exercise 8

1. part: $6

part	15
40	100

$$\begin{array}{r} 15 \\ \times 40 \\ \hline 600 \end{array}$$

$$100\overline{)600}^{6}$$

2. percent: 75%

6,000	%
8,000	100

$$\begin{array}{r} 6,000 \\ \times\ 100 \\ \hline 600,000 \end{array}$$

$$8,000\overline{)600,000}^{75}$$

3. part: $30

part	12
250	100

$$\begin{array}{r} 250 \\ \times 12 \\ \hline 500 \\ 250 \\ \hline 3,000 \end{array}$$

$$100\overline{)3,000}^{30}$$

4. whole: 215 pounds

172	80
whole	100

$$\begin{array}{r} 172 \\ \times\ 100 \\ \hline 17,200 \end{array}$$

$$80\overline{)17,200}^{215}$$

5. percent: 25%

150	%
600	100

$$\begin{array}{r} 150 \\ \times 100 \\ \hline 15,000 \end{array}$$

$$600\overline{)15,000}^{25}$$

6. part: $14.40

part	6
240	100

$$\begin{array}{r} 240 \\ \times\ 6 \\ \hline 1,440 \end{array}$$

$$100\overline{)1,440.00}^{14.40}$$

7. whole: 24 people

18	75
whole	100

$$\begin{array}{r} 18 \\ \times 100 \\ \hline 1,800 \end{array}$$

$$75\overline{)1,800}^{24}$$

8. percent: 30%

6	%
20	100

$$\begin{array}{r} 100 \\ \times 6 \\ \hline 600 \end{array}$$

$$20\overline{)600}^{30}$$

Percent Exercise 9

1. (3) $345.60

part	9
3,840	100

$$\begin{array}{r} 3,840 \\ \times\ 9 \\ \hline 34,560 \end{array}$$

$$100\overline{)34,560.00}^{345.60}$$

2. (2) 60%

90	%
150	100

$$\begin{array}{r} 90 \\ \times\ 100 \\ \hline 9,000 \end{array}$$

$$150\overline{)9,000}^{60}$$

3. (4) $325

First find how much she pays in rent per year.

part	25
15,600	100

$$\begin{array}{r} 15,600 \\ \times\ 25 \\ \hline 78\,000 \\ 312\,00 \\ \hline 390,000 \end{array}$$

$$100\overline{)390,000}^{3,900}$$

She pays $3,900 in one year. In one month she pays:

$$12\overline{)3,900}^{325}$$

4. (1) $42,000

2,100	5
whole	100

$$\begin{array}{r} 2,100 \\ \times\ 100 \\ \hline 210,000 \end{array}$$

$$5\overline{)210,000}^{42,000}$$

5. (4) 20%

The amount of markup is $90 − $75 = $15. This is the part.

15	%
75	100

$$\begin{array}{r} 15 \\ \times\ 100 \\ \hline 1,500 \end{array}$$

$$75\overline{)1,500}^{20}$$

6. (5) $1,440

First find the assessed value of the farm.

part	60
120,000	100

$$\begin{array}{r} 120,000 \\ \times\ \ \ \ \ 60 \\ \hline 7,200,000 \end{array}$$

$$100\overline{)7,200,000}^{72,000}$$

$72,000 is the assessed value for purposes of taxation. This is now the whole value; find the taxes that are the part.

part	2
72,000	100

$$\begin{array}{r} 72,000 \\ \times\ \ \ \ 2 \\ \hline 144,000 \end{array}$$

$$100\overline{)144,000}^{1,440}$$

7. (2) 12%

30	%
250	100

$$\begin{array}{r} 30 \\ \times 100 \\ \hline 3,000 \end{array}$$

$$250\overline{)3,000}^{12}$$

8. (1) 3,200

Since you know that 4,800 voters is 60% of the registered voters, you can find the total number of registered voters.

4,800	60
whole	100

$$\begin{array}{r} 4,800 \\ \times\ 100 \\ \hline 480,000 \end{array} \qquad \begin{array}{r} 8,000 \\ 60\overline{)480,000} \end{array}$$

There are 8,000 registered voters.
Now, find the number that did not go to the polls.

$$\begin{array}{r} 8,000 \quad \text{registered} \\ -4,800 \quad \text{went to polls} \\ \hline 3,200 \quad \text{did not go to polls} \end{array}$$

9. (4) $5,330

The total % discounted =
10% + 5% + 3% = 18%.
Find 18% of $65 to find the discount per chair.

part	18
65	100

$$\begin{array}{r} 65 \\ \times\ 18 \\ \hline 520 \\ 65 \\ \hline 1,170 \end{array} \qquad \begin{array}{r} 11.70 \ \text{discount} \\ 100\overline{)1,170.00} \end{array}$$

For one desk the school pays:
$$\begin{array}{r} \$65.00 \\ -11.70 \\ \hline \$53.30 \end{array}$$

For 100 desks the school pays:
$$\begin{array}{r} \$53.30 \\ \times\ \ 100 \\ \hline \$5,330.00 \end{array}$$

10. (1) 8%

The depreciation on the lathe is:
$$\begin{array}{r} \$3,600 \\ -3,312 \\ \hline \$\ \ 288 \end{array}$$

Use the $288 to find the depreciation rate on the lathe.

288	%
3,600	100

$$\begin{array}{r} 288 \\ \times\ 100 \\ \hline 28,800 \end{array} \qquad \begin{array}{r} 8 \\ 3,600\overline{)28,800} \end{array}$$

11. (5) $287.50

230	80
whole	100

$$\begin{array}{r} 230 \\ \times\ 100 \\ \hline 23,000 \end{array} \qquad \begin{array}{r} 287.50 \\ 80\overline{)23,000.00} \end{array}$$

12. (2) $19.80

First find the markup on a pair of shoes.

part	65
12	100

$$\begin{array}{r} 65 \\ \times 12 \\ \hline 130 \\ 65 \\ \hline 780 \end{array} \qquad \begin{array}{r} 7.80 \\ 100\overline{)780.00} \end{array}$$

The markup is $7.80.
The selling price is:
$$\begin{array}{r} \$12.00 \\ +7.80 \\ \hline \$19.80 \end{array}$$

13. (3) 12%

First find the additional amount he paid in interest:
$$\begin{array}{r} \$1,400 \\ -1,250 \\ \hline \$\ \ 150 \end{array}$$

The difference between the two amounts ($150) is the part. $1,250 is the whole or original amount of the loan. Find the percent.

150	%
1,250	100

$$\begin{array}{r} 150 \\ \times\ 100 \\ \hline 15,000 \end{array} \qquad \begin{array}{r} 12 \\ 1,250\overline{)15,000} \end{array}$$

14. (2) $852.80

The total % deductions =
9% + 5% + 4% = 18%.
Find the total amount that is deducted from his salary.

part	18
1,040	100

$$\begin{array}{r} 1,040 \\ \times\ \ 18 \\ \hline 8\ 320 \\ 10\ 40 \\ \hline 18,720 \end{array} \qquad \begin{array}{r} 187.20 \\ 100\overline{)18,720.00} \end{array}$$

His net salary is:
$$\begin{array}{r} \$1,040.00 \\ -187.20 \\ \hline \$\ \ 852.80 \end{array}$$

15. (4) 150

60	40
whole	100

$$\begin{array}{r} 60 \\ \times\ 100 \\ \hline 6,000 \end{array} \qquad \begin{array}{r} 150 \\ 40\overline{)6,000} \end{array}$$

16. (1) $2,070

First, find the down payment.

part	15
1,800	100

$$\begin{array}{r} 1,800 \\ \times\ \ 15 \\ \hline 9\ 000 \\ 18\ 00 \\ \hline 27,000 \end{array} \qquad \begin{array}{r} 270 \\ 100\overline{)27,000} \end{array}$$

Next, find the total monthly payments.
The monthly payments are:
$$\begin{array}{r} \$50 \\ \times 36 \\ \hline \$1,800 \end{array}$$

The total amount is:
$$\begin{array}{r} \$1,800 \ \text{payments} \\ +\ 270 \ \text{down} \\ \hline \$2,070 \end{array}$$

Percent Exercise 10

1. $7\frac{1}{7}\%$ increase

$$\begin{array}{r} 90 \\ -84 \\ \hline 6 \end{array}$$

6	%
84	100

$$\begin{array}{r} 100 \\ \times 6 \\ \hline 600 \end{array}$$

$$\begin{array}{r} 7\frac{1}{7}\% \\ 84\overline{)600} \\ \underline{588} \\ 12 \end{array}$$

3. 20% increase

$$\begin{array}{r} 12 \\ -10 \\ \hline 2 \end{array}$$

2	%
10	100

$$\begin{array}{r} 100 \\ \times 2 \\ \hline 200 \end{array}$$

$$10\overline{)200}$$

2. 25% decrease

$$\begin{array}{r} 1,200 \\ -900 \\ \hline 300 \end{array}$$

300	%
1200	100

$$\begin{array}{r} 300 \\ \times 100 \\ \hline 30,000 \end{array}$$

$$\begin{array}{r} 25\% \\ 1,200\overline{)30,000} \end{array}$$

4. $8\frac{1}{3}\%$ decrease

$$\begin{array}{r} 300 \\ -275 \\ \hline 25 \end{array}$$

25	%
300	100

$$\begin{array}{r} 25 \\ \times 100 \\ \hline 2,500 \end{array}$$

$$\begin{array}{r} 8\frac{1}{3}\% \\ 300\overline{)2,500} \end{array}$$

Percent Exercise 11

1. $375 \quad \dfrac{\overset{30}{\cancel{3000}}}{1} \times \dfrac{12.5}{\cancel{100}} = 375$

2. $842 \quad \dfrac{\overset{8}{\cancel{800}}}{1} \times \dfrac{5\frac{1}{4}}{\cancel{100}} =$

$$\dfrac{\overset{2}{\cancel{8}}}{1} \times \dfrac{21}{\cancel{4}} = 42 \qquad \begin{array}{r} 800 \text{ principal} \\ + 42 \text{ interest} \\ \hline \$842 \end{array}$$

3. $900 \quad \dfrac{\overset{50}{\cancel{5,000}}}{1} \times \dfrac{9}{\cancel{100}} \times \dfrac{2}{1} = 900$

4. $36 \quad$ 9 months $= \dfrac{9}{12} = \dfrac{3}{4}$ year

$$\dfrac{\overset{2}{\cancel{\overset{8}{\cancel{800}}}}}{1} \times \dfrac{6}{\cancel{100}} \times \dfrac{3}{\cancel{4}} = 36$$

5. $51.75 \quad$ 6 months $= \dfrac{6}{12} = \dfrac{1}{2}$

$$\overset{9}{\cancel{900}} \times \dfrac{11.5}{\cancel{100}} \times \dfrac{1}{2} = \dfrac{103.50}{2} = 51.75$$

6. $105 \quad$ *one year and six months* $= 1\frac{6}{12} = 1\frac{1}{2}$ *years*

$$\dfrac{500}{1} \times \dfrac{14}{100} \times 1\frac{1}{2} =$$

$$\dfrac{\overset{5}{\cancel{500}}}{1} \times \dfrac{\overset{7}{\cancel{14}}}{\cancel{100}} \times \dfrac{3}{2} = 105$$

7. $2,300
two years and six months $= 2\frac{6}{12} = 2\frac{1}{2}$ *years*
$$2,000 \times \dfrac{6}{100} \times 2\frac{1}{2} =$$

$$\overset{20}{\cancel{2,000}} \times \dfrac{\overset{3}{\cancel{6}}}{\cancel{100}} \times \dfrac{5}{2} = \qquad \begin{array}{r} \$2,000 \text{ principal} \\ + 300 \text{ interest} \\ \hline \$2,300 \end{array}$$

$$\dfrac{600}{2} = 300$$

8. $1,095 \quad$ *one year and 8 months* $=$
$$1\frac{8}{12} = 1\frac{2}{3} \text{ years}$$
$$\dfrac{900}{1} \times \dfrac{13}{100} \times 1\frac{2}{3} =$$

$$\dfrac{\overset{3}{\cancel{\overset{9}{\cancel{900}}}}}{1} \times \dfrac{13}{\cancel{100}} \times \dfrac{5}{3} = 195$$

$$\begin{array}{r} \$ \ 900 \text{ principal} \\ + \ 195 \text{ interest} \\ \hline \$1,095 \end{array}$$

9. $933.33 \quad$ *two years and 4 months* $=$
$$2\frac{4}{12} = 2\frac{1}{3} \text{ years}$$

$$\dfrac{4,000}{1} \times \dfrac{10}{100} \times 2\frac{1}{3} =$$

$$\dfrac{\overset{40}{\cancel{4,000}}}{1} \times \dfrac{10}{\cancel{100}} \times \dfrac{7}{3} = \dfrac{2800}{3} =$$

$933.33\frac{1}{3}$ *to the nearest penny* $=$
$933.33

10. $1,332.75 \quad$ *nine months* $= \dfrac{9}{12} = \dfrac{3}{4}$ *year*

$$\dfrac{\overset{3}{\cancel{\overset{12}{\cancel{1,200}}}}}{1} \times \dfrac{14.75}{\cancel{100}} \times \dfrac{3}{\cancel{4}} = 132.75$$

$$\begin{array}{r} \$1,200.00 \text{ principal} \\ + 132.75 \text{ interest} \\ \hline \$1,332.75 \end{array}$$

Percent Test

1. (4) 95% $.95 = .95 = 95\%$

2. (4) .032 $3.2\% = 03.2 = .032$

3. (1) $62\frac{1}{2}\%$ $\frac{5}{\underset{2}{8}} \times \frac{\overset{25}{\cancel{100}\%}}{1} = \frac{125}{2} = 62\frac{1}{2}\%$

4. (2) $\frac{3}{8}$ $\frac{37\frac{1}{2}}{100} = 37\frac{1}{2} \div 100$

   $= \frac{\overset{3}{\cancel{75}}}{2} \times \frac{1}{\underset{4}{\cancel{100}}} = \frac{3}{8}$

5. (2) 9.72

part	2.7
360	100

$\begin{array}{r} 360 \\ \times 2.7 \\ \hline 972.0 \end{array}$ $\begin{array}{r} 9.72 \\ 100\overline{)972.00} \end{array}$

6. (1) 60%

21	%
35	100

$\begin{array}{r} 21 \\ \times 100 \\ \hline 2,100 \end{array}$ $\begin{array}{r} 60 \\ 35\overline{)2,100} \end{array}$

7. (4) 375

75	20
whole	100

$\begin{array}{r} 75 \\ \times 100 \\ \hline 7,500 \end{array}$ $\begin{array}{r} 375 \\ 20\overline{)7,500} \end{array}$

8. (5) 32

48	150
whole	100

$\begin{array}{r} 48 \\ \times 100 \\ \hline 4,800 \end{array}$ $\begin{array}{r} 32 \\ 150\overline{)4,800} \end{array}$

9. (1) $12\frac{1}{2}\%$

   The part is the change in price:
   $\$1.20 - \$1.05 = \$.15$. The whole is the original amount, $1.20.

.15	%
1.20	100

$\begin{array}{r} .15 \\ \times 100 \\ \hline 15.00 \end{array}$

$\begin{array}{r} 12\frac{6}{12} = 12\frac{1}{2} \\ 1.2\overline{)15.0} \end{array}$

10. (5) $2,300

$115	5
whole	100

$\begin{array}{r} \$115 \\ \times 100 \\ \hline \$11,500 \end{array}$ $\begin{array}{r} \$2,300 \\ 5\overline{)\$11,500} \end{array}$

11. (4) $296.80
    First find the discount.

part	20
$350	100

$\begin{array}{r} \$350 \\ \times 20 \\ \hline \$7,000 \end{array}$ $\begin{array}{r} \$70 \\ 100\overline{)\$7,000} \end{array}$

   The discount is $70.
   The selling price is: $\begin{array}{r} \$350 \\ -70 \\ \hline \$280 \end{array}$

   Now find the tax.

part	6
$280	100

$\begin{array}{r} \$280 \\ \times 6 \\ \hline \$1,680 \end{array}$ $\begin{array}{r} \$16.80 \\ 100\overline{)\$1,680.00} \end{array}$

   The final price is: $\begin{array}{r} \$280.00 \\ +16.80 \; tax \\ \hline \$296.80 \end{array}$

12. (2) $8.00
    First find the amount marked up.

part	20
10	100

$\begin{array}{r} 20 \\ \times 10 \\ \hline 200 \end{array}$ $\begin{array}{r} 2 \\ 100\overline{)200} \end{array}$

   Next find the price the store will charge.
   $\begin{array}{r} 10 \\ +2 \\ \hline \$12 \end{array}$

   Then find the amount discounted.
   $\frac{\overset{4}{\cancel{12}}}{1} \times \frac{1}{\underset{1}{\cancel{3}}} = \$4$

   Finally, subtract to find the sale price.
   $\begin{array}{r} 12 \\ -4 \\ \hline \$8 \end{array}$

13. (4) $83\frac{1}{3}\%$

1,000	%
1,200	100

$\begin{array}{r} 1,000 \\ \times 100 \\ \hline 100,000 \end{array}$

$\begin{array}{r} 83\frac{1}{3} \\ 1,200\overline{)100,000} \\ \underline{9,600} \\ 4,000 \\ \underline{3,600} \\ 400 \end{array}$

14. (2) 10%

First find the additional amount she was paid:

$$\begin{array}{r} \$935 \\ -\ 850 \\ \hline \$\ 85 \end{array}$$

$85 is the part. $850 is the whole or original amount. Find the percent.

$85	%
$850	100

$$\begin{array}{r} 85 \\ \times\ 100 \\ \hline 8,500 \end{array} \qquad \begin{array}{r} 10 \\ 850\overline{)8,500} \end{array}$$

15. (2) $27

$$9 \text{ months} = \frac{9}{12} = \frac{3}{4} \text{ year}$$

$$\frac{\$\overset{9}{\cancel{450}}}{1} \times \frac{\overset{2}{\cancel{8}}}{\underset{2}{\cancel{100}}} \times \frac{3}{\underset{1}{\cancel{4}}} = \frac{\$54}{2} = \$27$$

Mixed Review

1. (2) 403.019

2. (5) 385.7

3. (3) $3.77

$$\begin{array}{r} 2.69 \\ 1.4 \\ \hline 1076 \\ 269 \\ \hline 3.766 = 3.77 \end{array}$$

4. (5) $9\frac{1}{2}$ lb.

$$\begin{array}{r} 2\frac{1}{4} = 2\frac{1}{4} \\ 3\frac{1}{2} = 3\frac{2}{4} \\ +3\frac{3}{4} = 3\frac{3}{4} \\ \hline 8\frac{6}{4} = 9\frac{2}{4} = 9\frac{1}{2} \end{array}$$

5. (3) $175

$$\frac{\overset{35}{\cancel{280}}}{1} \times \frac{3}{\cancel{8}} = \$105$$

$$\begin{array}{r} 280 \\ -105 \\ \hline 175 \end{array}$$

6. (2) 75%

$$\begin{array}{r} 14 \\ -8 \\ \hline \$6 \text{ markup} \end{array}$$

6	%
8	100

$$\begin{array}{r} 100 \\ \times\ 6 \\ \hline 600 \end{array} \qquad \begin{array}{r} 75 \\ 8\overline{)600} \\ 56 \\ \hline 40 \\ 40 \end{array}$$

7. (2) $5,280

$$8,000 \times \frac{12}{100} \times 5\frac{1}{2} =$$

$$\frac{\overset{80}{\cancel{8,000}}}{1} \times \frac{\overset{6}{\cancel{12}}}{\underset{1}{\cancel{100}}} \times \frac{11}{\underset{1}{\cancel{2}}} = 5,280$$

8. (4) $.40

$$\begin{array}{r} .40 \\ 11.5\overline{)4.60} \\ 4\ 60 \end{array}$$

9. (1) $\frac{5}{8}$ ft.

$$\begin{array}{r} 6\frac{1}{2} = 6\frac{4}{8} \\ 2\frac{3}{4} = 2\frac{6}{8} \\ +\frac{1}{8} = \frac{1}{8} \\ \hline 8\frac{11}{8} = 9\frac{3}{8} \end{array}$$

$$\begin{array}{r} 10 = 9\frac{8}{8} \\ -9\frac{3}{8} = 9\frac{3}{8} \\ \hline \frac{5}{8} \end{array}$$

10. (2) $3 + 9 + 2(6.50)$

RATIO AND PROPORTION

GRAPHS AND TABLES

MEASUREMENT

Ratio and Proportion

Ratio and proportion are two of the most useful tools you will learn in mathematics. You will be able to use the skills you develop in this section on many different types of math problems.

RATIO

In mathematics, we often compare numbers. Subtracting is one method of comparison. For example, suppose a man is 28 years old and his daughter is 7 years old. We can say that the man is 21 years older than his daughter.

Another way to compare numbers is by division. A **ratio** is a comparison of numbers by division. For example, if the man is 28 years old and his daughter is 7 years old, the man is 4 times as old as his daughter. The ratio of the man's age to his daughter's age is said to be 28 to 7 or 4 to 1.

Ratios can be written three ways: with the word *to*, with a colon (:), or as a fraction. Like fractions, ratios should be reduced. There are three ways of writing the ratio of the man's age to his daughter's age. Notice that each form has been reduced.

$$28 \text{ to } 7 = 4 \text{ to } 1$$
$$28 : 7 = 4 : 1$$
$$\frac{28}{7} = \frac{4}{1}$$

Notice that in a ratio, unlike a fraction, you keep a 1 in the denominator. This is because you are comparing two numbers.

The numbers in a ratio must be written in the order that the problem asks for. Look at these examples carefully.

EXAMPLE 1: Evelyn earns $600 a month. She pays $120 in rent. What is the ratio of her rent to her income?

Solution: Make a ratio with the rent on top (numerator) and the monthly income on the bottom (denominator). Then reduce.

$$\frac{\text{rent}}{\text{income}} = \frac{\$120}{\$600} = \frac{1}{5} \text{ or } \mathbf{1:5}$$

EXAMPLE 2: Simplify the ratio $\frac{1}{4} : \frac{2}{3}$.

   Solution: Ratio is a comparison of numbers by division. Rewrite this as a division problem and solve.

$$\frac{1}{4} : \frac{2}{3} = \frac{1}{4} \div \frac{2}{3} = \frac{1}{4} \times \frac{3}{2} = \frac{3}{8} \text{ or } \textbf{3:8}$$

EXAMPLE 3: On a test of 20 questions, Maceo got 2 questions wrong. What is the ratio of the number he got right to the number he got wrong?

①
total questions	20
number wrong	−2
number right	18

② $\dfrac{\text{right}}{\text{wrong}} \ \dfrac{18}{2} = \dfrac{9}{1}$ or **9:1**

   Step 1. Subtract the number of questions Maceo got wrong from the total number of questions.

   Step 2. Make a ratio with the number right on top (numerator) and the number wrong on the bottom (denominator). Then reduce.

RATIO AND PROPORTION EXERCISE 1

Express each of the following ratios in reduced form.

1. Simplify the ratio 12:15.

2. In a GED class with 20 students, there are 12 women. What is the ratio of the number of women to the total number of students?

3. In the class in problem 2, what is the ratio of the number of men to the total number of students?

4. For the class in problem 2, what is the ratio of the number of men to the number of women?

5. In a factory with 150 workers, 105 workers belong to the union. What is the ratio of the number of workers who do not belong to the union to the total number of workers in the factory?

6. Out of a total city budget of $18,000,000, $3,000,000 is spent for education. What is the ratio of the amount spent for education to the amount not spent on education?

7. Simplify the ratio $\frac{1}{5} : \frac{2}{7}$.

8. On a test with 50 questions, there were 15 fraction problems and 5 decimal problems. What is the ratio of the total number of fraction and decimal problems to the number of problems on the test?

9. In a day care center with 80 children, there are 48 girls. What is the ratio of the number of boys to the number of girls?

10. Express in simplest form the ratio $\frac{3}{4}$ to $\frac{5}{8}$.

11. In a season, a basketball team won 62 games and lost 20. What was the ratio of the number of games the team won to the total number of games they played?

12. Another basketball team played the same number of games as the team in problem 11 and lost 41. What was the ratio of the number of games that team won to the number of games they played?

Answers and solutions start on page 182.

PROPORTION

A **proportion** is a statement involving two equal ratios. The statement 2:4 = 1:2 is a proportion. Written with fractions, the proportion is $\frac{2}{4} = \frac{1}{2}$. A proportion is another use of equivalent fractions.

The **cross products** in a proportion are always equal. A cross product is the product of a number in one upper corner multiplied by the number in the opposite lower corner. For the proportion $\frac{2}{4} = \frac{1}{2}$, one cross product is $2 \times 2 = 4$. The other cross product is $4 \times 1 = 4$. Both cross products equal 4.

You have already used proportion to solve problems. The percent box (page 146) is a type of proportion. The relationship $\frac{part}{whole} = \frac{percent}{100}$ is a proportion. In percent box problems, you had to find the part, the whole, or the percent. With ordinary proportion problems, a letter stands for the missing **element** or **term.** Any letter can be used to stand for an unknown number, but in most cases, x is used to represent an unknown.

Rules for Solving a Proportion

1. Multiply the two numbers that are "diagonal to" each other. This operation is called **"cross multiplying."**
2. Divide the product by the remaining number in the proportion.

Notice that a letter stands for the missing number in the following proportions.

EXAMPLE 1: Find the missing term in $\frac{x}{8} = \frac{9}{12}$

①
$$\begin{array}{r} 8 \\ \times 9 \\ \hline 72 \end{array}$$

②
$$12 \overline{)72}^{\;6}$$

Step 1. Multiply the two numbers that are diagonal to each other, 8 and 9.

Step 2. Divide the product, 72, by the remaining number in the proportion, 12.
The value of x is **6.**

EXAMPLE 2: Solve for c in $\frac{3}{7} = \frac{8}{c}$

①
$$\begin{array}{r} 7 \\ \times 8 \\ \hline 56 \end{array}$$

②
$$3 \overline{)56}^{\;18\frac{2}{3}}$$

Step 1. Cross multiply the two numbers that are diagonal to each other, 7 and 8.

Step 2. Divide by the remaining number, 3.
The value of c is **$18\frac{2}{3}$.**

EXAMPLE 3: Solve for y in the proportion $5:y = 2:7$

①
$$\frac{5}{y} = \frac{2}{7}$$

②
$$\begin{array}{r} 5 \\ \times 7 \\ \hline 35 \end{array}$$

③
$$2 \overline{)35}^{\;17\frac{1}{2}}$$

Step 1. Rewrite the proportion in fractional form. Notice that the first term on each side of the equal sign becomes a numerator.

Step 2. Cross multiply the numbers that are diagonal to each other, 5 and 7.

Step 3. Divide 35 by the other number in the proportion, 2. The value of y is **$17\frac{1}{2}$.**

RATIO AND PROPORTION EXERCISE 2 ————————

Solve for the missing term in each proportion.

1. $\frac{m}{6} = \frac{10}{15}$ 　　　　**2.** $\frac{3}{a} = \frac{5}{6}$ 　　　　**3.** $\frac{4}{9} = \frac{y}{3}$

4. $\dfrac{8}{7} = \dfrac{4}{x}$ 7. $\dfrac{2}{11} = \dfrac{4}{p}$ 10. $3:7 = 4:y$

5. $\dfrac{1}{3} = \dfrac{s}{5}$ 8. $c:8 = 9:2$ 11. $15:40 = x:60$

6. $\dfrac{3}{6} = \dfrac{w}{5}$ 9. $4:e = 6:8$ 12. $30:a = 12:16$

Answers and solutions start on page 182.

PROPORTION WORD PROBLEMS

You saw earlier how the proportional method can be a useful tool for solving percent problems. In fact, proportions have many practical applications. Throughout the rest of this book, you will have several opportunities to use proportions. Read through the next examples carefully. The key to using proportions is setting up the problem carefully. Notice how corresponding parts of a proportion are set up. Although x is generally used to represent an unknown, you may want to use the first letter of the quantity you are looking for.

EXAMPLE 1:   If 12 yards of lumber cost \$40, how much does 30 yards cost?

① $\dfrac{\text{yards}}{\text{cost}} \dfrac{12}{40} = \dfrac{30}{c}$ ② $\begin{array}{r} 40 \\ \times 30 \\ \hline 1{,}200 \end{array}$ ③ $\begin{array}{r} \$100 \\ 12\overline{)1{,}200} \end{array}$

Step 1.   Write two ratios with corresponding numerators and denominators. Yards are on the top in each ratio. Cost is on the bottom. Notice that c stands for the cost you are asked to find.

Step 2.   Multiply the two numbers that are diagonal to each other, 30 and 40.

Step 3.   Divide 1,200 by the other number in the proportion, 12. The cost of 30 yards of lumber is **\$100**.

EXAMPLE 2:   The ratio of the number of men to the number of women working in a certain hospital is 2:3. If 480 women work in the hospital, how many men work there?

① $\dfrac{\text{men}}{\text{women}} \dfrac{2}{3} = \dfrac{m}{480}$ ② $\begin{array}{r} 480 \\ \times 2 \\ \hline 960 \end{array}$ ③ $\begin{array}{r} 320 \text{ men} \\ 3\overline{)960} \end{array}$

Step 1. Make a proportion. The top number in each ratio refers to men. The bottom number refers to women. Notice that *m* stands for the unknown number of men.

Step 2. Multiply the numbers that are diagonal to each other, 2 and 480.

Step 3. Divide 960 by the other number in the proportion, 3. **320 men** work in the hospital. Remember, it is essential that you decide which category goes on the top of each ratio and which goes on the bottom. Keep these the same on <u>both</u> sides of the proportion.

RATIO AND PROPORTION EXERCISE 3

Solve each problem with a proportion.

1. If 6 feet of wire cost $3.40, how much does 9 feet of wire cost?

2. The scale on a map says that 2 inches = 150 miles. If two cities are actually 325 miles apart, how far apart will they be on the map?

3. If 30 chickens lay 75 eggs a month, how many eggs will 50 chickens lay in a month?

4. How long will it take to go 1,200 miles if a plane travels 450 miles in 2 hours?

5. To make a certain color of paint, the ratio of blue paint to white paint is 5:2. How many gallons of blue paint are required to mix with 14 gallons of white paint in order to make the color?

6. If apples cost 90¢ a dozen, how much do 8 apples cost?

7. A snapshot that was 3 inches wide and 5 inches long was enlarged to be 12 inches long. How wide is the enlargement?

8. If a worker can make 16 parts in 2 hours, how long does he need to make 100 parts?

9. A recipe calls for two cups of sugar for every three cups of flour. If a cook is using 12 cups of flour, how many cups of sugar does he need?

10. On a certain bar graph, a line $2\frac{1}{2}$ inches long represents 75 degrees. How many inches of bar are required to represent 110 degrees?

Answers and solutions start on page 183.

ANSWERS AND SOLUTIONS

Ratio and Proportion Exercise 1

1. 4:5 $\frac{12}{15} = \frac{4}{5} = 4{:}5$

2. 3:5 $\frac{\text{women}}{\text{total}} = \frac{12}{20} = \frac{3}{5} = 3{:}5$

3. 2:5 $\frac{\text{men}}{\text{total}} = \frac{20-12}{20} = \frac{8}{20} = \frac{2}{5} = 2{:}5$

4. 2:3 $\frac{\text{men}}{\text{women}} = \frac{8}{12} = \frac{2}{3} = 2{:}3$

5. 3:10 $\frac{\text{not in union}}{\text{total workers}} = \frac{150-105}{150} = \frac{45}{150} = $
$\frac{9}{30} = \frac{3}{10}$ or 3:10

6. 1:5 $\frac{\$ \text{ for education}}{\$ \text{ not for education}}$
$= \frac{\$3,000,000}{\$18,000,000 - \$3,000,000}$
$= \frac{\$3,000,000}{\$15,000,000} = \frac{1}{5}$ or 1:5

7. 7:10 $\frac{1}{5} \div \frac{2}{7} = \frac{1}{5} \times \frac{7}{2} = \frac{7}{10}$ or 7:10

8. 2:5 $\frac{\text{fractions and decimals}}{\text{total number of problems}} =$
$\frac{15+5}{50} = \frac{20}{50} = \frac{2}{5}$ or 2:5

9. 2:3 $\frac{\text{boys}}{\text{girls}} = \frac{80-48}{48} = \frac{32}{48} = \frac{2}{3}$ or 2:3

10. 6:5 $\frac{3}{4} \div \frac{5}{8} = \frac{3}{4} \times \frac{8}{5} = \frac{6}{5}$ or 6:5

11. 31:41 $\frac{\text{games won}}{\text{games played}} = \frac{62}{62+20} = \frac{62}{82} =$
$\frac{31}{41}$ or 31:41

12. 1:2 $\frac{\text{games won}}{\text{games played}} = \frac{82-41}{82} = \frac{41}{82} =$
$\frac{1}{2}$ or 1:2

Ratio and Proportion Exercise 2

1. 4
$$\begin{array}{r} 10 \\ \times 6 \\ \hline 60 \end{array} \qquad 15\overline{)60}\,^{4}$$

2. $3\frac{3}{5}$
$$\begin{array}{r} 3 \\ \times 6 \\ \hline 18 \end{array} \qquad 5\overline{)18}\,^{3\frac{3}{5}}$$

3. $1\frac{1}{3}$
$$\begin{array}{r} 4 \\ \times 3 \\ \hline 12 \end{array} \qquad 9\overline{)12}\,^{1\frac{3}{9}} = 1\frac{1}{3}$$

4. $3\frac{1}{2}$
$$\begin{array}{r} 7 \\ \times 4 \\ \hline 28 \end{array} \qquad 8\overline{)28}\,^{3\frac{4}{8}} = 3\frac{1}{2}$$

5. $1\frac{2}{3}$
$$\begin{array}{r} 1 \\ \times 5 \\ \hline 5 \end{array} \qquad 3\overline{)5}\,^{1\frac{2}{3}}$$

6. $2\frac{1}{2}$
$$\begin{array}{r} 3 \\ \times 5 \\ \hline 15 \end{array} \qquad 6\overline{)15}\,^{2\frac{3}{6}} = 2\frac{1}{2}$$

7. 22
$$\begin{array}{r} 11 \\ \times 4 \\ \hline 44 \end{array} \qquad 2\overline{)44}\,^{22}$$

8. 36 $\frac{c}{8} = \frac{9}{2}$
$$\begin{array}{r} 8 \\ \times 9 \\ \hline 72 \end{array} \qquad 2\overline{)72}\,^{36}$$

9. $5\frac{1}{3}$ $\frac{4}{e} = \frac{6}{8}$
$$\begin{array}{r} 4 \\ \times 8 \\ \hline 32 \end{array} \qquad 6\overline{)32}\,^{5\frac{2}{6}} = 5\frac{1}{3}$$

10. $9\frac{1}{3}$ $\frac{3}{7} = \frac{4}{y}$
$$\begin{array}{r} 7 \\ \times 4 \\ \hline 28 \end{array} \qquad 3\overline{)28}\,^{9\frac{1}{3}}$$

11. $22\frac{1}{2}$ $\frac{15}{40} = \frac{x}{60}$
$$\begin{array}{r} 15 \\ \times 60 \\ \hline 900 \end{array}$$
$$40\overline{)900}\,^{22\frac{20}{40}} = 22\frac{1}{2}$$
$$\begin{array}{r} 80 \\ \hline 100 \\ 80 \\ \hline 20 \end{array}$$

12. 40 $\frac{30}{a} = \frac{12}{16}$
$$\begin{array}{r} 16 \\ \times 30 \\ \hline 480 \end{array} \qquad 12\overline{)480}\,^{40}$$

Ratio and Proportion Exercise 3

1. $5.10
$$\frac{feet}{\$} \qquad \frac{6}{3.40} = \frac{9}{x}$$

$$\begin{array}{r} 3.40 \\ \times\ 9 \\ \hline 30.60 \end{array} \qquad 6\overline{)\ 30.60}^{\,5.10}$$

2. $4\frac{1}{3}$ inches
$$\frac{inches}{miles} \qquad \frac{2}{150} = \frac{x}{325}$$

$$\begin{array}{r} 325 \\ \times\ 2 \\ \hline 650 \end{array} \qquad 150\overline{)\ 650}^{\,4\frac{50}{150}} = 4\frac{1}{3}$$
$$\qquad \qquad \frac{600}{50}$$

3. 125 eggs
$$\frac{chickens}{eggs} \qquad \frac{30}{75} = \frac{50}{x}$$

$$\begin{array}{r} 75 \\ \times 50 \\ \hline 3,750 \end{array} \qquad 30\overline{)\ 3,750}^{\,125}$$

4. $5\frac{1}{3}$ hours
$$\frac{miles}{hours} \qquad \frac{450}{2} = \frac{1,200}{x}$$

$$\begin{array}{r} 1,200 \\ \times\ \ 2 \\ \hline 2,400 \end{array} \qquad 450\overline{)\ 2,400}^{\,5\frac{150}{450}} = 5\frac{1}{3}$$
$$\qquad \qquad \frac{2,250}{150}$$

5. 35 gallons
$$\frac{blue}{white} \qquad \frac{5}{2} = \frac{x}{14}$$

$$\begin{array}{r} 14 \\ \times\ 5 \\ \hline 70 \end{array} \qquad 2\overline{)\ 70}^{\,35}$$

6. 60¢
$$\frac{cost}{number} \qquad \frac{90}{12} = \frac{x}{8}$$

$$\begin{array}{r} 90 \\ \times\ 8 \\ \hline 720 \end{array} \qquad 12\overline{)\ 720}^{\,60}$$

7. $7\frac{1}{5}$ inches
$$\frac{width}{length} \qquad \frac{3}{5} = \frac{x}{12}$$

$$\begin{array}{r} 12 \\ \times\ 3 \\ \hline 36 \end{array} \qquad 5\overline{)\ 36}^{\,7\frac{1}{5}}$$

8. $12\frac{1}{2}$ hours
$$\frac{parts}{hours} \qquad \frac{16}{2} = \frac{100}{x}$$

$$\begin{array}{r} 100 \\ \times\ 2 \\ \hline 200 \end{array} \qquad 16\overline{)\ 200}^{\,12\frac{8}{16}} = 12\frac{1}{2}$$
$$\qquad \qquad \frac{16}{40}$$
$$\qquad \qquad \frac{32}{8}$$

9. 8 cups
$$\frac{sugar}{flour} \qquad \frac{2}{3} = \frac{x}{12}$$

$$\begin{array}{r} 12 \\ \times\ 2 \\ \hline 24 \end{array} \qquad 3\overline{)\ 24}^{\,8}$$

10. $3\frac{2}{3}$ inches
$$\frac{inches}{degrees} \qquad \frac{2\frac{1}{2}}{75} = \frac{x}{110}$$

$$\frac{5}{\underset{1}{\cancel{2}}} \times \frac{\overset{55}{\cancel{110}}}{1} = 275 \qquad 75\overline{)\ 275}^{\,3\frac{50}{75}} = 3\frac{2}{3}$$
$$\qquad \qquad \frac{225}{50}$$

Graphs and Tables

TYPES OF GRAPHS AND TYPES OF QUESTIONS

A **graph** is a way of organizing numbers visually. From a graph, you can make a quick comparision of numbers.

Consider the following information: according to the 1980 census, the population of the metropolitan area of Chicago was 7,102,328. The population of Detroit was 4,352,762. The population of Los Angeles was 7,477,657. The population of New York was 9,119,737. The population of Philadelphia was 4,716,818. The population of San Francisco was 3,252,721.

The information in the last paragraph is precise, but it is hard to compare the sizes of the cities quickly. A table is a clearer way of presenting the information. The table below (Figure 1) contains the same information. The cities are listed in order of size with the largest city first.

Figure 1

POPULATION OF THE LARGEST METROPOLITAN AREAS IN THE U.S.	
New York City	9,119,737
Los Angeles	7,477,657
Chicago	7,102,328
Philadelphia	4,716,818
Detroit	4,352,762
San Francisco	3,252,721
Source: 1980 Census	

Another way of organizing this information is on a graph. In the following graph, called a pictograph (Figure 2), each symbol of a small person represents 1,000,000 people. The partial figures represent $\frac{1}{2}$ of 1,000,000, or 500,000 people. The information in a pictograph is not as exact as in a table.

184

To read the population of a metropolitan area from the pictograph, count the number of figures. Then multiply the number by 1,000,000. For example, Chicago has seven figures. The population of Chicago is:

$$7 \times 1,000,000 = \text{approximately } 7,000,000 \text{ people.}$$

Figure 2
APPROXIMATE POPULATION OF THE LARGEST METROPOLITAN AREAS IN THE U.S.

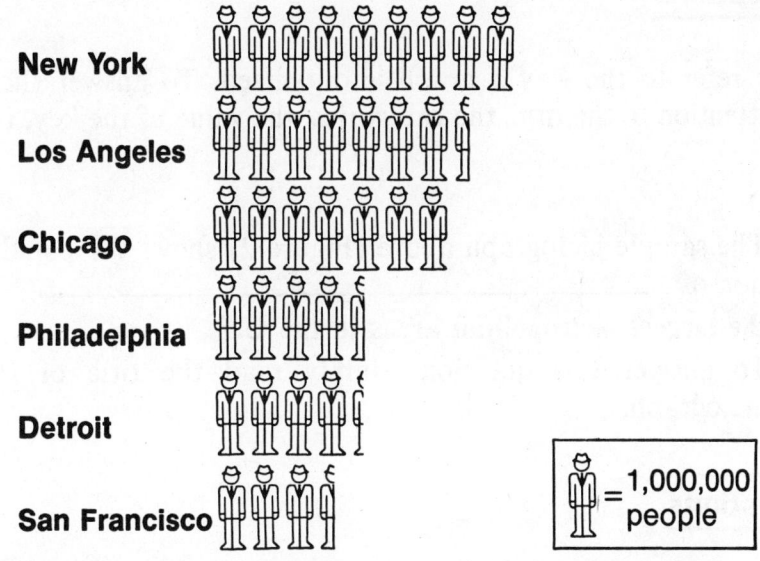

A bar graph is a more precise way to show the population information. In the graph below (Figure 3), the population of each metropolitan area is represented by the length of a bar extending across the graph. The horizontal axis that runs along the bottom of the graph represents millions of people. The bar for New York extends beyond the 9 million line. This

Figure 3
POPULATION OF THE LARGEST METROPOLITAN AREAS IN THE U.S.

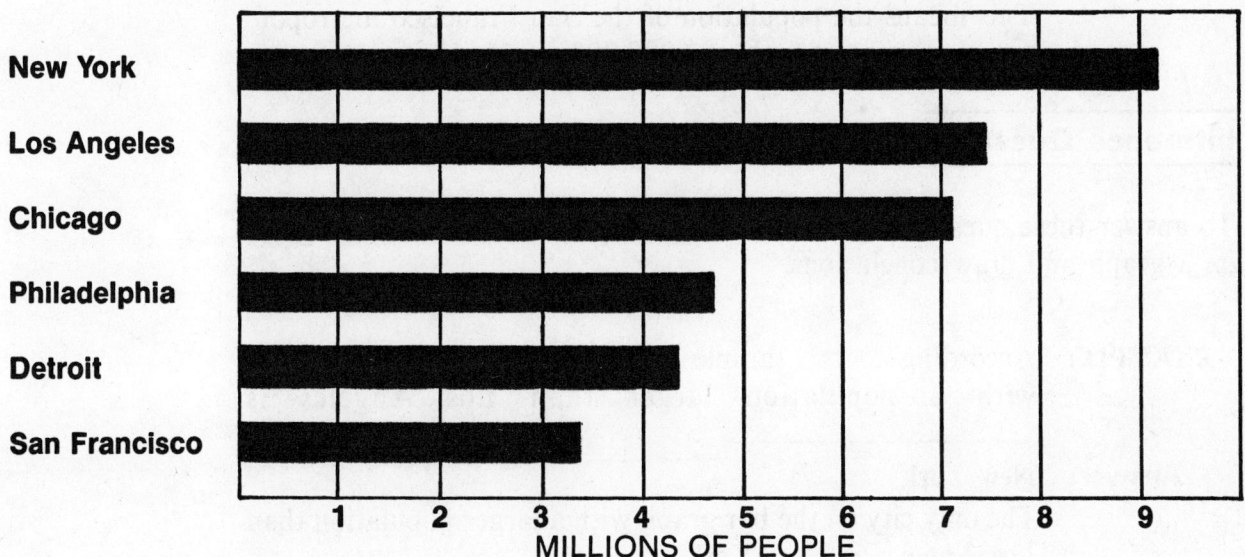

means that the population of New York is more than 9,000,000. The bar for Philadelphia goes farther beyond the 4 million line than does the bar for Detroit. This shows that Philadelphia's population is greater than Detroit's.

There are three common types of questions about graphs. Think about these as you work through this section of the book.

Scanning Questions

These questions refer to the way a graph is organized. To answer these questions, pay attention to the title, the categories, the value of the key, the source, etc.

> **EXAMPLE:**   The sample pictograph above (Figure 2) shows the population of _____.
>
> **Answer:**   **the largest metropolitan areas in the U.S.**
>   To answer this question, simply read the title of the pictograph.

Reading Questions

To answer these questions, you must find specific information on a graph.

> **EXAMPLE:**   According to the sample bar graph above (Figure 3), the population of the metropolitan area of San Francisco was less than 3,000,000.
>   True or False.
>
> **Answer:**   **False**
>   The bar for San Francisco goes beyond the 3,000,000 line. This means the population of the San Francisco metropolitan area was <u>more</u> than 3,000,000.

Inference Questions

To answer these questions, you must compare pieces of information given on a graph and draw conclusions.

> **EXAMPLE:**   According to the sample bar graph (Figure 3), the city with a population larger than Los Angeles is
>   _____.
>
> **Answer:**   **New York**
>   The only city on the bar graph with a larger population than Los Angeles is New York.

PICTOGRAPHS

You have already seen that a pictograph uses small pictures or symbols to represent information. A pictograph must have a key that tells what each symbol stands for.

Figure 4 is an example of a pictograph.

Figure 4

COMPARISON OF PER CAPITA PERSONAL INCOME FOR NINE METROPOLITAN AREAS
(to the nearest $500)

Atlanta, GA	$$$$$$$$$
Baltimore, MD	$$$$$$$$$
Birmingham, AL	$$$$$$$$$
Chicago, IL	$$$$$$$$$$$$
Honolulu, HI	$$$$$$$$$$$
Salt Lake City, UT	$$$$$$$$$
San Antonio, TX	$$$$$$$$$
San Jose, CA	$$$$$$$$$$$$
Washington, DC	$$$$$$$$$$$$

$\boxed{\$ = \$1,000}$

EXAMPLE 1: What was the per capita income for residents of San Antonio, Texas?

Solution: There are $8\frac{1}{2}$ symbols in the row labeled San Antonio. Each symbol represents $1,000. Compute the value of the symbols.

$\$1,000 \times 8 = \$8,000$
$\$1,000 \times \frac{1}{2} = \underline{+\ \ \ \ 500}$
$\$8,500$ per capita income

EXAMPLE 2: Between which two cities is there a larger difference in per capita income: Chicago and Birmingham or San Jose and Honolulu?

Solution: Chicago's per capita income is shown by $11\frac{1}{2}$ symbols and Birmingham's by 9 symbols, a difference of $2\frac{1}{2}$ symbols, or $2,500.

San Jose's per capita income is shown by 12 symbols and Honolulu's is shown by $10\frac{1}{2}$ symbols. The difference is $1\frac{1}{2}$ symbols, or $1,500.

There is a greater difference between Chicago's and Birmingham's per capita incomes.

GRAPH EXERCISE 1 ─────────────────────────────────

Use Figure 5 to answer the questions below. For each question, choose the best answer.

Figure 5
RESIDENTIAL TELEPHONES

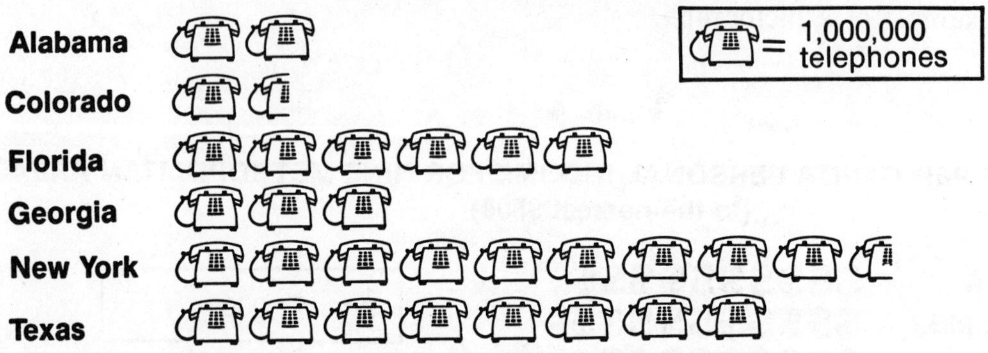

1. This graph tells
   (1) the population of certain states in the U.S.
   (2) the number of commercial telephones in certain states in the U.S.
   (3) the number of residential telephones in every state in the U.S.
   (4) the number of residential telephones in certain states in the U.S.
   (5) the number of residential telephones in the world

2. Each symbol in the pictograph represents
   (1) 1,000,000 people
   (2) 100,000 commercial telephones
   (3) 1,000,000 commercial telephones
   (4). 100,000 residential telephones
   (5) 1,000,000 residential telephones

3. How many residential phones are there in the state of Colorado?
   (1) 1,500,000   (4) 2,000,000
   (2) 150,000   (5) 1.5
   (3) 15,000,000

4. How many residential phones are there in the state of Texas?
   (1) 80,000   (4) 8,000,000
   (2) 800   (5) 80,000,000
   (3) 8

5. How many more residential phones are there in New York than in Florida?
   (1) 3,000,000   (4) 6,500,000
   (2) 3,500,000   (5) 9,500,000
   (3) 6,000,000

6. What is the ratio of the number of residential telephones in Georgia to the number of residential telephones in Alabama?
   (1) 3:5   (4) 2:3
   (2) 2:5   (5) 3:4
   (3) 3:2

7. The number of residential telephones in Texas is equal to the combined number of residential phones in which two states?
   (1) Alabama and Georgia   (4) Alabama and Florida
   (2) Colorado and New York   (5) Colorado and Florida
   (3) Florida and Georgia

8. Based on the pictograph, which of the following statements is true?
   (1) There are fewer residential telephones in Texas than in Florida.
   (2) There are three times as many residential telephones in New York as in Colorado.
   (3) There are more residential telephones in Colorado than in Texas.
   (4) There are twice as many residential telephones in Florida as in Georgia.
   (5) There are twice as many residential telephones in Georgia as in Florida.

Answers and solutions start on page 206.

CIRCLE GRAPHS

A circle graph shows the parts of a whole. The pie-shaped pieces show the comparative sizes of the parts. Often, the parts are expressed in percents.

In Figure 6 below, you can see that 8% of homes in the United States receive 1–4 television stations. The slice that represents 11 or more television stations is much larger, as 32% of U.S. homes receive that number of stations. Notice that all the slices of the "pie" added together equal 100%.

Figure 6
NUMBER OF STATIONS RECEIVED BY U.S. HOMES

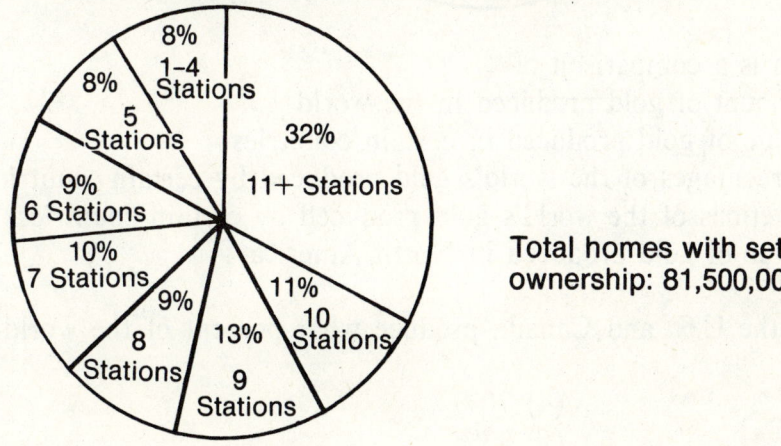

Total homes with set ownership: 81,500,000

EXAMPLE 1:   What percent of homes receive 8 stations?

    Solution: Look around the graph until you find the segment containing the label "8 stations." This segment shows that **9%** of U.S. homes receive 8 stations.

EXAMPLE 2:   Approximately how many homes receive 7 stations?

    Solution: The graph indicates that 10% of the homes receive 7 stations. Find 10% of the number of total homes.

    Step 1.   Find the total number of homes with sets; this is indicated below the graph as 81,500,000 homes.

    Step 2.   To find 10% of 81,500,000, either use the percent box or change 10% to $\frac{1}{10}$ and multiply as shown below.

$$\overset{8,150,000}{\cancel{81,500,000}} \times \frac{1}{\underset{1}{\cancel{10}}} = \textbf{8,150,000 homes receive 7 stations}$$

GRAPH EXERCISE 2

Use Figure 7 to answer the questions below. For each question, choose the best answer.

Figure 7
WORLD GOLD PRODUCTION

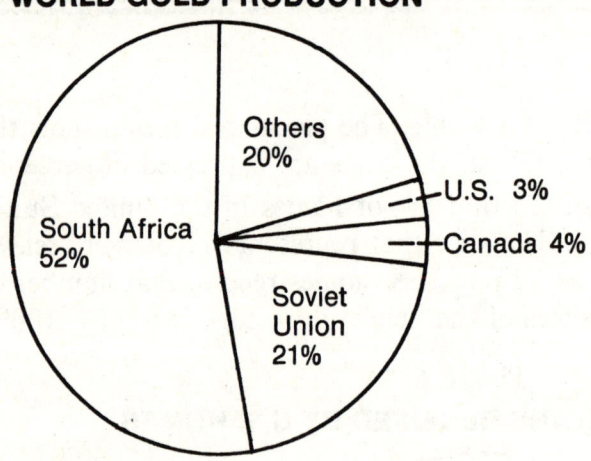

1. This graph is a comparison of
    (1) the amount of gold produced in the world
    (2) the value of gold produced in certain countries
    (3) the percentages of the world's gold produced by certain countries
    (4) the fractions of the world's gold produced by certain countries
    (5) the value of gold produced in North America

2. Together, the U.S. and Canada produce what percent of the world's gold?
    (1) 3%   (4) 70%
    (2) 4%   (5) 93%
    (3) 7%

3. What percent of the world's gold is produced outside South Africa?
   (1) 20%   (4) 41%
   (2) 27%   (5) 48%
   (3) 52%

4. The combined gold production of the U.S. and Canada is what fraction of the gold production of the Soviet Union?
   (1) $\frac{1}{3}$ (4) $\frac{3}{4}$
   (2) $\frac{1}{2}$ (5) $1\frac{1}{2}$
   (3) $\frac{2}{3}$

5. The Soviet Union produces approximately what fraction of the world's gold?
   (1) $\frac{1}{4}$ (4) $\frac{1}{10}$
   (2) $\frac{1}{5}$ (5) $\frac{1}{20}$
   (3) $\frac{1}{8}$

6. What is the ratio of gold production in Canada to gold production in South Africa?
   (1) 3:5   (4) 1:14
   (2) 3:4   (5) 5:3
   (3) 1:13

7. Which of the following statements is true?
   (1) The U.S. produces most of the gold in the world.
   (2) The Soviet Union produces more than half the world's gold.
   (3) The U.S. produces more gold than Canada.
   (4) South Africa produces more than half the gold in the world.
   (5) South Africa produces about three-fourths of the gold in the world.

8. In a year, the total world gold production is about 40,000,000 ounces. Approximately how many ounces of gold does Canada produce?
   (1) 16,000,000   (4) 160,000
   (2) 1,600,000   (5) 40,000,000
   (3) 4,000,000

9. Which of the following statements is false?
   (1) Canada produces more gold than the U.S.
   (2) The combined gold production of the Soviet Union, Canada, and the U.S. is greater than the gold production of South Africa.
   (3) The Soviet Union produces about seven times as much gold as the U.S.
   (4) South Africa produces more than twice as much gold as the Soviet Union.
   (5) The gold produced in all the countries referred to as "others" is less than one-fourth of the total world production.

Answers and solutions start on page 206.

Some circle graphs represent a total amount of money. The parts of the pie represent an amount of cents or dollars. In Figure 8, each part represents the section (in cents) of a town's budget. The total is 100 cents, or $1.

Notice that the number of cents is equivalent to a percent. For example, 10¢ out of a dollar is the same as 10% of the total.

GRAPH EXERCISE 3 ─────────────────

Use Figure 8 to answer the questions below. For each question, choose the best answer.

Figure 8

A DOLLAR OF CENTERVILLE'S OPERATING BUDGET

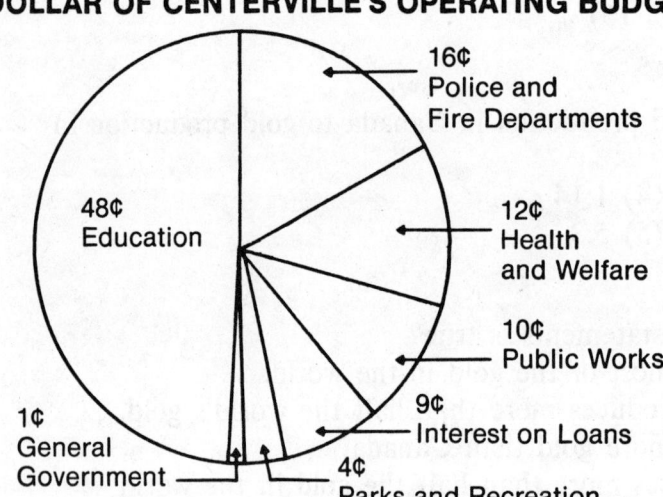

1. This circle graph shows
    (1) the amount of money the town of Centerville spends in a year
    (2) how much more the town of Centerville will spend this year than last year
    (3) how much more money the town of Centerville spends for some services than for others
    (4) the difference between the amounts spent for certain services and the percentages spent for others
    (5) the portion of every dollar in Centerville's budget that goes to each budget category

2. Out of every dollar in the budget, public works and loan interest represent how much money?
    (1) 9¢   (4) 27¢
    (2) 10¢   (5) 81¢
    (3) 19¢

3. The amount spent for the police and fire departments is how many times the amount spent on parks and recreation?
   (1) 3
   (2) 4
   (3) 12
   (4) 16
   (5) 20

4. The amount spent for education is approximately what fraction of the whole budget?
   (1) $\frac{1}{8}$
   (2) $\frac{1}{4}$
   (3) $\frac{1}{16}$
   (4) $\frac{1}{2}$
   (5) $\frac{1}{12}$

5. The amount of the Centerville budget spent on health and welfare and parks and recreation is equal to the amount spent on what category?
   (1) Police and fire departments
   (2) Education
   (3) Public works
   (4) Loan interest
   (5) General government

6. General government uses what percent of Centerville's budget?
   (1) 100%
   (2) 99%
   (3) 10%
   (4) .01%
   (5) 1%

7. The actual yearly budget for the police and fire departments in Centerville is $2,400,000. Find the yearly budget amount for public works.
   (1) $2,600,000
   (2) $3,200,000
   (3) $1,500,000
   (4) $1,600,000
   (5) $1,200,000

8. The actual amount of Centerville's budget for a year is $15,000,000. How much money does the town spend in a year on health and welfare?
   (1) $1,200,000
   (2) $1,800,000
   (3) $12,000,000
   (4) $120,000
   (5) $13,200,000

9. Which of the following statements is true?
   (1) Centerville spends twice as much on loan interest as on public works.
   (2) Centerville spends more for general government than for public works.
   (3) Centerville spends four times as much on education as on health and welfare.
   (4) Centerville spends about the same amount on education as on loan interest.
   (5) Centerville spends three times as much for parks and recreation as for general government.

Answers and solutions start on page 206.

BAR GRAPHS

You have already seen a bar graph at the beginning of this section. The populations of the six largest metropolitan areas in the U.S. were shown in Figure 3.

Bar graphs are more complicated than pictographs or circle graphs. The value of the bars depends on reading two sets of information. One set of information goes up and down along the **vertical axis.** The other set of information goes across along the **horizontal axis.**

Figure 9

NOON TEMPERATURES DURING A WEEK IN APRIL

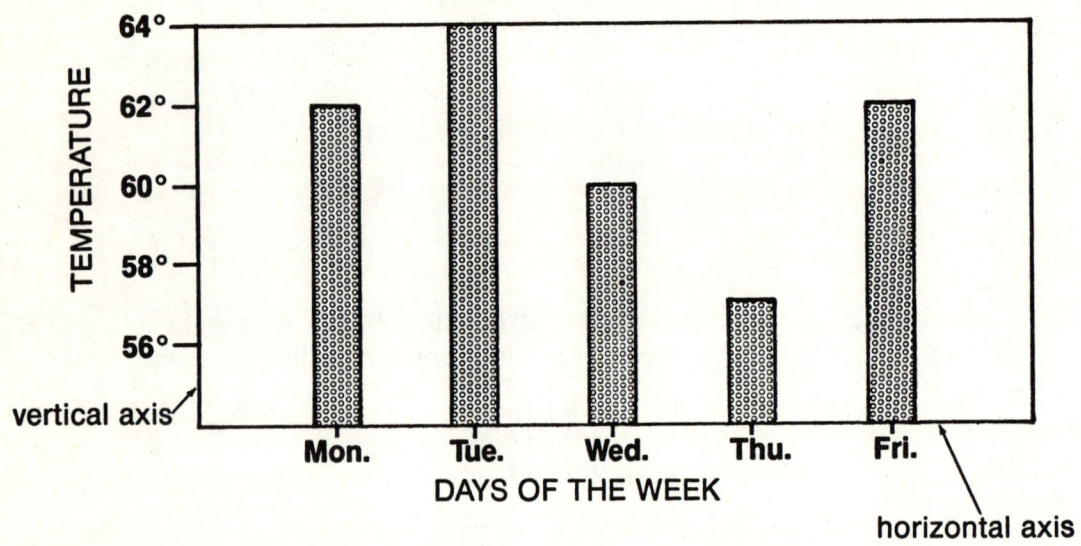

The vertical axis measures the temperature in degrees. The horizontal axis lists the days of the week.

EXAMPLE 1: What was the noon temperature on Wednesday?

    Solution: Find Wednesday on the horizontal axis. Follow the bar above "Wed." to the top. Read the temperature straight across on the axis to the left. The bar stops at the line labeled "60°." **The temperature at noon on Wednesday was 60°.**

EXAMPLE 2: On what day was the noon temperature 57°?

    Solution: Look halfway between 56° and 58° to find 57° on the vertical axis. Look across until you find a bar that rises to this point. The bar labeled "Thu." ends halfway between 56° and 58°. **The noon temperature was 57° on Thursday.**

━━━━━━━━━━━━━━━━

Use Figure 10 to answer the questions below. For each question, choose the best answer.

Figure 10
DISTANCE NEEDED TO STOP A CAR TRAVELING AT A CERTAIN SPEED

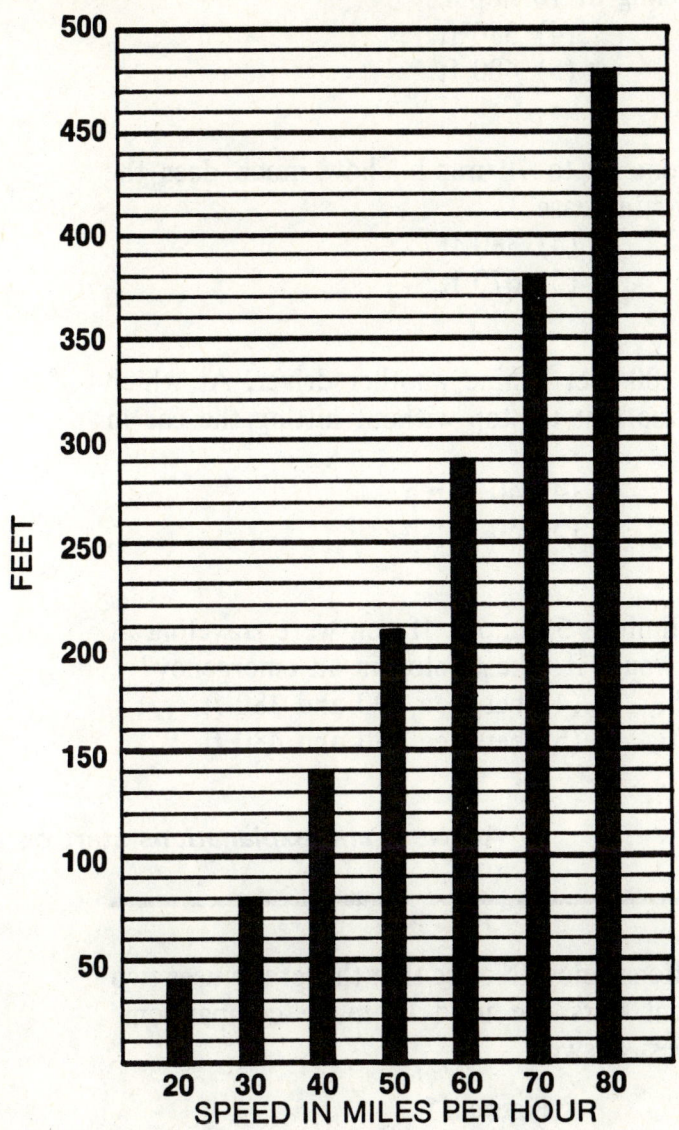

1. If a car is traveling at a speed of 40 m.p.h., what is the stopping distance for the car?
   (1) 100 ft.   (4) 150 ft.
   (2) 140 ft.   (5) 70 ft.
   (3) 80 ft.

2. A car with a stopping distance of 290 feet is traveling at a speed of
   (1) 50 m.p.h.   (4) 70 m.p.h.
   (2) 55 m.p.h.   (5) 290 m.p.h.
   (3) 60 m.p.h.

3. How much faster is a car with a stopping distance of 210 feet traveling than a car with a stopping distance of 80 feet?
   (1) 20 m.p.h.   (4) 80 m.p.h.
   (2) 30 m.p.h.   (5) 130 m.p.h.
   (3) 50 m.p.h.

4. The distance needed to stop a car traveling at 80 m.p.h. is how much greater than for a car traveling at 70 m.p.h.?
   (1) 100 ft.   (4) 380 ft.
   (2) 120 ft.   (5) 480 ft.
   (3) 210 ft.

5. When a car accelerates from 60 to 70 m.p.h., how much does the stopping distance for the car increase?
   (1) 45 ft.   (4) 380 ft.
   (2) 90 ft.   (5) 670 ft.
   (3) 290 ft.

6. Suppose you were driving 200 feet behind another driver. At which speed would you be going too fast to stop without hitting the car in front?
   (1) 10 m.p.h.   (4) 40 m.p.h.
   (2) 20 m.p.h.   (5) 50 m.p.h.
   (3) 30 m.p.h.

7. The maximum legal speed limit is 55 m.p.h. If you were traveling at that speed, how many feet would it take to stop in an emergency?
   (1) between 140 and 210 ft.   (4) between 290 and 380 ft.
   (2) between 250 and 290 ft.   (5) between 380 and 480 ft.
   (3) between 210 and 290 ft.

Answers and explanations start on page 207.

GRAPH EXERCISE 5 ——————————————————————————

Use figure 11 to answer the questions below. Notice that this graph uses two different kinds of bars. Different bars are used to make comparisons. For each question, choose the best answer.

Figure 11

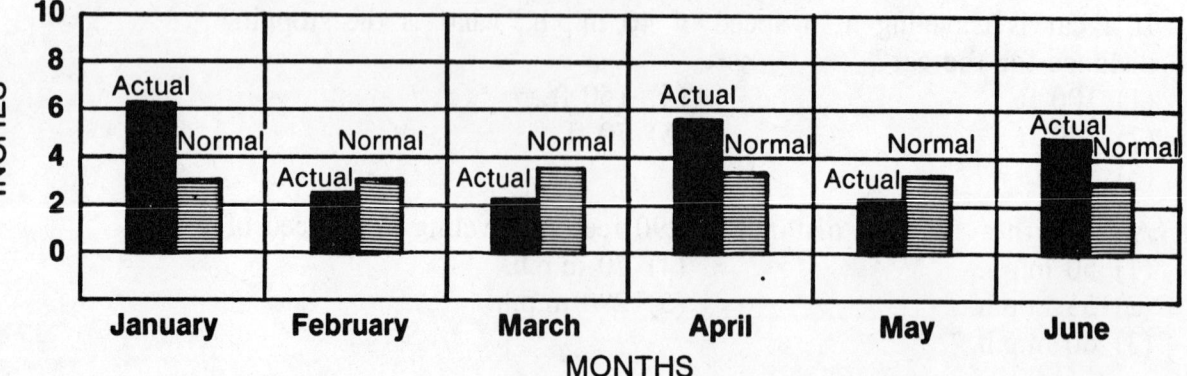

PRECIPITATION IN INCHES FOR A SIX-MONTH PERIOD IN NEW YORK CITY

1. Complete this statement: The bar graph in Figure 11 shows
    (1) the actual and normal temperature averages for a six-month period in New York City
    (2) the actual and normal precipitation for a year in New York City
    (3) the actual and normal precipitation for a six-month period in New York City
    (4) the daily precipitation for a year in New York City
    (5) the difference between actual and normal temperatures for six months in New York City

2. According to the graph, the normal precipitation for each month shown is
    (1) 2 inches
    (2) a little over 4 inches
    (3) about half the actual amount
    (4) about twice the actual amount
    (5) between 2 and 4 inches

3. For which months was the actual precipitation less than the normal precipitation?
    (1) February, March, April   (4) January, March, April
    (2) February, March, May   (5) March, May, June
    (3) January, April, June

4. For what month was actual precipitation about twice normal precipitation?
    (1) January   (4) April
    (2) February   (5) June
    (3) March

5. For the six months shown, which month had the highest normal precipitation?
    (1) January   (4) April
    (2) February   (5) June
    (3) March

6. For which month were the actual and normal precipitation most nearly the same?
    (1) January   (4) April
    (2) February   (5) June
    (3) March

7. Between which two consecutive months did actual precipitation levels change the most?
    (1) May and June   (4) February and March
    (2) April and May   (5) January and February
    (3) March and April

Answers and solutions start on page 207.

LINE GRAPHS

A **line graph,** like a bar graph, has information on both a vertical axis and a horizontal axis. The values that you read are displayed along a continuous line. Every point along the line has a value. To tell the value of a point, read across to the vertical axis at the side and down to the horizontal axis across the bottom.

Figure 12 is a line graph that shows the speed of a moving object.

Figure 12

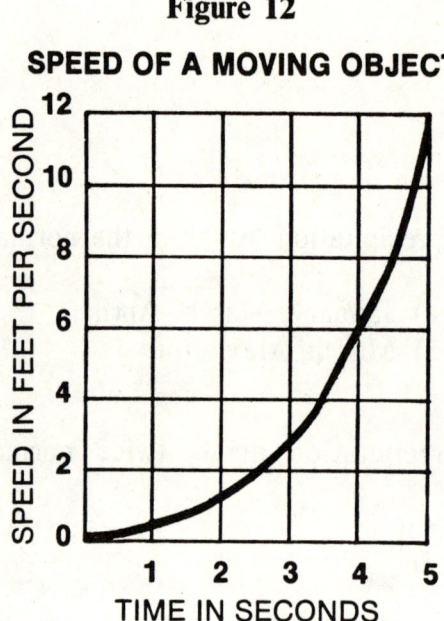

SPEED OF A MOVING OBJECT

EXAMPLE 1: Find the speed of the moving object at 4 seconds.

Solution: Find 4 on the axis labeled "Time in Seconds." Follow the line labeled "4" up until it meets the curved line. Read the speed on the vertical axis directly to the left. The point on the vertical axis is 6. Therefore, at 4 seconds, the object is moving at a speed of **6 feet per second.**

EXAMPLE 2: How many seconds does it take the object to reach a speed of 8 feet per second?

Solution: Find 8 on the vertical axis labeled "Speed in Feet per Second." Follow the line labeled "8" directly across to the curved line. Read the number of seconds directly below on the horizontal axis. This point on the horizontal axis is about halfway between 4 and 5. The object reaches a speed of 8 feet per second in about $4\frac{1}{2}$ **seconds.**

GRAPH EXERCISE 6

Use Figure 13 to answer the questions below. For each question, choose the best answer.

Figure 13
AVERAGE PRICE OF 1,000 BOARD FEET OF TIMBER

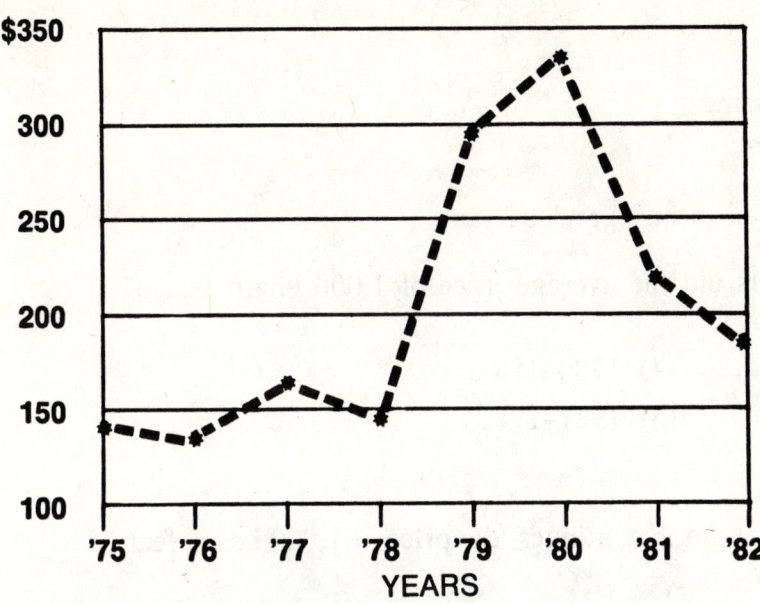

1. The graph in Figure 13 shows
   (1) timber production from 1975 to 1982
   (2) the change in the construction industry from 1975 to 1982
   (3) the change in the average price of 1,000 board feet of timber from 1975 to 1982
   (4) the change in timber construction contracts from 1975 to 1982
   (5) none of the above

2. In what year shown on the graph did the price of timber reach its lowest level?
   (1) 1975   (4) 1980
   (2) 1976   (5) 1981
   (3) 1978

3. During what year did the price of 1,000 board feet of timber first reach $200?
   (1) 1975   (4) 1979
   (2) 1977   (5) 1980
   (3) 1978

4. For the years shown on the graph, the highest price for 1,000 board feet of timber was about

(1) $210   (4) $300

(2) $240   (5) $330

(3) $270

5. From the beginning of 1978 to the beginning of 1979, the price of 1,000 board feet of timber

(1) nearly doubled
(2) stayed about the same
(3) dropped slightly
(4) was cut in half
(5) decreased more than in any previous year

6. Between what two years did the average price of 1,000 board feet of timber rise the most?

(1) 1976–1977   (4) 1980–1981

(2) 1978–1979   (5) 1981–1982

(3) 1979–1980

7. If the trend shown on the graph continues, the price of 1,000 board feet of timber will probably

(1) soon double   (4) continue to drop

(2) continue to rise   (5) rise to $350

(3) stay about the same

Answers and solutions start on page 207.

GRAPH EXERCISE 7 ━━━━━━━━━━━━━

Refer to Figure 14 to answer the questions on page 201. Notice that in this graph, there are two lines. One is solid. The other is broken. For each question, choose the best answer.

Figure 14

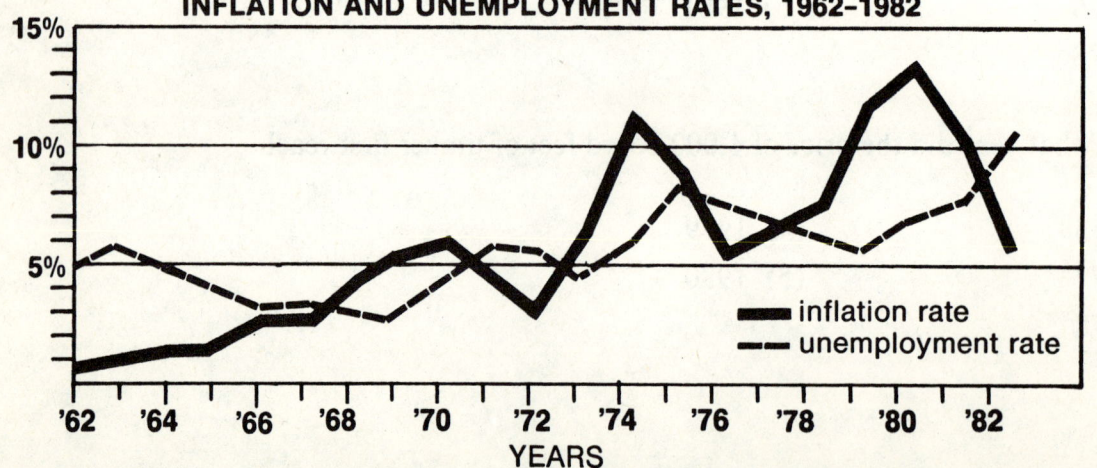

1. In Figure 14, both the inflation rate and the unemployment rate are measured in
   (1) dollars
   (2) number of people
   (3) years
   (4) percents
   (5) fractions

2. In 1962, the rate of inflation was close to
   (1) 1%
   (2) 3%
   (3) 5%
   (4) 6%
   (5) 8%

3. Inflation reached its highest rate in
   (1) 1964
   (2) 1972
   (3) 1974
   (4) 1980
   (5) 1982

4. Unemployment was at its lowest rate in
   (1) 1965
   (2) 1969
   (3) 1971
   (4) 1973
   (5) 1979

5. In 1974, the unemployment rate was close to
   (1) 4%
   (2) 6%
   (3) 8%
   (4) 11%
   (5) 12%

6. For what three-year period was the inflation rate over 10%?
   (1) 1966–68
   (2) 1972–74
   (3) 1977–79
   (4) 1979–81
   (5) 1978–80

7. From 1962 to 1967
   (1) the inflation rate and the unemployment rate were about the same
   (2) the unemployment rate was higher than the inflation rate
   (3) the inflation rate was higher than the unemployment rate
   (4) the inflation rate dropped and the unemployment rate rose
   (5) none of the above

8. In 1962, the unemployment rate was how many points higher than the inflation rate?
   (1) 1%
   (2) 4%
   (3) 5%
   (4) 6%
   (5) 10%

Answers and solutions start on page 208.

TABLES

You have already seen a table at the beginning of this section. The populations of the six largest metropolitan areas in the U.S. were shown in a table in Figure 1. A **table** displays facts in columns and rows for purposes of comparison and interpretation.

Use Figure 15 to see how to read a table. Look carefully at the title of the table and at the headings over all of the columns.

Figure 15
NORMAL TEMPERATURES, HIGHS, LOWS

State	Station	Normal Temperature January Max.	Min.	July Max.	Min.	Extreme Temperature Highest	Lowest
Alabama	Mobile	61	41	91	73	104	7
Alaska	Juneau	29	18	64	48	90	−22
California	San Francisco	55	41	71	54	106	20
Florida	Jacksonville	65	45	90	72	105	12
Iowa	Dubuque	26	9	82	61	99	−28
Massachusetts	Boston	36	23	81	65	102	−12
New Hampshire	Concord	31	10	33	57	102	−37
North Dakota	Bismarck	19	−3	84	57	109	−44
Oklahoma	Oklahoma City	48	26	93	70	110	−4
Tennessee	Nashville	48	29	90	69	107	−15
Utah	Salt Lake City	37	19	93	61	107	−30
Wisconsin	Madison	25	8	81	59	104	−37
Wisconsin	Milwaukee	27	11	80	59	101	−24

Source: World Almanac

EXAMPLE 1: What was the highest extreme temperature recorded at the station at Milwaukee, Wisconsin?

  Solution: Find the correct row for Milwaukee, Wisconsin. There are two entries for Wisconsin, one for Madison and one for Milwaukee. Follow the row for "Wisconsin . . . Milwaukee" to the right until you come to the column under "Extreme Temperature: Highest." The highest extreme temperature for Milwaukee, Wisconsin, was **101°**.

EXAMPLE 2: Of the cities listed, which had the highest normal temperature for January?

  Solution: Skim down the column labeled "Normal Temperature: January: Max." The highest temperature in that column is 65°. Look to the left, to the column under "States . . . Stations" to see which city this is for. **Jacksonville, Florida** had the highest normal temperature for January.

TABLES EXERCISE 8 ━━━━━━━━━━━━━━━━━━━━━━━━━━━━━━

Use the table in Figure 16 to answer the questions below. For each question, choose the best answer.

Figure 16

SAFETY RECORD: U.S. SCHEDULED AIR CARRIERS, 1976–1981

Year	Accidents	Fatalities	Miles Flown
1976	22	38	2,319,967,000
1977	21	78	2,418,652,000
1978	21	160	2,520,165,000
1979	24	351	2,736,129,000
1980	15	0	2,890,000,000
1981	24	4	2,695,000,000

Source: National Transportation Safety Board

1. Based on the table, which year had the fewest accidents?
   (1) 1977
   (2) 1978
   (3) 1979
   (4) 1980
   (5) 1981

2. Which year had the highest number of fatalities?
   (1) 1976
   (2) 1977
   (3) 1978
   (4) 1979
   (5) 1980

3. How many fatalities occurred in the year in which the greatest number of miles were flown?
   (1) 4
   (2) 0
   (3) 351
   (4) 160
   (5) 78

4. The number of fatalities in 1979 was
   (1) about half the number of fatalities in 1980
   (2) about the same as the number of fatalities in 1978
   (3) about half the number of fatalities in 1978
   (4) only 3 more than the number of fatalities in 1978
   (5) more than twice the number of fatalities in 1978

5. Which two years had the most accidents?
   (1) 1976 and 1977
   (2) 1977 and 1981
   (3) 1978 and 1979
   (4) 1979 and 1980
   (5) 1979 and 1981

6. The number of accidents in 1980 was what percent of the number of accidents in 1979?
   (1) $37\frac{1}{2}$
   (2) 50%
   (3) $62\frac{1}{2}\%$
   (4) 75%
   (5) 160%

Answers and solutions start on page 208.

TABLES EXERCISE 9

Use Figure 17 to answer the questions below. For each question, choose the best answer.

Figure 17

PRODUCTION OF ELECTRICITY IN THE U.S. BY SOURCE
(Amounts include both privately owned and publicly owned utilities)

Calendar Year	Net production, million kwh	Percentage produced by source					
		Coal	Oil	Gas	Nuclear	Hydro	Other*
1971	1,612,593	44.3	13.6	23.2	2.4	16.5	0.05
1974	1,867,103	44.5	16.0	17.2	6.1	16.1	0.1
1976	2,037,775	46.4	15.7	14.4	9.4	13.9	0.2
1977	2,124,580	46.4	16.8	14.4	11.8	10.4	0.2
1978	2,206,515	44.2	16.5	13.8	12.5	12.7	0.2
1979	2,247,372	47.8	13.5	14.7	11.4	12.4	0.2
1980	2,286,439	50.8	10.7	15.1	10.9	12.0	0.2
1981	2,294,812	52.4	9.0	15.1	11.9	11.4	0.2

*Includes electricity produced from geothermal power, wood, and waste.

Source: World Almanac

1. Which category is *not* a source of electricity?
(1) coal
(2) oil
(3) gas
(4) nuclear power
(5) kwh

2. In 1971, what percent of electricity was produced by gas?
(1) 44.3%
(2) 13.6%
(3) 23.2%
(4) 2.4%
(5) 16.5%

3. Nuclear power accounted for 12.5% of the production of electricity in the year
(1) 1971
(2) 1974
(3) 1976
(4) 1977
(5) 1978

4. In what year did oil make its largest contribution percentagewise to electrical output?
(1) 1974
(2) 1976
(3) 1977
(4) 1978
(5) 1979

5. By how much did the percent of nuclear power's contribution to electrical output rise between 1971 and 1981?

   (1) 2.4%
   (2) 9.5%
   (3) 11.9%
   (4) 15.1%
   (5) 23.2%

6. How did the percent of electrical power produced by gas change from 1971 to 1981?

   (1) increased 2.4%
   (2) increased 8.1%
   (3) decreased 8.1%
   (4) increased 23.2%
   (5) decreased 23.2%

7. In 1981, the difference in the percent of electrical production between nuclear and hydro power was

   (1) $\frac{1}{2}$%
   (2) 5%
   (3) 11.4%
   (4) 11.9%
   (5) 23.3%

Answers and solutions start on page 208.

ANSWERS AND SOLUTIONS

Graph Exercise 1

1. (4) the number of residential telephones in certain states in the U.S.

2. (5) 1,000,000 residential telephones

3. (1) 1,500,000
$$\begin{array}{r} 1.5 \\ \times\ 1,000,000 \\ \hline 1,500,000 \end{array}$$

4. (4) 8,000,000
$$\begin{array}{r} 8 \\ \times 1,000,000 \\ \hline 8,000,000 \end{array}$$

5. (2) 3,500,000
New York has $9\frac{1}{2} \times 1,000,000 = \quad 9,500,000$
Florida has $6 \times 1,000,000 \quad = -6,000,000$
The difference is $\qquad\qquad 3,500,000$

6. (3) 3:2
Georgia has 3,000,000 telephones.
Alabama has 2,000,000 telephones.
3,000,000:2,000,000 = 3:2

7. (4) Alabama and Florida
Alabama has $2 \times 1,000,000 = \quad 2,000,000$
Florida has $6 \times 1,000,000 \quad = +6,000,000$
The number of phones $\qquad 8,000,000$
in Texas is the sum

8. (4) There are twice as many telephones in Florida as in Georgia.
Florida has 6,000,000 telephones.
Georgia has 3,000,000 telephones.
$$3,000,000 \overline{)\,6,000,000\,}^{\ \ 2}$$

Graph Exercise 2

1. (3) the percentage of gold produced by certain countries.
Choices (1), (2), and (5) are wrong because the graph is not concerned with amounts and values of gold. Choice (4) is not right because percentages, not fractions, are given on the graph.

2. (3) 7%
The U.S. produces $\quad 3\%$
Canada produces $\quad +4\%$
The total is $\qquad\ \ 7\%$

3. (5) 48%
Total production $\quad = \quad 100\%$
South Africa's production $\quad = -52\%$
The difference is $\qquad\ \ 48\%$

4. (1) $\frac{1}{3}$
Together, the U.S. and Canada produce 7%. The Soviet Union produces 21%.
$\frac{7\%}{21\%} = \frac{1}{3}$

5. (2) $\frac{1}{5}$
The Soviet Union produces 21%. Total world production is 100%. 21% can be rounded off to 20% for ease of comparison.
$\frac{20\%}{100\%} = \frac{1}{5}$

6. (3) 1:13
Canada: S. Africa
$4:52 = 1:13$

7. (4) South Africa produces more than half the gold in the world, 52%.

8. (2) 1,600,000
Find 4% of 40,000,000.
$4\% = .04$
$$\begin{array}{r} 40,000,000 \\ \times \qquad .04 \\ \hline 1,600,000.00 \end{array}$$

9. (2) Statement (2) is false. In fact, the combined gold production of the Soviet Union, Canada, and the U.S. (28%) is less than the gold production of South Africa (52%).

Graph Exercise 3

1. (5) The portion of every dollar in Centerville's budget that goes to each budget category.
The other choices refer to actual total dollar amounts. The amount of the budget is not given on the graph.

2. (3) 19¢
public works = $\quad 10¢$
loan interest = $\quad +9¢$
total $\quad = \qquad 19¢$

3. (2) 4

$$4¢\overline{)16¢}^{\,4}$$

4. (4) $\frac{1}{2}$ 48¢ is approximately 50¢.

$$\frac{\$\ .50}{\$1.00} = \frac{1}{2}$$

5. (1) Police and fire departments

6. (5) 1%

General government uses 1¢ out of every dollar.

$\frac{1}{100}$ is exactly 1%.

7. (3) $1,500,000

$$\frac{\%\ \text{police \& fire}}{\%\ \text{public works}} = \frac{\text{amount police \& fire}}{\text{amount public works}}$$

$$\frac{16}{10} = \frac{2,400,000}{x}$$

$$\begin{array}{r} 2,400,000 \\ \times\ \ \ \ \ \ \ 10 \\ \hline 24,000,000 \end{array}$$

$$16\overline{)24,000,000}^{\,1,500,000}$$
$$\underline{16}$$
$$80$$
$$\underline{80}$$
$$0$$

8. (2) $1,800,000

12¢ out of a dollar = $\frac{12}{100}$ = 12%

12% = .12

$$\begin{array}{r} \$15,000,000 \\ \times\ \ \ \ \ \ \ \ .12 \\ \hline 300\ 000\ 00 \\ \underline{1\ 500\ 000\ 0} \\ \$1,800,000.00 \end{array}$$

9. (3) Centerville spends four times as much on education as on health and welfare.

$$12¢\overline{)48¢}^{\,4}$$

Graph Exercise 4

1. (2) 140 ft.

2. (3) 60 m.p.h.

3. (1) 20 m.p.h.

$$\begin{array}{r} 210\ \text{feet} =\ \ 50\ m.p.h. \\ 80\ \text{feet} = \underline{-30\ m.p.h.} \\ 20\ m.p.h. \end{array}$$

4. (1) 100 ft.

$$\begin{array}{r} 80\ m.p.h. =\ \ 480\ ft. \\ 70\ m.p.h. = \underline{-380\ ft.} \\ 100\ ft. \end{array}$$

5. (2) 90 ft.

$$\begin{array}{r} 70 =\ \ 380\ ft. \\ 60 = \underline{-290\ ft.} \\ 90\ ft. \end{array}$$

6. (5) 50 m.p.h.

7. (3) between 210 and 290 ft.

Graph Exercise 5

1. (3) the actual and normal precipitation for a six-month period in New York City.

2. (5) between 2 and 4 inches.

3. (2) February, March, May

4. (1) January

Actual precipitation was a little over 6 inches.

Normal precipitation was about 3 inches.

5. (3) March

The normal precipitation is almost the same each month, but March is slightly closer to the 4-inch line than the others.

6. (2) February

7. (5) January and February

The actual precipitation declined about 4 inches between January and February.

Graph Exercise 6

1. (3) the change in the average price of 1,000 board feet of timber from 1975 to 1982

2. (2) 1976

3. (3) 1978

4. (5) $330

5. (1) nearly doubled

The price rose from about $150 in 1978 to about $300 in 1979.

6. (2) The line increases the most dramatically from 1978–1979.

7. (4) continue to drop

Graph Exercise 7

1. (4) percents
2. (1) 1%
3. (4) 1980
4. (2) 1969
5. (2) 6%
6. (4) 1979–1981

7. (2) the unemployment rate was higher than the inflation rate.
8. (2) 1962 unemployment =   5%
   1962 inflation   = −1%

   4% greater

Tables Exercise 8

1. (4) 1980
2. (4) 1979
3. (2) 0
   There were no fatalities in the year that the greatest number of miles were flown, 1980.
4. (5) more than twice the number of fatalities in 1978.
   351 is more than twice 160.

5. (5) 1979 and 1981
6. (3) $62\frac{1}{2}$%
   There were 15 accidents in 1980 and 24 accidents in 1979.
   $\frac{15}{24} = \frac{5}{8} = 62\frac{1}{2}$%

Tables Exercise 9

1. (5) kwh
2. (3) 23.2%
3. (5) 1978
4. (3) 1977
5. (2) 9.5%   11.9%
              −2.4%
               9.5%

6. (3) decreased 8.1%
   Percentage went from 23.2 to 15.1
   23.2%
   −15.1%
    8.1%
7. (1) $\frac{1}{2}$%   11.9%
              −11.4%
              .5% = $\frac{1}{2}$%

Measurement

STANDARD MEASUREMENT SYSTEM

Below are some **standard units** of measurement of length, time, liquid measure, and weight. These are the units commonly used in the U.S. Next to the units are other units that they are equal to. Notice the abbreviations in parentheses.

Take the time now to memorize any of the units and their equivalencies that you do not already know. You will need to remember these for problems in this section.

Measures of Length

1 foot (ft.) = 12 inches (in.)
1 yard (yd.) = 36 inches
1 yard = 3 feet
1 mile (mi.) = 5,280 feet
1 mile = 1,760 yards

Measures of Time

1 minute (min.) = 60 seconds (sec.)
1 hour (hr.) = 60 minutes
1 day = 24 hours
1 week (wk.) = 7 days
1 year (yr.) = 365 days

Liquid Measures

1 pint (pt.) = 16 ounces (oz.)
1 cup = 8 ounces
1 pint = 2 cups
1 quart (qt.) = 2 pints
1 gallon (gal.) = 4 quarts

Measures of Weight

1 pound (lb.) = 16 ounces (oz.)
1 ton (t.) = 2,000 pounds

You can use proportions to change from one unit of measurement to another. This process is called **conversion.**

Rules for Changing Units of Measurement

1. Make a proportion. The ratio on the left side should represent the equivalency between the two units. For example, if a problem asks you to change pounds to ounces, the left side of the proportion should be written $\frac{1 \text{ lb.}}{16 \text{ oz.}}$. The ratio on the right side should include the number of units given in the problem as well as the letter x to represent the missing unit. Remember that when you set up a proportion, you must put the units in the same order on both sides of the equal sign.

2. Solve the proportion. First, cross multiply.

3. Divide by the remaining number.

EXAMPLE 1:   Change 5 pounds into ounces.

① $\dfrac{1 \text{ lb.}}{16 \text{ oz.}} = \dfrac{5 \text{ lb.}}{x}$ ② $\begin{array}{r} 16 \\ \times 5 \\ \hline 80 \end{array}$ ③ $\begin{array}{r} 80 \text{ oz.} \\ 1\overline{)80} \end{array}$

Step 1.   Make a proportion. On the left is the ratio of 1 lb. to 16 oz. On the right, put 5 in the pounds' position and x under it to represent the number of ounces you must find.

Step 2.   Cross multiply the numbers that are diagonal to each other, 16 and 5.

Step 3.   Divide by the remaining number in the proportion, 1. **5 pounds is equal to 80 ounces.**

EXAMPLE 2:   10 quarts is equal to how many gallons?

① $\dfrac{1 \text{ gal.}}{4 \text{ qt.}} = \dfrac{x}{10 \text{ qt.}}$ ② $\begin{array}{r} 10 \\ \times 1 \\ \hline 10 \end{array}$ ③ $\begin{array}{r} 2\frac{2}{4} = 2\frac{1}{2} \text{ gal.} \\ 4\overline{)10} \end{array}$

Step 1.   Make a proportion. On the left is the ratio of 1 gal. to 4 qt. On the right, put x in place of gallons and the 10 (as given in the problem) in the quarts place.

Step 2. Multiply the numbers that are diagonal to each other, 1 and 10.

Step 3. Divide by the other number in the proportion, 4, and reduce the remainder.

10 quarts is equal to $2\frac{1}{2}$ gallons.

Although there are other ways to make conversions, the proportional method gives you a single way to change from either larger to smaller or from smaller to larger units of measurement.

MEASUREMENT EXERCISE 1

Change each measurement to the new unit indicated.

1. 40 ounces = _____ pounds

2. 3 tons = _____ pounds

3. 6 feet = _____ yards

4. 23 days = _____ weeks

5. 14 ounces = _____ pound

6. 6 gallons = _____ quarts

7. 2640 feet = _____ mile

8. 100 inches = _____ yards

9. 20 ounces = _____ cups

10. 30 quarts = _____ pints

11. 3 miles = _____ feet

12. 2 yards = _____ feet

13. 130 quarts = _____ gallons

14. 3 days = _____ hours

15. 45 minutes = _____ hour

16. 10,560 feet = _____ miles

17. 4 pints = _____ ounces

18. 200 minutes = _____ hours

19. 100 feet = _____ inches

20. 500 pounds = _____ ton

Answers and solutions start on page 221.

METRIC MEASUREMENTS

The metric system is used in most countries outside the U.S. Gradually, this system is becoming more popular in the U.S.

The basic unit of weight in the metric system is the *gram*. A gram is about $\frac{1}{450}$ of a pound.

The basic unit of length in the metric system is the *meter*. A meter is just a little longer than one yard.

The basic unit of liquid measure in the metric system is the *liter*. A liter is about the same size as a quart.

Since you may not be familiar with metric measurement, here are a few hints to help you remember the relationships between metric units.

Prefix	Meaning
milli-	$\frac{1}{1,000}$
centi-	$\frac{1}{100}$
deci-	$\frac{1}{10}$
kilo-	1,000

EXAMPLE 1: One kilometer is _____ meters. Since "kilo" means 1,000, 1 km = 1,000 m.

One millimeter is _____ meter. Since "milli" means $\frac{1}{1,000}$, 1 mm = $\frac{1}{1,000}$ m.

Below are the most common metric measurements. Take the time now to memorize these units before you go on.

Measures of Length
1 meter (m) = 1,000 millimeters (mm)
1 meter = 100 centimeters (cm)
1 kilometer (km) = 1,000 meters

Measures of Weight
1 gram (g) = 1,000 milligrams (mg)
1 gram = 100 centigrams (cg)
1 kilogram (kg) = 1,000 grams

Liquid Measures
1 liter (1) = 1,000 milliliters (ml)
1 liter = 100 centiliters (cl)

To change from one metric unit to another, use a proportion as you did with standard measurements.

EXAMPLE 2: Change 5 meters into centimeters.

① $\dfrac{1 \text{ m}}{100 \text{ cm}} = \dfrac{5 \text{ m}}{x}$

② $\begin{array}{r} 100 \\ \times\ 5 \\ \hline 500 \end{array}$

③ $\begin{array}{r} 500 \text{ cm} \\ 1\overline{)500} \end{array}$

Step 1. Make a proportion. On the left is the ratio of 1 m to 100 cm. This is the equivalency from the two types of units you are working with. On the right, put 5 in the meters' place and x in place of centimeters.

Step 2. Cross multiply the numbers that are diagonal to each other, 100 and 5.

Step 3. Divide by the other number in the proportion, 1. **5 meters is equal to 500 centimeters.**

EXAMPLE 3: 6,400 grams equals how many kilograms?

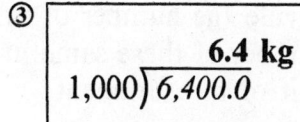

Step 1. Make a proportion. On the left is the ratio of 1 kg to 1,000 g. On the right, put x in place of the kilograms and 6,400 in the grams' place.

Step 2. Multiply the numbers that are diagonal to each other, 1 and 6,400.

Step 3. Divide by the other number in the proportion, 1. **6,400 grams is equal to 6.4 kilograms.**

Although the last answer could be written as $6\frac{2}{5}$ kg, decimals, not fractions, are generally used with the metric system.

MEASUREMENT EXERCISE 2 ━━━━━━━━━━━━━━━━━━━

Change each measurement into the new unit indicated.

1. 915 milliliters = ___ liters

2. 50 centimeters = ___ meter

3. 2.4 kilometers = ___ meters

4. 10 grams = ___ milligrams

5. .5 liter = ___ centiliters

6. 200 meters = ___ kilometer

7. 125 centigrams = ___ grams

8. 3 kilograms = ___ grams

9. 4.8 liters = ___ milliliters

10. 8,750 millimeters = ___ meters

Answers and solutions start on page 222.

SIMPLIFYING MEASUREMENTS

When you add and multiply measurements, you will sometimes have to **simplify** your answers.

Suppose the answer to a problem is 15 inches. This can be changed to feet and inches since 15 inches is more than one foot. This is similar to changing improper fractions to mixed numbers.

Rule for Simplifying Measurements

1. Divide the number of units in your answer by the total number of these same units contained in the next larger unit of measurement.

2. Write the remainder in terms of the smaller unit.

EXAMPLE 1: Change 15 inches to feet and inches.

$$\begin{array}{r} \textbf{1 ft. 3 in.} \\ 12\overline{)\,15} \\ \underline{12} \\ 3 \end{array}$$

Step 1. Divide 15 inches by 12, the number of inches in one foot.

Step 2. Write the remainder in inches.
15 inches equal 1 ft. 3 in.

EXAMPLE 2: Simplify 150 minutes.

$$\begin{array}{r} \textbf{2 hr. 30 min. or } 2\tfrac{1}{2}\textbf{ hr.} \\ 60\overline{)\,150} \\ \underline{120} \\ 30 \end{array}$$

Step 1. Divide 150 minutes by 60, the number of minutes in an hour.

Step 2. Write the remainder in minutes.
150 minutes equal 2 hr. 30 min. or $2\tfrac{1}{2}$ hours.

Depending upon the problem, you may find the remainder either in terms of the smaller unit or in fractional or decimal form. The following exercise asks for the remainder expressed in terms of the smaller unit.

MEASUREMENTS EXERCISE 3

Solve each problem.

1. Change 19 inches to feet and inches.

2. Change 75 seconds to minutes and seconds.

3. Change 36 ounces to pounds and ounces.

4. Change 25 feet to yards and feet.

5. Change 200 minutes to hours and minutes.

6. Change 4,500 pounds to tons and pounds.

7. Change 15 quarts to gallons and quarts.

8. Change 265 centimeters to meters and centimeters.

9. Change 10,560 feet to miles and feet.

10. Change 3,180 grams to kilograms and grams.

11. Change 9 pints to quarts and pints.

12. Change 40 days to weeks and days.

Answers and solutions start on page 222.

ARITHMETIC OPERATIONS AND MEASUREMENTS

Adding, subtracting, multiplying, and dividing measurements require special carrying and borrowing steps. Generally, units of measure must be changed into the type of unit in the column to the right when borrowing or carrying. The following are examples of the basic operations with measurements. Read through each example carefully.

EXAMPLE 1: Find the sum of 2 ft. 8 in., 4 ft. 11 in., and 1 ft. 9 in.

①
```
  2 ft.  8 in.
  4 ft. 11 in.
+ 1 ft.  9 in.
  7 ft. 28 in.
```

②
```
         2 ft. 4 in.
    12) 28
        24
         4
```

③
```
   7 ft.
+  2 ft. 4 in.
   9 ft. 4 in.
```

Step 1. Line up feet under feet and inches under inches. Add the feet and inches separately.

Step 2. Since 28 inches is more than one foot, you must simplify. Change 28 inches to feet and inches. Divide 28 by 12, the number of inches in a foot.

Step 3. Add the 2 ft. 4 in. from Step 2 to the 7 ft. you got in Step 1.

EXAMPLE 2: Subtract 5 kilometers 575 meters from 10 kilometers.

①
| 10 km |
| − 5 km 575 m |

②
| 10 km = 9 km 1,000 m |

③
| $\overset{9\ 9\ 1}{9\ \text{km}\ 1,\cancel{000}\ \text{m}}$ |
| −5 km   575 m |
| **4 km   425 m** |

Step 1. Put 10 kilometers on the top. Line up kilometers under kilometers.

Step 2. Notice that there is nothing to subtract 575 meters from. Borrow 1 km from 10 km and rewrite it as 9 km 1,000 m. This can be done because there are 1,000 meters in a kilometer.

Step 3. Subtract 575 m from 1,000 m. Subtract 5 km from the remaining 9 km.

EXAMPLE 3: Take 1 hr. 40 min. from 3 hr. 35 min.

①
| 3 hr. 35 min. |
| −1 hr. 40 min. |

②
| 60 min. + 35 min. = 95 min. |

③
| 2 hr. 95 min. |
| −1 hr. 40 min. |
| **1 hr. 55 min.** |

Step 1. Put 3 hr. 35 min. on top. Line up hours under hours and minutes under minutes.

Step 2. Since you cannot subtract 40 min. from 35 min., borrow 1 hour from 3 hours. Change the hour to minutes: 1 hr. = 60 min. Then add the 60 min. you borrowed to the 35 min. in the original problem.

Step 3. Subtract 40 min. from 95 min., and subtract 1 hr. from 2 hr.

EXAMPLE 4: Multiply 5 lb. 10 oz. by 3.

①
| 5 lb. 10 oz. |
| × 3 |
| 15 lb. 30 oz. |

②
| 1 lb. 14 oz. |
| 16) 30 |
| 16 |
| 14 |

③
| 15 lb. |
| + 1 lb. 14 oz. |
| 16 lb. 14 oz. |

Step 1. Multiply the pounds and ounces separately.

Step 2. Since 30 ounces is more than a pound, you must simplify. Change 30 ounces to pounds and ounces. Divide 30 by 16, the number of ounces in a pound.

Step 3. Add the 15 lb. from Step 1 to the 1 lb. 14 oz. from Step 2.

EXAMPLE 5:   Divide 6 liters 55 milliliters by 5.

①
$$
\begin{array}{r}
1 \\
5\overline{)6\ \ell\ \ 55\ m\ell} \\
\underline{-5} \\
1\ \ell
\end{array}
$$

② $1\ \ell = 1,000\ m\ell$

③
$$
\begin{array}{rr}
1\ \ell & 211\ m\ell \\
5\overline{)6\ \ell} & 55\ m\ell \\
\underline{-5} & \\
1\ \ell = & +1,000\ m\ell \\
& 1,055\ m\ell \\
& \underline{-1,055\ m\ell} \\
& 0
\end{array}
$$

Step 1.   Divide 5 into 6 liters. Then multiply the quotient by the divisor and subtract.

Step 2.   Change the remainder (1 ℓ) into milliliters. 1 ℓ = 1,000 mℓ.

Step 3.   Add 1,000 mℓ remainder to the 55 mℓ already in the problem. Divide 5 into the new total (1,055 mℓ), multiply the milliliters in the quotient by 5, and subtract.

MEASUREMENTS EXERCISE 4

Solve each problem and simplify your answer.

1. Find the sum of 4 ft. 9 in. and 2 ft. 4 in.

2. Subtract 1 min. 35 sec. from 4 min. 20 sec.

3. Multiply 8 yd. 2 ft. by 5.

4. Add 2 liters 750 milliliters and 490 milliliters.

5. Divide 10 min. 15 sec. by 3.

6. Subtract 540 meters from 2.5 kilometers.

7. Multiply 2 yd. 5 ft. 8 in. by 3.

8. Find the sum of 2 tons 500 lb., 1 ton 900 lb., and 8 tons 1,200 lb.

9. Multiply 2 kilometers 450 meters by 3.

10. Add 4 gal. 2 qts. to 3 gal. 3 qts.

11. Subtract 1 day 12 hr. 30 min. from 3 days 10 hr. 15 min.

12. Divide 2 kilograms by 5.

13. Divide 8 yd. 2 ft. 8 in. by 5.

14. Subtract 200 centiliters from 2 liters.

15. Divide 4 meters by 16.

Answers and solutions start on page 223.

MEASUREMENT WORD PROBLEMS

The next problems give you a chance to use your knowledge of measurements in practical situations. These problems use the basic operations with whole numbers as well as with decimals and fractions. Read each problem carefully before you start to solve it. Also, look at the answer choices before you solve each problem. Sometimes, you can tell from the answers whether remainders in your answer should be in unit, fraction, or decimal form.

MEASUREMENTS EXERCISE 5

Circle the number of the correct answer.

1. Maria bought 4.5 yards of material for $21.11. To the nearest cent, what was the price of one yard of the material?

(1) $4.50
(2) $4.58
(3) $4.69
(4) $4.74
(5) $5.80

2. For a picnic, the Chung family bought 6 lb. 5 oz. of hamburger, 4 lb. 10 oz. of sausage, and 7 lb. 3 oz. of ribs. Find the total weight of their purchases.

(1) 18 lb. 2 oz.
(2) 17 lb. 2 oz.
(3) 18 lb. 10 oz.
(4) 17 lb. 10 oz.
(5) 18 lb. 6 oz.

3. Flight 121 leaves Newark at 10:20 a.m. and arrives in Los Angeles at 1:05 p.m. The time difference between Newark and Los Angeles is three hours. (When it is 8 a.m. in Newark, it is 5 a.m. in Los Angeles.) How long is the flight from Newark to Los Angeles?

(1) 5 hr. 25 min.
(2) 5 hr. 45 min.
(3) 3 hr. 45 min.
(4) 2 hr. 45 min.
(5) 4 hr. 15 min.

4. City workers in Centerville participated in a blood donation campaign for three days. On Monday, they gave 29.3 liters of blood; on Tuesday, they gave 36.5 liters; and on Wednesday, they gave 42.2 liters. What was the average daily amount of blood donated by the workers?

(1) 18 ℓ
(2) 36 ℓ
(3) 29.3 ℓ
(4) 36.5 ℓ
(5) 54 ℓ

5. Mr. Asaoka mailed three packages weighing .65 kg, 1.2 kg, and 3.15 kg respectively. What was the total weight of the packages?

(1) 1.67 kg
(2) .33 kg
(3) 3.3 kg
(4) 5 kg
(5) 10.85 kg

6. Fred bought 5 cans of tomato paste, each weighing 14 ounces. What was the total weight of the cans?

(1) 4 lb. 7 oz.
(2) 4 lb. 6 oz.
(3) 3 lb. 10 oz.
(4) 7 lb. 6 oz.
(5) 5 lb. 14 oz.

7. Mae is filling quart jars with her homemade applesauce. She has made a total of $3\frac{1}{2}$ gallons of applesauce. How many quart jars can she fill?

(1) $3\frac{1}{2}$
(2) 7
(3) 8
(4) 10
(5) 14

8. There are three doors into the Alvarados' house. The measurement around each door is 18 ft. 8 in. Weatherstripping costs $.45 a foot. How much will it cost to buy weatherstripping for all three doors?

(1) $8.40
(2) $16.80
(3) $24.30
(4) $25.20
(5) $25.65

9. Rolando walked a total distance of 3 kilometers. He divided his trip into 4 equal segments, with a stop between each segment of the trip. What distance did he walk between each stop?

(1) 12 km
(2) 750 km
(3) 750 m
(4) 1 km 35 km
(5) 7 km

10. James wants to make 5 liters of punch. He needs 350 milliliters of tonic water for each liter of punch. What is the total amount of tonic water he needs?

(1) 1 ℓ 500 mℓ
(2) 17 mℓ
(3) 1750 ℓ
(4) 70 mℓ
(5) 1 ℓ 750 mℓ

11. 12,600 seconds is equal to how many hours?

(1) $3\frac{1}{2}$
(2) 3
(3) $2\frac{1}{6}$
(4) 2
(5) 21

12. John uses 3 gallons 1 quart of chemicals when he fertilizes his lawn. He puts fertilizer on his lawn four times a year. How much fertilizer does he use in a year?

(1) 16 gal.
(2) 15 gal.
(3) 14 gal.
(4) 13 gal.
(5) 12 gal. 1 qt.

13. From a cable 4 meters long, Italo cut a piece 1.8 meters long. How long was the remaining piece?

(1) 3.8 m
(2) 2.8 m
(3) 2.2 m
(4) 2.4 m
(5) 3.2 m

14. 40 people ran in a 10.5 kilometer race on Saturday, and all of them finished. A local business donated a penny to charity for every meter each participant ran that day. How much money was donated to charity?

(1) $20
(2) $4,000
(3) $42,000
(4) $4,200
(5) $420

15. Yolanda uses 8 oz. of chocolate and 2 cups of sugar for every batch of fudge she makes. How many pints of this mixture does she use to make a batch of fudge?

(1) 3
(2) 20
(3) 4
(4) $1\frac{1}{2}$
(5) 12

16. Sam walks $1\frac{1}{2}$ miles to work every day and walks the same distance home. If Sam works a 5-day week, how many miles does he walk in 12 weeks?

(1) 15
(2) 90
(3) 180
(4) 60
(5) 190

17. Sandra ran in a 10 km race. If 1 km = .62 mile, how far did Sandra run?

(1) .62 mile
(2) 6.2 miles
(3) 62 miles
(4) $\frac{1}{6}$ mile
(5) $\frac{3}{5}$ mile

18. The corner gas station now sells gas by the liter. If Luis wants to buy the equivalent of 10 gallons of gas, about how many liters will he buy? (1 liter = 1.06 quarts)

(1) 4 liters
(2) 10 liters
(3) 40 liters
(4) 50 liters
(5) 400 liters

Answers and solutions start on page 223.

ANSWERS AND SOLUTIONS

Measurements Exercise 1

1. $2\frac{1}{2}$ lb. $\quad \dfrac{1\ lb.}{16\ oz.} = \dfrac{x}{40\ oz.} \quad \begin{array}{r} 40 \\ \times 1 \\ \hline 40 \end{array} \quad \begin{array}{r} 2\frac{8}{16} = 2\frac{1}{2} \\ 16\overline{)40} \\ \underline{32} \\ 8 \end{array}$

2. 6,000 lb. $\dfrac{1\ T}{2,000\ lb.} = \dfrac{3\ T.}{x} \quad \begin{array}{r} 2,000 \\ \times\ 3 \\ \hline 6,000 \end{array} \quad \begin{array}{r} 6,000 \\ 1\overline{)6,000} \end{array}$

3. 2 yd. $\quad \dfrac{1\ yd.}{3\ ft} = \dfrac{x}{6\ ft.} \quad \begin{array}{r} 6 \\ \times 1 \\ \hline 6 \end{array} \quad \begin{array}{r} 2 \\ 3\overline{)6} \end{array}$

4. $3\frac{2}{7}$ wk $\dfrac{1\ wk.}{7\ da.} = \dfrac{x}{23\ da.} \quad \begin{array}{r} 23 \\ \times 1 \\ \hline 23 \end{array} \quad \begin{array}{r} 3\frac{2}{7} \\ 7\overline{)23} \end{array}$

5. $\frac{7}{8}$ lb. $\quad \dfrac{1\ lb.}{16\ oz.} = \dfrac{x}{14\ oz.} \quad \begin{array}{r} 14 \\ \times 1 \\ \hline 14 \end{array} \quad \dfrac{14}{16} = \dfrac{7}{8}$

Since 14 is less than 16, you can put the 14 over 16 and reduce.

6. 24 qt. $\quad \dfrac{1\ gal.}{4\ qt.} = \dfrac{6\ gal.}{x} \quad \begin{array}{r} 6 \\ \times 4 \\ \hline 24 \end{array} \quad \begin{array}{r} 24 \\ 1\overline{)24} \end{array}$

7. $\frac{1}{2}$ mi. $\quad \dfrac{1\ mi.}{5,280\ ft.} = \dfrac{x}{2,640\ ft.}$

$\begin{array}{r} 2,640 \\ \times\ 1 \\ \hline 2\ 640 \end{array} \quad \dfrac{2,640}{5,280} = \dfrac{1}{2}$

8. $2\frac{7}{9}$ yd. $\dfrac{1\ yd.}{36\ in.} = \dfrac{x}{100\ in.}$

$\begin{array}{r} 100 \\ \times 1 \\ \hline 100 \end{array} \quad \begin{array}{r} 2\frac{28}{36} = 2\frac{7}{9} \\ 36\overline{)100} \\ \underline{72} \\ 28 \end{array}$

9. $2\frac{1}{2}$ cups $\dfrac{1\ cup}{8\ oz.} = \dfrac{x}{20\ oz.}$

$\begin{array}{r} 20 \\ \times 1 \\ \hline 20 \end{array} \quad \begin{array}{r} 2\frac{4}{8} = 2\frac{1}{2} \\ 8\overline{)20} \\ \underline{16} \\ 4 \end{array}$

10. 60 pt. $\quad \dfrac{1\ qt.}{2\ pt.} = \dfrac{30\ qt.}{x} \quad \begin{array}{r} 30 \\ \times 2 \\ \hline 60 \end{array} \quad \begin{array}{r} 60 \\ 1\overline{)60} \end{array}$

11. 15,840 ft. $\dfrac{1\ mi.}{5,280\ ft.} = \dfrac{3\ mi.}{x}$

$\begin{array}{r} 5,280 \\ \times\ 3 \\ \hline 15,840 \end{array} \quad \begin{array}{r} 15,840 \\ 1\overline{)15,840} \end{array}$

12. 6 ft. $\quad \dfrac{1\ yd.}{3\ ft.} = \dfrac{2\ yd.}{x} \quad \begin{array}{r} 3 \\ \times 2 \\ \hline 6 \end{array} \quad \begin{array}{r} 6 \\ 1\overline{)6} \end{array}$

13. $32\frac{1}{2}$ gal. $\dfrac{1\ gal.}{4\ qt.} = \dfrac{x}{130\ qt.}$

$\begin{array}{r} 130 \\ \times 1 \\ \hline 130 \end{array} \quad \begin{array}{r} 32\frac{2}{4} = 32\frac{1}{2} \\ 4\overline{)130} \end{array}$

14. 72 hr. $\quad \dfrac{1\ da.}{24\ hr.} = \dfrac{3\ da.}{x} \quad \begin{array}{r} 24 \\ \times 3 \\ \hline 72 \end{array} \quad \begin{array}{r} 72 \\ 1\overline{)72} \end{array}$

15. $\frac{3}{4}$ hr. $\quad \dfrac{1\ hr.}{60\ min.} = \dfrac{x}{45\ min.} \quad \begin{array}{r} 45 \\ \times 1 \\ \hline 45 \end{array} \quad \dfrac{45}{60} = \dfrac{3}{4}$

16. 2 mi. $\quad \dfrac{1\ mi.}{5,280\ ft.} = \dfrac{x}{10,560\ ft.}$

$\begin{array}{r} 10,560 \\ \times 1 \\ \hline 10,560 \end{array} \quad \begin{array}{r} 2 \\ 5,280\overline{)10,560} \end{array}$

17. 64 oz. $\quad \dfrac{1\ pt.}{16\ oz.} = \dfrac{4\ pt.}{x} \quad \begin{array}{r} 16 \\ \times 4 \\ \hline 64 \end{array} \quad \begin{array}{r} 64 \\ 1\overline{)64} \end{array}$

18. $3\frac{1}{3}$ hr. $\quad \dfrac{60\ min.}{1\ hr.} = \dfrac{200}{x}$

$\begin{array}{r} 200 \\ \times 1 \\ \hline 200 \end{array} \quad \begin{array}{r} 3\frac{20}{60} = 3\frac{1}{3} \\ 60\overline{)200} \\ \underline{180} \\ 20 \end{array}$

19. 1,200 in. $\dfrac{1\ ft.}{12\ in.} = \dfrac{100}{x}$

$\begin{array}{r} 12 \\ \times 100 \\ \hline 1,200 \end{array} \quad \begin{array}{r} 1,200 \\ 1\overline{)1,200} \end{array}$

20. $\frac{1}{4}$ T. $\quad \dfrac{1\ T.}{2,000\ lb.} = \dfrac{x}{500\ lb.}$

$\begin{array}{r} 500 \\ \times 1 \\ \hline 500 \end{array} \quad \dfrac{500}{2,000} = \dfrac{1}{4}$

Measurements Exercise 2

1. .915 ℓ

$$\frac{1\ \ell}{1,000\ m\ell} = \frac{x}{915\ m\ell}$$

$$\begin{array}{r} 915 \\ \times 1 \\ \hline 915 \end{array} \qquad \frac{915}{1,000\ m\ell} = .915$$

2. .5 m

$$\frac{1\ m}{100\ cm} = \frac{x}{50\ cm}$$

$$\begin{array}{r} 50 \\ \times 1 \\ \hline 50 \end{array} \qquad \frac{50}{100} = .5$$

3. 2,400 m

$$\frac{1\ km}{1,000\ m} = \frac{2.4\ km}{x}$$

$$\begin{array}{r} 2.4 \\ \times 1,000 \\ \hline 2,400.0 \end{array} \qquad 1\overline{)2,400}$$

4. 10,000 mg

$$\frac{1\ g}{1,000\ mg} = \frac{10g}{x}$$

$$\begin{array}{r} 1,000 \\ \times\ \ 10 \\ \hline 10,000 \end{array} \qquad 1\overline{)10,000}$$

5. 50 cℓ

$$\frac{1\ \ell}{100\ c\ell} = \frac{.5\ \ell}{x}$$

$$\begin{array}{r} 100 \\ \times .5 \\ \hline 50.0 \end{array} \qquad 1\overline{)50}\ \ ^{50}$$

6. .2 km

$$\frac{1\ km}{1,000\ m} = \frac{x}{200\ m}$$

$$\begin{array}{r} 200 \\ \times 1 \\ \hline 200 \end{array} \qquad \frac{200}{1,000} = .2$$

7. 1.25 g

$$\frac{1\ g}{100\ cg} = \frac{x}{125\ cg}$$

$$\begin{array}{r} 125 \\ \times 1 \\ \hline 125 \end{array} \qquad 100\overline{)125.00}\ ^{1.25}$$

8. 3,000 g

$$\frac{1\ kg}{1,000\ g} = \frac{3\ kg}{x}$$

$$\begin{array}{r} 1,000 \\ \times\ \ 3 \\ \hline 3,000 \end{array} \qquad 1\overline{)3,000}\ ^{3,000}$$

9. 4,800 mℓ

$$\frac{1\ \ell}{1,000\ m\ell} = \frac{4.8\ \ell}{x}$$

$$\begin{array}{r} 4.8 \\ \times 1,000 \\ \hline 4,800.0 \end{array} \qquad 1\overline{)4,800}\ ^{4,800}$$

10. 8.75 m

$$\frac{1\ m}{1,000\ m} = \frac{x}{8,750\ mm}$$

$$\begin{array}{r} 8,750 \\ \times\ \ 1 \\ \hline 8,750 \end{array} \qquad 1,000\overline{)8,750.00}\ ^{8.75}$$

Measurements Exercise 3

1. 1 ft. 7 in.

$$12\overline{)19}\ ^{1\ ft.\ 7\ in.}$$
$$\begin{array}{r} 12 \\ \hline 7 \end{array}$$

2. 1 min. 15 sec.

$$60\overline{)75}\ ^{1\ min.\ 15\ sec.}$$
$$\begin{array}{r} 60 \\ \hline 15 \end{array}$$

3. 2 lb. 4 oz.

$$16\overline{)36}\ ^{2\ lb.\ 4\ oz.}$$
$$\begin{array}{r} 32 \\ \hline 4 \end{array}$$

4. 8 yd. 1 ft.

$$3\overline{)25}\ ^{8\ yd.\ 1\ ft.}$$
$$\begin{array}{r} 24 \\ \hline 1 \end{array}$$

5. 3 hr. 20 min.

$$60\overline{)200}\ ^{3\ hr.\ 20\ min.}$$
$$\begin{array}{r} 180 \\ \hline 20 \end{array}$$

6. 2 T. 500 lb.

$$2,000\overline{)4,500}\ ^{2\ T.\ 500\ lb.}$$
$$\begin{array}{r} 4,000 \\ \hline 500 \end{array}$$

7. 3 gal. 3 qt.

$$4\overline{)15}\ ^{3\ gal.\ 3\ qt.}$$
$$\begin{array}{r} 12 \\ \hline 3 \end{array}$$

8. 2 m 65 cm

$$100\overline{)265}\ ^{2\ m\ 65\ cm}$$
$$\begin{array}{r} 200 \\ \hline 65 \end{array}$$

9. 2 mi.

$$5,280\overline{)10,560}\ ^{2\ mi.}$$
$$\begin{array}{r} 10,560 \end{array}$$

10. 3 kg 180 g

$$1,000\overline{)3,180}\ ^{3\ kg\ 180\ g}$$
$$\begin{array}{r} 3,000 \\ \hline 180 \end{array}$$

11. 4 qt. 1 pt.

$$2\overline{)9}\ ^{4\ qt.\ 1\ pt.}$$
$$\begin{array}{r} 8 \\ \hline 1 \end{array}$$

12. 5 wk. 5 days

$$7\overline{)40}\ ^{5\ wk.\ 5\ days}$$
$$\begin{array}{r} 35 \\ \hline 5 \end{array}$$

Measurement Exercise 4

1. 7 ft. 1 in. 　*4 ft.　9 in.*
　　　　　　　+2 ft.　4 in.
　　　　　　　6 ft. 13 in. = 7 ft. 1 in.

2. 2 min. 45 sec.　*4 min. 20 sec. =　3 min. 80 sec.*
　　　　　　　　−1 min. 35 sec.　−1 min. 35 sec.
　　　　　　　　　　　　　　　2 min. 45 sec.

3. 43 yd. 1 ft.　　*8 yd.　2 ft.*
　　　　　　×　　　5
　　　　　40 yd. 10 ft. = 43 yd. 1 ft.

4. 3 ℓ 240 mℓ　　*2 ℓ　750 mℓ*
　　　　　　×　　　490 mℓ
　　　　　2 ℓ 1240 mℓ = 3 ℓ 240 mℓ

5. 3 min. 25 sec.　　*3 min.　　25 sec.*
　　　　　　3) 10 min.　　15 sec.
　　　　　　　9
　　　　　1 min. = + 60 sec.
　　　　　　　　　75
　　　　　　　　　75

6. 1 km 960 m　　*2.5 km =　2500m*
　　　　−　540m　−540m
　　　　　　　1960m　= 1km 960m
　　　　　　　　　or 1.96 km

7. 11 yd. 2 ft.　*2 yd.　5 ft.　8 in.*
　　　　×　　　　　3
　　　6 yd. 15 ft. 24 in. = 6 yd. 17 ft.
　　　　　　　　= 11 yd. 2 ft.

8. 12 tons 600 lb.　　*2 tons　500 lb.*
　　　　　　　1 ton　900 lb.
　　　　　+ 8 tons 1,200 lb.
　　　　　11 tons 2,600 lb.
　　　　　= 12 tons 600 lb.

9. 7 km　350 m　　*2 km　450 m*
　　　　　×　　　3
　　6 km 1,350 m = 7 km 350 m

10. 8 gal. 1 qt.　　*4 gal. 2 qt.*
　　　　　　+3 gal. 3 qt.
　　　　7 gal. 5 qt. = 8 gal. 1 qt.

11. 1 day 21 hr. 45 min.　　*3 days 10 hr. 15 min.*
　　　　　　　　　−1 day　12 hr. 30 min.

　　　　　　　　= *3 days　9 hr. 75 min.*
　　　　　　　　−1 day　12 hr. 30 min.

　　　　　　　　= *2 days 33 hr. 75 min.*
　　　　　　　　−1 day　12 hr. 30 min.
　　　　　　　　　1 day　21 hr. 45 min.

12. 400 g　　　　　*400 g*
　　　　　5) 2 kg
　　　　　= 2,000 g
　　　　　− 2,000
　　　　　　　0

13. 1 yd. 2 ft. 4 in.
　　　1 yd.　　2 ft.　　　4 in.
　5) 8 yd.　　2 ft.　　　8 in.
　　5
　3 yd. = +9 ft.
　　　11 ft.
　　　10
　　1 ft. = +12 in.
　　　　　20
　　　　　20

14. 0　　　*2 ℓ　　=　200 cℓ*
　　　−　200 cℓ　−200 cℓ
　　　　　　　　　0

15. 25 cm　　　　*25 cm*
　　　　16) 4m
　　　　= 400 cm
　　　　− 32
　　　　　80
　　　　　80
　　　　　　0

Measurement Exercise 5

1. (3) $4.69　　　*$　4.691*
　　　　　4.5,) $21.1,100
　　　　　18 0
　　　　　3 1 1
　　　　　2 7 0
　　　　　4 10
　　　　　4 05
　　　　　　50
　　　　　　45
　　　$4.691 to the nearest cent = $4.69

2. (1) 18 lb. 2 oz.　　*6 lb.　5 oz.*
　　　　　　　4 lb. 10 oz.
　　　　　+ 7 lb.　3 oz.
　　　　17 lb. 18 oz. = 18 lb. 2 oz.

3. (2) 5 hr. 45 min.

First, find the Newark departure in Los Angeles time (3 hours earlier):

　10 hr.　20 min. (10:20)
　− 3 hr.　　　　　or
　7 hr.　20 min. (7:20 departure in LA time)

Change 1:05 p.m. (LA arrival) to hours and minutes:

12 hr. (noon) + 1 hr. 5 minutes = 13 hr. 5 min.

　13 hr.　5 min. (1:05 p.m.) =　12 hr. 65 min.
　− 7 hr. 20 min. (7:20 a.m.) = −　7 hr. 20 min.
　　　　　　　　　　　5 hr. 45 min.

Exercise 5 cont'd.

4. (2) 36 ℓ

$$\begin{array}{r} 29.3\,ℓ \\ 36.5 \\ +42.2 \\ \hline 108.0\,ℓ \end{array}$$

$$\begin{array}{r} 36\,ℓ \\ 3\overline{)108} \end{array}$$

5. (4) 5 kg

$$\begin{array}{r} .65\,kg \\ 1.2 \\ +3.15 \\ \hline 5.00\,kg \end{array}$$

6. (2) 4 lb. 6 oz.

$$\begin{array}{r} 14\,oz. \\ \times\ 5 \\ \hline 70\,oz. = 4\,lb.\,6\,oz. \end{array}$$

7. (5) 14

$$\dfrac{1\,gal.}{4\,qt.} = \dfrac{3\frac{1}{2}\,gal.}{x}$$

1 gal. = 4 qt.

$$\begin{array}{r} 4 \\ \times 3\frac{1}{2} \\ \hline 2 \\ 12 \\ \hline 14 \end{array}$$

$$\begin{array}{r} 14 \\ 1\overline{)14} \\ 14 \end{array}$$

8. (4) $25.20

$$\begin{array}{r} 18\,ft.\ 8\,in. \\ \times\ 3\ (doors) \\ \hline 54\,ft.\ 24\,in. = 56\,ft. \end{array}$$

$$\begin{array}{r} 56 \\ \times .45 \\ \hline 280 \\ 224 \\ \hline \$25.20 \end{array}$$

9. (3) 750m

$$4\overline{)3\,km}$$

$$\begin{array}{r} 750\,m \\ = 3000\,m \\ 28 \\ \hline 20 \\ 20 \\ \hline 00 \\ 00 \end{array}$$

10. (5) 1 ℓ 750 mℓ

$$\begin{array}{r} 350\,mℓ \\ \times\ 5 \\ \hline 1,750\,mℓ = 1\,ℓ\,750\,mℓ \end{array}$$

11 (1) $3\frac{1}{2}$ hr

First find how many minutes in 12,600 seconds.

$$\begin{array}{r} 210\ min. \\ 60\overline{)12,600} \\ 12\ 0 \\ \hline 60 \\ 60 \\ \hline 00 \end{array}$$

Then find how many hours in 210 minutes.

$$3\frac{30}{60} = 3\frac{1}{2}$$

$$\begin{array}{r} 60\overline{)210} \\ 180 \\ \hline 30 \end{array}$$

12. (4) 13 gal.

$$\begin{array}{r} 3\,gal.\ 1\,qt. \\ \times\ 4 \\ \hline 12\,gal.\ 4\,qt. = 13\,gal. \end{array}$$

13. (3) 2.2 m

$$\begin{array}{r} 4.0\,m \\ -1.8 \\ \hline 2.2\,m \end{array}$$

14. (4) $4,200

$$\begin{array}{r} 10.5\,km \\ \times\ 40\ people \\ \hline 420\,km\ total \end{array}$$

Find how many meters in 420 kilometers.

$$\dfrac{1\,km}{1,000\,m} = \dfrac{420\,km}{m}$$

$$\begin{array}{r} 420 \\ \times\ 1,000 \\ \hline 420,000\,m \end{array}$$

Then find what a penny per meter would earn.

$$\begin{array}{r} 420,000 \\ \times\ .01\ (penny) \\ \hline \$4,200.00 \end{array}$$

15. (4) $1\frac{1}{2}$

$$\begin{array}{r} 8\,oz. = \frac{1}{2}\,pint \\ +2\,cups = 1\,pint \\ \hline 1\frac{1}{2}\,pints \end{array}$$

16. (3) 180

$$\begin{array}{r} 1\frac{1}{2}\,miles \\ \times\ 2 \\ \hline 3\,miles\ daily \end{array}$$

$$\begin{array}{r} 3\,miles \\ \times 5\,days \\ \hline 15\,miles\ weekly \end{array}$$

$$\begin{array}{r} 15\,miles \\ \times 12\,weeks \\ \hline 30 \\ 150 \\ \hline 180\,miles\ in\ 12\,weeks \end{array}$$

17. (2) 6.2 miles

$$\dfrac{1\,km}{.62\,mile} = \dfrac{10\,km}{x\,miles}$$

$$\begin{array}{r} .62 \\ \times 10\,km \\ \hline 6.20\,miles \end{array}$$

18. (3) 40 liters

First, find out approximately how many liters are in a quart.

1 quart = approx. 1 liter
1 gallon = 4 quarts

Therefore, there are approximately 4 liters in a gallon.

$$\begin{array}{r} 10\,gallons \\ \times 4\,liters \\ \hline 40\,liters \end{array}$$

MIXED REVIEW

These problems give you a chance to test the skills you learned in the ratio and proportion, graphs, and measurements sections of this book. For each problem fill in the circle that corresponds to the correct answer.

1. The ratio 18:24 is equal to which of the following?
   (1) 3:4
   (2) 2:3
   (3) 9:10
   (4) 1:2
   (5) 4:3

   1 ① ② ③ ④ ⑤

2. Simplify the ratio $\frac{2}{3} : \frac{5}{6}$.
   (1) 4:5
   (2) 2:5
   (3) 3:5
   (4) 5:3
   (5) 1:3

   2 ① ② ③ ④ ⑤

3. Consolidated Utilities' softball team won 12 games and lost 8 last season. What was the ratio of the number of games the team won to the number of games they played?
   (1) 3:2
   (2) 2:5
   (3) 3:5
   (4) 5:3
   (5) 5:2

   3 ① ② ③ ④ ⑤

4. Find s in $\frac{s}{15} = \frac{6}{20}$.
   (1) $2\frac{1}{4}$
   (2) $4\frac{1}{2}$
   (3) $6\frac{1}{2}$
   (4) 9
   (5) 18

   4 ① ② ③ ④ ⑤

5. What is the value of x in 5:12 = x:84?
   (1) $17\frac{1}{2}$
   (2) 35
   (3) 42
   (4) 9
   (5) 18

   5 ① ② ③ ④ ⑤

6. Pat worked 6 overtime hours for $38.40. At the same rate, how much will she make for 10 hours of overtime work?

6 ① ② ③ ④ ⑤

   (1) $32
   (2) $51.20
   (3) $64
   (4) $76.80
   (5) $96

7. The ratio of the number of people who voted to the total number of registered voters was 3:8 in the last election in Midvale. Out of 38,000 registered voters, how many people voted?

7 ① ② ③ ④ ⑤

   (1) 10,133
   (2) 14,250
   (3) 23,750
   (4) 28,500
   (5) 47,500

8. For every $5 in her budget, Mrs. Murphy spends $2 on food. Her weekly budget is $215. How much does she spend each week on food?

8 ① ② ③ ④ ⑤

   (1) $43
   (2) $54
   (3) $68
   (4) $86
   (5) $129

Use the following table for problems 9 and 10.

Average Farm Wages (in dollars per hour)			
Method of pay:	1978	1979	1980
Paid by piece-rate	3.76	4.07	4.61
Paid by hour	3.08	3.38	3.63
Paid cash wages	3.22	3.58	3.82

9. In 1980, what was the difference in an hour's pay for someone paid by piece-rate and someone paid by the hour?

9 ① ② ③ ④ ⑤

   (1) $1.02
   (2) $.19
   (3) $.68
   (4) $.02
   (5) $.98

10. For a 40-hour week in 1978, how much did the average farm   10 ① ② ③ ④ ⑤
worker make if he was paid by the hour?
- (1) $123.20
- (2) $128.80
- (3) $135.20
- (4) $145.20
- (5) $150.40

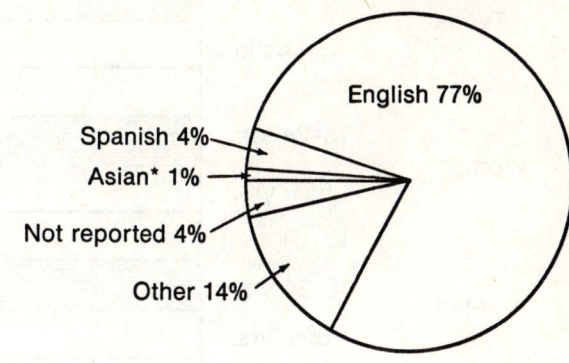

**MOTHER TONGUE OF
AMERICANS
(% of population)**

English 77%

Spanish 4%

Asian* 1%

Not reported 4%

Other 14%

Use the graph at the right for
problems 11 and 12.

*includes Chinese, Japanese,
and Philippine

11. The fraction of Americans who reported that their mother tongue   11 ① ② ③ ④ ⑤
is English is about
- (1) $\frac{1}{7}$
- (2) $\frac{3}{5}$
- (3) $\frac{1}{4}$
- (4) $\frac{2}{3}$
- (5) $\frac{3}{4}$

12. The number of Americans who reported Spanish as their mother   12 ① ② ③ ④ ⑤
tongue is about how many times the number who reported an
Asian language as their mother tongue?
- (1) 1
- (2) 2
- (3) 3
- (4) 4
- (5) 5

Use the graph below for problems 13 through 17.

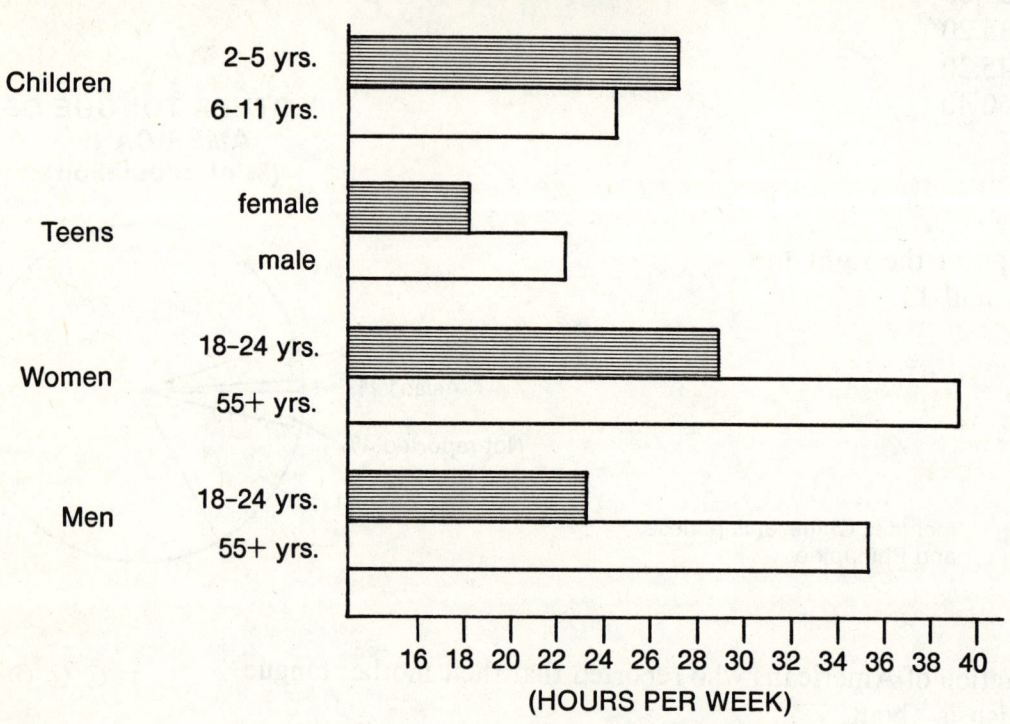

AVERAGE TELEVISION
VIEWING TIME

13. Women ages 18 to 24 watch television an average of how many   13 ① ② ③ ④ ⑤
hours per week?
(1) 9
(2) 19
(3) 29
(4) 39
(5) 40

14. Which category shown on the graph watches television the least?   14 ① ② ③ ④ ⑤
(1) male teens
(2) female teens
(3) children 6-11
(4) women 18-24
(5) men 55+

15. Which groups shown on the graph watch television more than 30   15 ① ② ③ ④ ⑤
hours a week?
(1) teens
(2) men & women 18-24
(3) children
(4) men & women 55+
(5) none of these

16. Teen males watch television about how many hours more per week than teen females?

    16 ① ② ③ ④ ⑤

    (1) 4 hrs.
    (2) 6 hrs.
    (3) 8 hrs.
    (4) 18 hrs.
    (5) 22 hrs.

17. Which of the following statements is false?

    17 ① ② ③ ④ ⑤

    (1) Children ages 2 to 5 watch television more than children ages 6 to 11.
    (2) Women over 55 years old watch television an average of nearly 40 hours a week.
    (3) Of children, teens, adult women, and adult men, the group that watches the most television is adult men.
    (4) Male teens and men, ages 18 to 24, watch about the same number of hours of television per week.
    (5) Except for female teens, every group shown on the graph watches television at least 20 hours per week.

18. Change $3\frac{1}{2}$ tons to pounds.

    18 ① ② ③ ④ ⑤

    (1) 14,000 lbs.
    (2) 7,000 lbs.
    (3) 3,500 lbs.
    (4) 700 lbs.
    (5) 350 lbs.

19. 250 inches is approximately how many yards?

    19 ① ② ③ ④ ⑤

    (1) 7
    (2) 6
    (3) 5
    (4) 4
    (5) 3

20. Write 1,950 meters as kilometers.

    20 ① ② ③ ④ ⑤

    (1) .0195 km
    (2) 1950 km
    (3) 19.5 km
    (4) 1.95 km
    (5) .195 km

21. What is the sum of 8 ft. 9 in., 6 ft. 5 in., and 4 ft. 6 in.?

    21 ① ② ③ ④ ⑤

    (1) 19 ft. 8 in.
    (2) 18 ft. 10 in.
    (3) 20 ft. 8 in.
    (4) 19 ft. 10 in.
    (5) 18 ft. 8 in.

22. A truck weighing 1 ton 850 pounds was filled with cargo weighing 2 tons 1,500 pounds. What was the combined weight of the truck and the cargo?

22 ① ② ③ ④ ⑤

(1) 5T. 350 lbs.
(2) 4T. 350 lbs.
(3) 3T. 350 lbs.
(4) 4T. 650 lbs.
(5) 3T. 650 lbs.

23. The trip from New York to Montreal by bus takes 7 hours 25 minutes. The same trip by train takes 9 hours 10 minutes. The bus takes how much less time than the train?

23 ① ② ③ ④ ⑤

(1) 2 hr. 15 min.
(2) 1 hr. 15 min.
(3) 1 hr. 45 min.
(4) 2 hr. 30 min.
(5) 2 hr. 45 min.

24. Nick welded together 3 pieces of pipe each 2 feet 5 inches long. Find the total length of the welded pipe.

24 ① ② ③ ④ ⑤

(1) 7 ft. 5 in.
(2) 7 ft. 3 in.
(3) 6 ft. 5 in.
(4) 6 ft. 3 in.
(5) 7 ft. 9 in.

25. Andrew cut a piece of electric cable 43 ft. 4 in. long into 8 equal pieces. How long was each piece?

25 ① ② ③ ④ ⑤

(1) 5 ft. 5 in.
(2) 5 ft. $\frac{1}{2}$ in.
(3) 5 ft. 1 in.
(4) 5 ft. 7 in.
(5) 5 ft. 8 in.

Answers and solutions start on page 231.

Mixed Review

1. (1) 3:4 $\frac{18}{24} = \frac{3}{4}$ or 3:4

2. (1) 4:5

$$\frac{2}{3} \div \frac{5}{6} = \frac{2}{\cancel{3}_1} \times \frac{\cancel{6}^2}{5} = \frac{4}{5} \text{ or } 4:5$$

3. (3) 3:5

games won   = 12
games lost   = +8
games played = 20

$\frac{won}{played} = \frac{12}{20} = \frac{3}{5}$ or 3:5

4. (2) $4\frac{1}{2}$

$\begin{array}{r} 15 \\ \times 6 \\ \hline 90 \end{array}$ $20\overline{)90}$ $\begin{array}{r} 80 \\ \hline 10 \end{array}$ $4\frac{10}{20} = 4\frac{1}{2}$

5. (2) 35

$\frac{5}{12} = \frac{x}{84}$ $\begin{array}{r} 84 \\ \times 5 \\ \hline 420 \end{array}$ · $12\overline{)420}$ $\begin{array}{r} 35 \\ \hline 36 \\ 60 \\ 60 \end{array}$

6. (3) $64 $\frac{hours}{\$}$ $\frac{6}{38.40} = \frac{10}{x}$

$\begin{array}{r} 38.40 \\ \times 10 \\ \hline 384.00 \end{array}$ $6\overline{)384}$ 64

7. (2) 14,250

$\frac{voters}{total}$ $\frac{3}{8} = \frac{x}{38,000}$

$\begin{array}{r} 38,000 \\ \times 3 \\ \hline 114,000 \end{array}$ $8\overline{)114,000}$ $14,250$

8. (4) $86 $\frac{food}{total}$ $\frac{2}{5} = \frac{x}{215}$

$\begin{array}{r} 215 \\ \times 2 \\ \hline 430 \end{array}$ $5\overline{)430}$ 86

9. (5) $.98 $\begin{array}{r} \$4.61 \\ -3.63 \\ \hline \$ .98 \end{array}$

10. (1) $123.20 $\begin{array}{r} \$3.08 \\ \times 40 \\ \hline \$123.20 \end{array}$

11. (5) $\frac{3}{4}$ 77% is close to 75%.

$\frac{75}{100} = \frac{3}{4}$

12. (4) 4   Asian languages = 1%
Spanish languages = 4%

$1\overline{)4}$ 4

13. (3) 29

14. (2) female teens

15. (4) men & women 55+

16. (1) 4 hr.

Teen males watch $22\frac{1}{2}$ hr.
Teen females watch $-18\frac{1}{2}$
difference $\overline{4}$ hr.

17. (3) Of children, teens, adult women, and adult men, the group that watches the most television is adult men.

This is false because adult women watch the most television.

18. (2) 7,000 lb. $\frac{1 \text{ T.}}{2,000 \text{ lb.}} = \frac{3\frac{1}{2}}{x}$

$3\frac{1}{2} \times 2,000 = \frac{7}{\cancel{2}_1} \times \frac{\cancel{2,000}^{1,000}}{1} = 7,000$

19. (1) 7 $\frac{1 \text{ yd.}}{36 \text{ in.}} = \frac{x}{250}$

$\begin{array}{r} 250 \\ \times 1 \\ \hline 250 \end{array}$ $36\overline{)250.0}$ 6.9 to the nearest whole number = 7.
$\begin{array}{r} 216 \\ \hline 34\,0 \\ 32\,4 \end{array}$

20. (4) 1:95 km $\frac{1 \text{ km}}{1,000 \text{ m}} = \frac{x}{1,950}$

$\begin{array}{r} 1,950 \\ \times 1 \\ \hline 1,950 \end{array}$ $1,000\overline{)1,950.00}$ 1.95 km

21. (1) 19 ft. 8 in.

$\begin{array}{r} 8 \text{ ft. } 9 \text{ in.} \\ 6 \text{ ft. } 5 \text{ in.} \\ + 4 \text{ ft. } 6 \text{ in.} \\ \hline 18 \text{ ft. } 20 \text{ in.} = 19 \text{ ft. } 8 \text{ in.} \end{array}$

22. (2) 4 T. 350 lb.

$\begin{array}{r} 1 \text{ T. } 850 \text{ lb.} \\ +2 \text{ T. } 1,500 \text{ lb.} \\ \hline 3 \text{ T. } 2,350 \text{ lb.} = 4 \text{ T. } 350 \text{ lb.} \end{array}$

23. (3) 1 hr. 45 min.

$\begin{array}{r} 9 \text{ hr. } 10 \text{ min.} = 8 \text{ hr. } 70 \text{ min.} \\ -7 \text{ hr. } 25 \text{ min.} = -7 \text{ hr. } 25 \text{ min.} \\ \hline 1 \text{ hr. } 45 \text{ min.} \end{array}$

Mixed Review, cont'd.

24.   (2) 7 ft. 3 in.

$$2 \text{ ft.} \quad 5 \text{ in.}$$
$$\underline{\times \qquad 3}$$
$$6 \text{ ft. } 15 \text{ in.} = 7 \text{ ft. } 3 \text{ in.}$$

25.   (1) 5 ft. 5 in.

$$\begin{array}{r} 5 \text{ ft.} \qquad 5 \text{ in.} \\ 8\overline{) \,43 \text{ ft.} \qquad 4 \text{ in.}} \\ \underline{40} \\ 3 \text{ ft.} = 36 \text{ in.} \\ \underline{+\ 4} \\ 40 \text{ in.} \\ \underline{40} \end{array}$$

INTRODUCTION TO GEOMETRY AND ALGEBRA

GEOMETRY

ALGEBRA

Introduction To Geometry and Algebra

In this section, you will learn skills that you can apply in the geometry and algebra sections that follow. These skills are an extension of the basic arithmetic skills you have already developed.

POWERS

5^2 is read "five to the second power." The 5 is called the **base**. The 2 is called the **exponent**. The exponent tells how many times to write the base in a multiplication problem.

Rule for Finding Powers

1. Write the base as many times as the exponent indicates.
2. Multiply.

EXAMPLE 1: What is 5^2?

$$5^2 = 5 \times 5 = \mathbf{25}$$

Solution: Write 5 two times and multiply.

A number to the second power is sometimes called the **square** of a number. 5 squared is 25.

EXAMPLE 2: Evaluate 4^3.

$$4^3 = \underbrace{4 \times 4}_{16} \times 4 = \mathbf{64.}$$

Solution: Write 4 three times and multiply. The first product is $4 \times 4 = 16$. The next product is $16 \times 4 = 64$.

A number to the third power is sometimes called the **cube** of a number. 4 cubed equals 64.

EXAMPLE 3: Find the value of 2^5.

$$2^5 = 2 \times 2 \times 2 \times 2 \times 2 = 32$$

Solution: Write 2 five times and multiply.

There are some special cases:

> ### 1 to any power equals 1.

EXAMPLE 4: What is 1^6?

$$1^6 = 1 \times 1 \times 1 \times 1 \times 1 \times 1 = \mathbf{1}$$

Solution: Write 1 six times and multiply. No matter how many times you multiply 1 by 1, you will get 1 as your answer.

> ### Any number to the first power is that number.

EXAMPLE 5: What is 7^1?

$$7^1 = 7$$

Solution: There is no multiplication in this problem. 7 to the first power indicates that you should simply write 7 one time.

> ### Any number to the 0 power is 1.

EXAMPLE 6: What is 8^0?

$$8^0 = \frac{8}{8} = 1.$$

Solution: The zero power actually means a number <u>divided</u> by itself.

You may use powers in combination with other operations. In the following examples, keep in mind that you find the value of each power before you add or subtract.

EXAMPLE 7: Simplify $3^2 + 2^4$.

①
$$3^2 = 3 \times 3 = 9$$
$$2^4 = 2 \times 2 \times 2 \times 2 = 16$$

②
$$9 + 16 = \mathbf{25}$$

Step 1. Find each power.
Step 2. Add the results.

EXAMPLE 8: Simplify $5^2 - 2^3 + 10^1$.

① | $5^2 = 25$
$2^3 = 8$
$10^1 = 10$ |

② | $\underbrace{25 - 8}_{17} + 10 = 27$ |

Step 1. Find each power.
Step 2. Combine the results from left to right.

EXAMPLE 9: Evaluate $3^4 - 6^0 + 3^2$.

① | $3^4 = 81$
$6^0 = 1$
$3^2 = 9$ |

② | $\underbrace{81 - 1}_{80} + 9 = 89$ |

Step 1. Find each power.
Step 2. Combine the results from left to right.

POWERS EXERCISE

Find the value of each of the following.

1. 6^2	**6.** $5^3 - 10^2$	**11.** 20^3
2. 1^3	**7.** 12^0	**12.** 13^1
3. 5^3	**8.** 25^2	**13.** 10^4
4. 3^5	**9.** $11^2 - 4^2$	**14.** $8^2 - 2^4$
5. $5^2 + 3^2$	**10.** $10^0 + 10^1 + 10^2$	**15.** $3^3 + 7^2$

Answers and solutions start on page 251.

SQUARE ROOTS

You have seen that finding a number to the second power is called squaring it. When you find a square root, you reverse that process. The **square root** of a number is another number, which multiplied by itself will give you the original number. The symbol $\sqrt{}$ means "the square root of."

EXAMPLE 1: Find $\sqrt{25}$
$\sqrt{25} = 5$

Solution: The number multiplied by itself that gives 25 is 5.

EXAMPLE 2: Find $\sqrt{144}$

$\sqrt{144} = 12$

Solution: The number multiplied by itself that gives 144 is 12.

Following are some common square roots. Take the time now to memorize these square roots.

$$\sqrt{1} = 1 \qquad \sqrt{36} = 6 \qquad \sqrt{121} = 11$$
$$\sqrt{4} = 2 \qquad \sqrt{49} = 7 \qquad \sqrt{144} = 12$$
$$\sqrt{9} = 3 \qquad \sqrt{64} = 8 \qquad \sqrt{169} = 13$$
$$\sqrt{16} = 4 \qquad \sqrt{81} = 9 \qquad \sqrt{196} = 14$$
$$\sqrt{25} = 5 \qquad \sqrt{100} = 10 \qquad \sqrt{225} = 15$$

You may need to find the square root of a larger number, and there are several methods for finding the square root of such numbers. One method is by averaging. On the GED, the square roots you will have to find will not be hard ones. This averaging method for finding a square root is also a good way to improve your estimating skills.

Rule for Finding Square Roots

Step 1. Guess. Choose a number that is fairly close to the square root of the nearest "round number."

Step 2. Divide. Divide the number you guessed into the number for which you are finding the square root.

Step 3. Average. Find the average of the number you guessed and the number you get when you divide.

Step 4. Check. Multiply the average by itself.

EXAMPLE 1: Find $\sqrt{484}$.

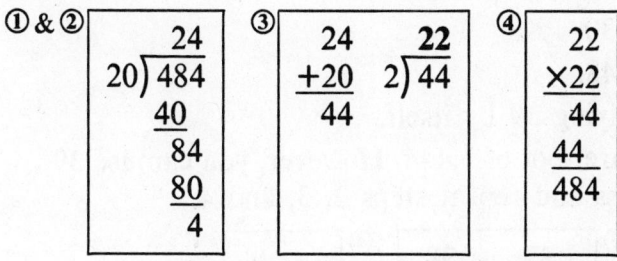

Step 1. Guess a "round" number that is fairly close to the correct answer. For example, 50 is a bad guess because 50 × 50 is 2,500. 20 is a good guess because 20 × 20 = 400, which is much closer to 484.

Step 2. Divide 20 into 484. Ignore the remainder.

Step 3. Then average the division answer, 24, with the guess, 20.

Step 4. Check by multiplying 22 by itself. 22 times 22 *is* 484. This means that **22 is the square root of 484.**

EXAMPLE 2: Find $\sqrt{841}$.

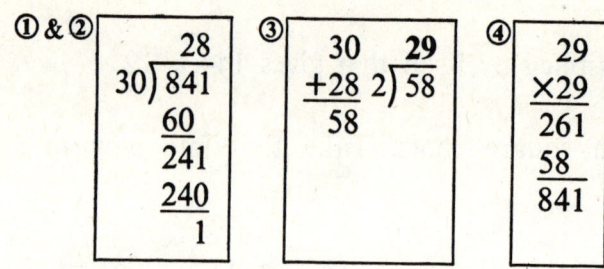

Step 1. Guess 30, because 30 × 30 = 900, which is close to 841. Notice that the guess can be larger than the final answer.

Step 2. Divide 841 by 30. Ignore the remainder.

Step 3. Then average 30 and 28.

Step 4. Check by multiplying 29 by itself.

Many students find it difficult to guess an answer. This takes practice and a willingness to experiment. In the solutions to the next exercise, the guesses are those the author finds reasonable. The guess must be reasonably close in order to work well. If the average is not the square root, use the average as a second guess. Then divide it into the number you are finding the square root of.

EXAMPLE 3: Find $\sqrt{1,444}$

① & ②
```
        48
30) 1,444
    1 20
     244
     240
       4
```

③
```
  30
 +48     39
  78   2) 78
```

④
```
    39
   ×39
   351
  1 17
  1,521
```

Step 1. Suppose you guess 30 because 30 × 30 = 900.

Step 2. Divide 1,444 by 30.

Step 3. Average 30 and 48.

Step 4. Check by multiplying 39 by itself.

39 is <u>not</u> the square root of 1,444. However, you can use 39 as your next guess and repeat steps 2, 3, and 4.

②
```
        37
39) 1,444
    1 17
     274
     273
       1
```

③
```
  39
 +37     38
  76   2) 76
```

④
```
    38
   ×38
   304
  1 14
  1,444
```

Step 2. Divide 1,444 by 39.

Step 3. Average 39 and 37.

Step 4.   Check by multiplying 38 by itself.

$$\sqrt{1{,}444} = 38.$$

SQUARE ROOTS EXERCISE

Find the following.

1. $\sqrt{196}$ 5. $\sqrt{361}$ 9. $\sqrt{4{,}624}$

2. $\sqrt{441}$ 6. $\sqrt{676}$ 10. $\sqrt{6{,}241}$

3. $\sqrt{1{,}024}$ 7. $\sqrt{1{,}849}$ 11. $\sqrt{7{,}396}$

4. $\sqrt{1{,}521}$ 8. $\sqrt{3{,}364}$ 12. $\sqrt{9{,}604}$

Answers and solutions start on page 251.

PROPERTIES OF NUMBERS

There are three properties or laws that you should know for the GED exam. These properties apply to addition and multiplication. You do not need to memorize these laws, but you should be able to understand them and to apply them to problems. These properties are based on the fact that we work with numbers in groups of twos.

The Commutative Property

For addition:

$$a + b = b + a$$

This means that the same two numbers added forward or backward will give you the same result.

EXAMPLE 1: $6 + 3 = 3 + 6$

$9 = 9$ Both sides equal 9.

For multiplication:

$$ab = ba$$

You can multiply the same two numbers forward or backward and the results will be the same.

EXAMPLE 2: $7 \times 4 = 4 \times 7$

$28 = 28$ Both sides equal 28.

This property is true for addition and multiplication, but it is not true for subtraction or division.

The Associative Property

For addition:

$$(a + b) + c = a + (b + c)$$

When you add three or more numbers, you can add any two at a time. The results will be the same.

EXAMPLE 3: $(2 + 5) + 6 = 2 + (5 + 6)$
$7 + 6 = 2 + 11$
$\mathbf{13 = 13}$ Both sides equal 13.

For multiplication:

$$(ab)\ c = a\ (bc)$$

When you multiply three numbers, you can multiply them in any order. The results will be the same.

EXAMPLE 4: $(2 \cdot 3)\ 4 = 2\ (3 \cdot 4)$
$(6)\ 4 = 2\ (12)$
$\mathbf{24 = 24}$ Both sides equal 24.

Again, this property is not true of subtraction and division.

The Distributive Property

$$a(b + c) = ab + ac$$
$$a(b - c) = ab - ac$$

The distributive property means that a number outside parentheses is multiplied times each number inside the parentheses. Their products are then combined according to the sign within the parentheses. This is true whether the numbers are separated by a plus (+) or a minus (−). You could say that the number outside the parentheses is being "distributed" to both numbers on the inside.

EXAMPLE 5: $2(6 + 3) =$
$2 \cdot 6 + 2 \cdot 3 =$
$12 + 6 = \mathbf{18}$

The purpose of showing you the distributive property is to help you understand why you get a particular solution. However, in your work later with algebra and geometry, many times you will solve for the quantity within the parentheses before multiplying it by the number on the outside.

EXAMPLE 6: Find $2(l + w)$ if $l = 6$ and $w = 3$.

$$2(l + w) =$$
$$2(6 + 3) =$$
$$2(9) = \mathbf{18}$$

Notice that Example 5 and Example 6 use the same numbers. The difference between the two is that in Example 5, you used the distributive law to multiply each number within parentheses by the number on the outside, while in Example 6, you solved for the quantity first and then multiplied. In both cases, you got the same answer.

In an algebraic statement, be sure to notice whether a single number or an entire quantity is being multiplied. For example, $3(4 + 2)$ will give a different answer than $3 \times 4 + 2$.

$$3(4 + 2) = \qquad\qquad 3 \times 4 + 2 =$$
$$3{\cdot}4 + 3{\cdot}2 = \qquad \text{BUT} \qquad 12 + 2 = \mathbf{14}$$
$$12 + 6 = \mathbf{18}$$

EXAMPLE 7: Which of the following is equal to $9(6 + 4)$?

    (1) $9 \times 6 + 4$
    (2) $9 + 6 \times 9 + 4$
    (3) $9 \times 6 + 9 \times 4$
    (4) $9 + 6 + 4$
    (5) $9 \times 6 \times 4$

  Answer: Choice (3) $\mathbf{9 \times 6 + 9 \times 4}$

  Solution: Choice (3) is correct. This is the distributive property. In choice (1), 9 is multiplied by only 6 and not 4. In choice (2), 9 is being added to 6 and to 4, and the quantities are being multiplied. In choice (4), there is no multiplication. In choice (5), there is no addition.

A good way to check your answer is to work out each answer choice. For the last example, the solution is:

$$9(6 + 4) = 9(10) = 90$$

Choice (3) is $9 \times 6 + 9 \times 4 = 54 + 36 = 90$. None of the other choices would give you 90.

You may see word problems on the GED that require you to show that you understand these properties of numbers.

EXAMPLE 8: On Saturday afternoon, 265 people were at the Uptown Movie Theatre. Saturday night, 304 people were there. Everyone paid $4 for a ticket. Which of the following represents the total value of the ticket sales that day?

(1) $4 \times 265 \times 304$
(2) $4(265 + 304)$
(3) $265(4 + 304)$
(4) $4 \times 265 + 304$
(5) $265 + 4(304)$

Answer: Choice (2) **$4(265 + 304)$**

Solution: Choice (2) is correct. It shows $4 multiplied by 265 people and by 304 people—the cost of each ticket multiplied by the total number of people.

PROPERTIES EXERCISE

Choose the correct solution to each problem.

1. Which of the following is equal to $(5 \cdot 3)4$?

(1) $(5 + 3)4$
(2) $5(3 \cdot 4)$
(3) $5(3 + 4)$
(4) $5 + 3 \cdot 4$
(5) $5 + 4 \cdot 3 + 4$

2. $(10 + 12) + 15$ can also be written as

(1) $10(12 + 15)$
(2) $(10 + 12)15$
(3) $10 \cdot 15 + 12 \cdot 15$
(4) $10 + 15 + 12 + 15$
(5) $10 + (12 + 15)$

3. Which of the following is equal to $16(14 + 27)$?

(1) $16 + 14 + 27$
(2) $14(16 + 27)$
(3) $27(16 + 14)$
(4) $16 \times 14 + 27$
(5) $(16 \times 14) + (16 \times 27$

4. During the month of February, Herb's Furniture sold 5 dinette sets at $200 apiece. In March, Herb sold 10 of the dinette sets. Which expression represents the total value of the merchandise sold?

(1) $200(5 + 10)$
(2) $5(200 + 10)$
(3) $10(200 + 5)$
(4) $200(5 \times 10)$
(5) $200(10 - 5)$

5. Which of the following is the same as $(9 + 7) + 4$?

(1) $4(9 + 7)$
(2) $9 + (7 + 4)$
(3) $9(7 + 4)$
(4) $7(9 + 4)$
(5) $4(9 + 7)$

6. Which of the following is equal to $\frac{1}{2}(15 - 6)$?

(1) $\frac{1}{2} \cdot 15 - 6$
(2) $15 - \frac{1}{2} \cdot 6$
(3) $\frac{1}{2} + 15 - 6$
(4) $(\frac{1}{2} \cdot 15) - (\frac{1}{2} \cdot 6)$
(5) $(\frac{1}{2} + 15) - (\frac{1}{2} + 6)$

7. Frank bought a shirt for $18 and a tie for $6. He had to pay 6% sales tax. Which of the following represents the sales tax he paid for these items?

(1) $18 + .06 \times 6$
(2) $.06 \times 18 \times 6$
(3) $6(.06 + 18)$
(4) $.06 \times 18 + 6$
(5) $.06(18 + 6)$

8. One week, Mark worked overtime for 3 hours on Wednesday and 4 hours on Thursday. For overtime, Mark makes $9 an hour. Which expression tells how much Mark made in overtime for that week?

(1) $4(3 + 9)$
(2) $3(4 + 9)$
(3) $9(3 + 4)$
(4) $(4 \times 3) + 9$
(5) $(4 \times 3) + (9 \times 3)$

9. Sharon drove for 4 hours at an average speed of 55 mph and then for 3 hours at an average speed of 40 mph. Which of the following expressions represents the total distance she drove?

(1) $7(55 + 40)$
(2) $95(3 + 4)$
(3) $(3 \times 55) + (4 \times 40)$
(4) $(4 \times 55) + (3 \times 40)$
(5) $40(3 + 4)$

10. All but 10 dozen out of 50 dozen eggs were sold at Marone's grocery store. Which expression represents how many eggs were sold?

(1) $12(50 - 10)$
(2) $50(12 - 10)$
(3) $10(50 - 12)$
(4) $(12 \cdot 50) - (10 \cdot 50)$
(5) $12(50 + 10)$

Answers and solutions start on page 252.

SUBSTITUTION

Substituting Numbers for Letters

In geometry and algebra, letters are often used in place of numbers. You have already used letters in place of numbers in proportion problems. You used the letter x to stand for the missing term in a proportion problem.

To substitute means to replace one thing with another. Substitution is a skill you will often use in algebra and geometry. For example, you will learn several formulas for perimeter, area, and volume when you study geometry. When you use these formulas in actual problems, you will replace the letters in the formula with numbers given in the problems.

The next examples illustrate the language of substitution. Watch for the signs in each example that tell you what operation to use.

Addition is easy to recognize.

EXAMPLE 1: What is the value of $a + b$ when a is 5 and b is 3?

$$a + b = 5 + 3 = 8$$

Step 1. Replace a with 5 and b with 3.

Step 2. Add 5 and 3.

Subtraction is also easy to recognize.

EXAMPLE 2: When $p = 10$ and $t = 4$, find the value of $p - t$.

$$p - t = 10 - 4 = 6$$

Step 1. Replace p with 10 and t with 4.

Step 2. Subtract 4 from 10.

Letters standing next to each other (with no signs between them) indicate multiplication.

EXAMPLE 3: If $y = 3$ and $z = 9$, what is the value of yz?

$$yz = 3 \times 9 = 27$$

Step 1. Replace y with 3 and z with 9.

Step 2. Multiply 3 by 9.

Another symbol for multiplication is a raised dot. We can write the last example another way:

$$yz = 3 \cdot 9 = 27$$

Remember, a number standing next to a letter (with no sign between them) also means multiplication. For example $3x$, when x is equal to 10, means the same thing as 3 times 10.

EXAMPLE 4. Let $m = 10$. Find the value of $\frac{1}{2}m$.

$$\frac{1}{2}m = \frac{1}{\cancel{2}} \cdot \frac{\cancel{10}^{\,5}}{1} = \mathbf{5}$$

Step 1. Replace m with 10.
Step 2. Cancel and multiply across.

The fraction bar is used for both dividing and reducing.

EXAMPLE 5. Find the value of $\frac{r}{s}$ when $r = 18$ and $s = 3$.

$$\frac{r}{s} = \frac{18}{3} = \mathbf{6}$$

Step 1. Replace r with 18 and s with 3.
Step 2. In this case, since the numerator is larger than the denominator, divide 18 by 3.

EXAMPLE 6: Find the value of $\frac{k}{m}$ when $k = 9$ and $m = 15$.

$$\frac{k}{m} = \frac{9 \div 3}{15 \div 3} = \mathbf{\frac{3}{5}}$$

Step 1. Replace k with 9 and m with 15.
Step 2. In this case, since the numerator is smaller than the denominator, you can reduce the fraction $\frac{9}{15}$ by 3.

Powers are easy to recognize.

EXAMPLE 7. Let $x = 4$. Find the value of x^3.

$$x^3 = 4^3 = 4 \cdot 4 \cdot 4 = \mathbf{64}$$

Step 1. Replace x with 4.
Step 2. Find 4 to the third power.

Square roots are indicated by the $\sqrt{}$ sign.

EXAMPLE 8. When $v = 225$, what is the value of $\sqrt{v}$?

$$\sqrt{v} = \sqrt{225} = \mathbf{15}$$

Step 1. Replace v with 225.
Step 2. Find the square root of 225.

SUBSTITUTION EXERCISE 1 ———————————————

Solve each problem.

1. When $c = 9$ and $d = 7$, what is the value of $c - d$?

2. If $r = 3$ and $s = 8$, what is rs?

3. When $u = 12$, what is the value of $5u$?

4. Let $f = 24$ and $g = 6$. Find the value of $\frac{g}{f}$.

5. Let $f = 24$ and $g = 6$. Find the value of $\frac{f}{g}$.

6. When $w = 24$, what is the value of $\frac{2}{3}w$?

7. If $a = 3$, $b = 4$, and $c = 5$, what is the value of abc?

8. What is the value of $\sqrt{m}$ when $m = 400$?

9. Find the value of $x - y$ when $x = 20$ and $y = 13$.

10. When $c = 36$ and $s = 12$, what is $\frac{c}{s}$?

11. What is d^3 when $d = 5$?

12. What is q^2 when $q = 8$?

13. If $a = 10$, $b = 3$, and $c = 6$, what is $a + b - c$?

14. What is mno when $m = 12$, $n = \frac{2}{3}$, and $o = \frac{1}{2}$?

Answers and solutions start on page 252.

——————————
Order of Operations
——————————

Suppose $a = 5$, $b = 3$, and $c = 4$. The expression $a + bc$ might be read two different ways. You might think that you should add 5 and 3 first, or you might think you should multiply 3 by 4 before adding. Only one way is correct.

In the expression $a + bc$, if you were to add first, you would get this <u>incorrect</u> answer:

$$\text{WRONG:} \quad a + bc = 5 + 3 \cdot 4 = 8 \cdot 4 = 32$$

The <u>correct</u> answer is found by first finding the value of 3 times 4 and then adding the result to the 5. Example 1 below illustrates this.

Mathematicians have agreed on a correct order to perform operations. Memorize the list below before you go on. Although a particular problem may not use all the steps below, this chart gives you an order to follow.

Order of Operations

1. Do operations on groupings of numbers (within parentheses or above or below a division bar)
2. Find powers and roots
3. Do multiplication or division
4. Do addition or subtraction

EXAMPLE 1: Find the value of $a + bc$ when $a = 5$, $b = 3$, and $c = 4$.

$$a + bc = 5 + 3 \cdot 4 = 5 + 12 = \mathbf{17}$$

Step 1. Substitute 5 for a, 3 for b, and 4 for c.

Step 2. This problem requires both addition and multiplication. According to the order of operations, you must multiply before you add. Multiply 3 by 4.

Step 3. Add 5 and 12.

EXAMPLE 2: What is the value of xy^2 when $x = 5$ and $y = 4$?

$$xy^2 = 5 \cdot 4^2 = 5 \cdot 16 = \mathbf{80}$$

Step 1. Replace x with 5 and y with 4.

Step 2. This problem involves multiplying two numbers and finding a power. Powers precede multiplication in the order of operations. Find 4 to the second power.

Step 3. Multiply 5 by 16.

The next two examples show different ways of grouping numbers with parentheses and with a fraction bar. Within each grouping, follow the same order of operations.

EXAMPLE 3: If $m = 5$ and $n = 2$, find the value of $(m + n)(m - n)$.

① $\boxed{(m + n)(m - n) = (5 + 2)(5 - 2)}$ ② & ③ $\boxed{7 \cdot 3 = \mathbf{21}}$

Step 1. Replace m with 5 and n with 2.

Step 2. Parentheses side-by-side indicate multiplication just as letters side-by-side do. First find the groupings of numbers within the parentheses. Add 5 and 2 in the first set of parentheses. Subtract 2 from 5 in the second set of parentheses.

Step 3. Since there is no sign between parentheses, multiply 7 by 3.

EXAMPLE 4:   If $b = 6$ and $c = 2$, find the value of $\dfrac{b+c}{c} + b$.

①
$$\frac{b+c}{c} + b = \frac{6+2}{2} + 6$$

②&③
$$\frac{8}{2} + 6 = 4 + 6 = 10$$

Step 1.   Replace b with 6 and c with 2.

This problem requires you to find a quantity that includes a division bar and then to add that result to a number.

Step 2.   First, find the value of the quantity. The fraction bar splits the quantity into two parts or groupings. Solve for each grouping separately. Add the 6 and 2 on top and divide by the 2 on the bottom.

Step 3.   Add that solution to the number to the right.

SUBSTITUTION EXERCISE 2

Solve each problem.

1.   Find $w^2 - v^2$ when $w = 5$ and $v = 3$.

2.   If $c = 4$, what is $2c^2$?

3.   What is the value of $3t - 10$ when $t = 12$?

4.   When $a = 10$ and $b = 9$, what is $3a - 2b$?

5.   If $x = 4$ and $y = 3$, what is the value of $\dfrac{xy}{x} + y^2$?

6.   If $a = 10$, find the value of $a^2 - 3$.

7.   If $a = 5$, $b = 7$, and $c = 6$, find $a^2 + 2b + 3bc$.

8.   What is $\dfrac{2}{3}\,sr^2$ when $s = 3$ and $r = 5$?

9.   If $a = 1$, $b = 2$, $c = 3$, and $d = 4$, what is $(a + b)\,(c + d)$?

10.   If $x = 5$, $y = 6$, and $z = 3$, find the value of $x\,(y - z)$.

11.   When $k = 9$, $p = 4$, and $r = 5$, what is the value of $\dfrac{k^2 + p}{r}$?

12.   When $m = 10$ and $n = 20$, what is $\dfrac{2\,m^2}{n}$?

13.   If $a = 2$, $b = 4$, $c = 6$, and $d = 8$, what is $cd - ab$?

14.   When $j = 20$ and $k = 30$, what is $\dfrac{2jk}{j + k}$?

15. Let $s = 5$ and $t = 4$. Find the value of $\dfrac{st - t^2}{2s}$.

16. If $r = 64$, what is the value of $\dfrac{\sqrt{r}}{r}$?

Answers and solutions start on page 253.

FORMULAS

A **formula** is a kind of mathematical instruction. In geometry, you will learn several formulas for finding perimeter, area, and volume. To **evaluate** a formula means to replace the letters on one side of the equal sign (=) with the numbers given in a problem. Then follow the order of operations rules to find your answer.

The next examples illustrate how to evaluate formulas.

EXAMPLE 1: Find A in the formula $A = ns^2$ when $n = 3$ and $s = 5$.

$$A = ns^2 = 3 \times 5^2 = 3 \times 25$$
$$A = 75$$

Step 1. Replace n with 3 and s with 5.
Step 2. According to the order of operations, find powers before multiplication. First find 5 to the second power.
Step 3. Multiply 25 by 3.

EXAMPLE 2: $P = 2(l + w)$. Find P when $l = 15$ and $w = 8$.

$$P = 2(l + w) = 2(15 + 8) = 2 \times 23$$
$$P = 46$$

Step 1. Replace l with 15 and w with 8.
Step 2. Perform any operation with the numbers inside parentheses first. In this formula, add the numbers inside the parentheses first. Add 15 and 8.
Step 3. There is no operation sign between the 2 and the parentheses. Multiply 23 by 2.

FORMULAS EXERCISE

Evaluate each of the following formulas.

1. Find the value of P in $P = A + B + C$ when $A = 5$, $B = 12$, and $C = 13$.

2. $F = ma$. What is F when $m = 900$ and $a = 2$?

3. $D = RT$. Find D when $R = 450$ and $T = 3.5$.

4. $1 = \dfrac{A}{w}$. Find 1 when $A = 56$ and $w = 4$.

5. What is the value of A in the formula $A = \frac{1}{2}bh$ when $b = 6$ and $h = 5$?

6. $F = \frac{9}{5}C + 32$. Find F when $C = 20$.

7. $I = PRT$. Find I when $P = 500$, $R = .06$, and $T = 3$.

8. Find the value of b in the formula $b = \dfrac{2A}{h}$ when $A = 24$ and $h = 6$.

9. $A = s^2$. Find A when $s = 11$.

10. $P = 2(l + w)$. Find P when $l = 10$ and $w = 8$.

11. $V = s^3$. Find V when $s = 4$.

12. What is the value of s in the formula $s = \frac{1}{2}gt^2$ when $g = 32$ and $t = 4$?

13. $C = \frac{5}{9}(F - 32)$. Find C when $F = 50$.

14. $A = \dfrac{h}{2}(b + c)$. Find A when $h = 10$, $b = 6$, and $c = 8$.

Answers and solutions start on page 253.

ANSWERS AND SOLUTIONS

Powers Exercise

1. 36 $6 \times 6 = 36$

2. 1 $1 \times 1 \times 1 = 1$

3. 125 $5 \times 5 \times 5 = 125$

4. 243 $3 \times 3 \times 3 \times 3 \times 3 = 243$

5. 34 $5 \times 5 = 25$
   $3 \times 3 = 9$
   $25 + 9 = 34$

6. 25 $5 \times 5 \times 5 = 125$
   $10 \times 10 = 100$
   $125 - 100 = 25$

7. 1   Any number to the zero power is 1.

8. 625 $25 \times 25 = 625$

9. 105 $11 \times 11 = 121$
   $4 \times 4 = 16$

10. 111 $1 + 10 + 100 = 111$

11. 8,000 $20 \times 20 \times 20 = 8,000$

12. 13   Any number to the first power is that number.

13. 10,000 $10 \times 10 \times 10 \times 10 = 10,000$

14. 48 $8 \times 8 = 64$
   $2 \times 2 \times 2 \times 2 = 16$
   $64 - 16 = 48$

15. 76 $3 \times 3 \times 3 = 27$
   $7 \times 7 = 49$
   $27 + 49 = 76$

Square Roots Exercise

1. 14
   Guess 12 because $12 \times 12 = 144$.

   $$12\overline{)196} \quad 16 \quad \begin{array}{r} 16 \\ +12 \\ \hline 28 \end{array} \quad 2\overline{)28}$$

   $$\begin{array}{r} 16 \\ 12\overline{)196} \\ \underline{12} \\ 76 \\ \underline{72} \\ 4 \end{array}$$

2. 21
   Guess 20 because $20 \times 20 = 400$.

   $$\begin{array}{r} 22 \\ 20\overline{)441} \\ \underline{40} \\ 41 \\ \underline{40} \\ 1 \end{array} \quad \begin{array}{r} 22 \\ +20 \\ \hline 42 \end{array} \quad \begin{array}{r} 21 \\ 2\overline{)42} \\ \underline{42} \end{array}$$

3. 32
   Guess 30 because $30 \times 30 = 900$.

   $$\begin{array}{r} 34 \\ 30\overline{)1,024} \\ \underline{90} \\ 124 \\ \underline{120} \\ 4 \end{array} \quad \begin{array}{r} 34 \\ +30 \\ \hline 64 \end{array} \quad \begin{array}{r} 32 \\ 2\overline{)64} \\ \underline{64} \end{array}$$

4. 39
   Guess 40 because $40 \times 40 = 1,600$.

   $$\begin{array}{r} 38 \\ 40\overline{)1,521} \\ \underline{120} \\ 321 \\ \underline{320} \\ 1 \end{array} \quad \begin{array}{r} 38 \\ +40 \\ \hline 78 \end{array} \quad \begin{array}{r} 39 \\ 2\overline{)78} \end{array}$$

5. 19
   Guess 20 because $20 \times 20 = 400$.

   $$\begin{array}{r} 18 \\ 20\overline{)361} \\ \underline{20} \\ 161 \\ \underline{160} \\ 1 \end{array} \quad \begin{array}{r} 18 \\ +20 \\ \hline 38 \end{array} \quad \begin{array}{r} 19 \\ 2\overline{)38} \\ \underline{38} \end{array}$$

6. 26
   Guess 30 because $30 \times 30 = 900$

   $$\begin{array}{r} 22 \\ 30\overline{)676} \\ \underline{60} \\ 76 \\ \underline{60} \\ 16 \end{array} \quad \begin{array}{r} 22 \\ +30 \\ \hline 52 \end{array} \quad \begin{array}{r} 26 \\ 2\overline{)52} \\ \underline{52} \end{array}$$

Square Roots Exercise cont'd

7. 43

 Guess 40 because 40 × 40 = 1,600.

```
        46            46            43
   40) 1,849         +40         2) 86
       160            86            86
       249
       240
         9
```

8. 58

 Guess 60 because 60 × 60 = 3,600.

```
        56            56            58
   60) 3,364         +60         2) 116
       300           116            10
       364                          16
       360                          16
         4
```

9. 68

 Guess 70 because 70 × 70 = 4,900.

```
        66            66            68
   70) 4,624         +70         2) 136
       420           136            12
       424                          16
       420                          16
         4
```

10. 79

 Guess 80 because 80 × 80 = 6,400.

```
        78            78            79
   80) 6,241         +80         2) 158
       560           158            14
       641                          18
       640                          18
         1
```

11. 86

 Guess 90 because 90 × 90 = 8,100

```
        82            82            86
   90) 7,396         +90         2) 172
       720           172            16
       196                          12
       180                          12
        16
```

12. 98

 Guess 100 because 100 × 100 = 10,000

```
         96            96            98
  100) 9,604         +100         2) 196
       900            196            18
       604                           16
       600                           16
         4
```

Properties Exercise

1. (2) $5 (3 \cdot 4)$

2. (5) $10 + (12 + 15)$

3. (5) $(16 \times 14) + (16 \times 27)$

4. (1) $200 (5 + 10)$

5. (2) $9 + (7 + 4)$

6. (4) $\frac{1}{2} \cdot 15 - \frac{1}{2} \cdot 6$

7. (5) $.06 (18 + 6)$

8. (3) $9 (3 + 4)$

9. (4) $(4 \times 55) + (3 \times 40)$

10. (1) $12 (50 - 10)$

Substitution Exercise 1

1. 2 $c - d = 9 - 7 = 2$

2. 24 $rs = 3 \cdot 8 = 24$

3. 60 $5u = 5 \cdot 12 = 60$

4. $\frac{1}{4}$ $\frac{g}{f} = \frac{6}{24} = \frac{1}{4}$

5. 4 $\frac{f}{g} = \frac{24}{6} = 4$

6. 16 $\frac{2}{3}w = \frac{2}{\cancel{3}} \times \frac{\overset{8}{\cancel{24}}}{1} = 16$

7. 60 $abc = 3 \times 4 \times 5 = 60$

8. 20 $\sqrt{m} = \sqrt{400} = 20$

9. 7 $x - y = 20 - 13 = 7$

10. 3 $\frac{c}{s} = \frac{36}{12} = 3$

11. 125 $d^3 = 5^3 = 5 \times 5 \times 5 = 125$

12. 64 $q^2 = 8^2 = 8 \times 8 = 64$

13. 7 $a + b - c =$
     $10 + 3 - 6 = 13 - 6 = 7$

14. 4 $mno = \frac{\overset{4}{\cancel{12}}}{1} \times \frac{2}{\underset{1}{\cancel{3}}} \times \frac{1}{\underset{1}{\cancel{2}}} = 4$

Substitution Exercise 2

1. 16 $\quad w^2 - v^2 =$
$5^2 - 3^2 = 25 - 9 = 16$

2. 32 $\quad 2c^2 =$
$2 \cdot 4^2 = 2 \cdot 16 = 32$

3. 26 $\quad 3t - 10 =$
$3 \cdot 12 - 10 = 36 - 10 = 26$

4. 12 $\quad 3a - 2b =$
$3 \cdot 10 - 2 \cdot 9 = 30 - 18 = 12$

5. 12 $\quad \dfrac{xy}{x} + y^2 =$
$\dfrac{4 \cdot 3}{4} + 3^2 = \dfrac{12}{4} + 9 = 3 + 9 = 12$

6. 97 $\quad a^2 - 3 =$
$10^2 - 3 = 100 - 3 = 97$

7. 165 $\quad a^2 + 2b + 3bc =$
$5^2 + 2 \cdot 7 + 3 \cdot 7 \cdot 6 =$
$25 + 14 + 126 = 165$

8. 50 $\quad \frac{2}{3} sr^2 =$
$\frac{2}{3} \cdot 3 \cdot 5^2 = \frac{2}{\cancel{3}} \cdot \cancel{3} \cdot 25 = 50$

9. 21 $\quad (a + b)(c + d) =$
$(1 + 2)(3 + 4) = 3 \cdot 7 = 21$

10. 15 $\quad x(y - z) =$
$5(6 - 3) = 5(3) = 15$

11. 17 $\quad \dfrac{k^2 + p}{r} =$
$\dfrac{9^2 + 4}{5} = \dfrac{81 + 4}{5} = \dfrac{85}{5} = 17$

12. 10 $\quad \dfrac{2m^2}{n} =$
$\dfrac{2 \cdot 10^2}{20} = \dfrac{2 \cdot 100}{20} = 10$

13. 40 $\quad cd - ab =$
$6 \cdot 8 - 2 \cdot 4 = 48 - 8 = 40$

14. 24 $\quad \dfrac{2jk}{j + k} =$
$\dfrac{2 \cdot 20 \cdot 30}{20 + 30} = \dfrac{1,200}{50} = 24$

15. $\frac{2}{5}$ $\quad \dfrac{st - t^2}{2s} =$
$\dfrac{5 \cdot 4 - 4^2}{2 \cdot 5} = \dfrac{20 - 16}{10} = \dfrac{4}{10} = \dfrac{2}{5}$

16. $\frac{1}{8}$ $\quad \dfrac{\sqrt{r}}{r} = \dfrac{\sqrt{64}}{64} = \dfrac{8}{64} = \dfrac{1}{8}$

Formulas Exercise

1. 30 $\quad P = A + B + C$
$P = 5 + 12 + 13 = 30$

2. 1,800 $\quad F = ma$
$F = 900 \times 2 = 1,800$

3. 1,575 $\quad D = RT$
$D = 450 \times 3.5 = 1,575$

4. 14 $\quad l = \dfrac{A}{w}$
$l = \dfrac{56}{4} = 14$

5. 15 $\quad A = \frac{1}{2}bh$
$A = \frac{1}{2} \times 6 \times 5 = \frac{1}{2} \times \frac{\cancel{6}^3}{1} \times \frac{5}{1} = 15$

6. 68 $\quad F = \frac{9}{5}C + 32$
$F = \frac{9}{5} \cdot 20 + 32$
$F = \frac{9}{\cancel{5}} \cdot \frac{\cancel{20}^4}{1} + 32 = 36 + 32 = 68$

7. 90 $\quad I = PRT$
$I = 500 \times .06 \times 3 = 90$

8. 8 $\quad b = \dfrac{2A}{h}$
$b = \dfrac{2 \times 24}{6} = \dfrac{48}{6} = 8$

9. 121 $\quad A = s^2$
$A = 11^2 = 11 \cdot 11 = 121$

10. 36 $\quad P = 2(l + w)$
$P = 2(10 + 8) = 2(18) = 36$

11. 64 $\quad V = s^3$
$V = 4^3 = 4 \cdot 4 \cdot 4 = 64$

12. 256 $\quad s = \frac{1}{2}gt^2$
$s = \frac{1}{2} \times 32 \times 4^2 = \frac{1}{\cancel{2}} \times \frac{\cancel{32}^{16}}{1} \times \frac{16}{1} = 256$

13. 10 $\quad C = \frac{5}{9}(F - 32)$
$C = \frac{5}{9}(50 - 32) = \frac{5}{\cancel{9}} \times \frac{\cancel{18}^2}{1} = 10$

14. 70 $\quad A = \frac{h}{2}(b + c)$
$A = \frac{10}{2}(6 + 8) = \frac{10}{\cancel{2}} \times \frac{\cancel{14}^7}{1} = 70$

Geometry

COMMON GEOMETRIC FIGURES

Geometry is the study of lines, angles, flat figures, and three-dimensional shapes. If you have done any carpentry or sewing, you are already familiar with many ideas in geometry.

To understand geometry, you must know several special terms. Following are some of the terms we will use in the rest of the geometry section of this book. Take the time now to learn the terms you do not already know.

An **angle** is a figure made of two lines extending from the same point. Figures A and B are examples of angles.

Fig. A   Fig. B

The size of an angle is not determined by the length of its lines, but by the size of the opening between the lines. The angle in Figure B is larger than the angle in Figure A because it is more open.

A **right angle** looks like a square corner. Figures C and D are both right angles. A small square inside the angle, as in Figure C, is often used to indicate a right angle. There are many examples of right angles around us. The top and side of a door form a right angle. The corners of this page are right angles. You will learn about other types of angles later in this section.

Fig. C   Fig. D

Parallel lines are lines that run in the same direction. They will never cross no matter how far they are extended. Train tracks are examples of parallel lines. Figures E, F, and G each show pairs of parallel lines. In Figure E, the two parallel lines are **vertical**. Vertical means running straight up and down. The parallel lines in Figure F are **horizontal**. Horizontal means running straight across. The parallel lines in Figure G are neither vertical nor horizontal.

Fig. E   Fig. F   Fig. G

Perpendicular lines are lines that cross or meet to form right angles. Figure H is an example of perpendicular lines. The wood or metal supports of a window are usually perpendicular lines.

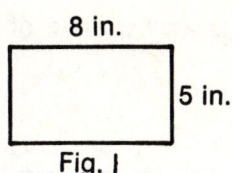

Fig. H

Three common flat shapes are the rectangle, the square, and the triangle. These three figures are examples of **plane figures.** A plane figure is completely flat.

A **rectangle** is a four-sided figure with four right angles. The sides across from each other are parallel and equal in measurement. Figure I is a rectangle. The two short sides each measure 5 inches. The two long sides each measure 8 inches. A short side and a long side of a rectangle are perpendicular to each other. Rectangles are very common figures; most walls, floors, and windows are rectangular.

8 in.

5 in.

Fig. I

A **square** is a four-sided figure with four right angles and four equal sides. The sides across from each other are parallel. The only difference between a square and a rectangle is that the sides of a square all have the same measurement. Figure J is a square. Each side measures 3 feet.

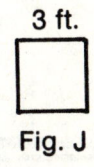

3 ft.

Fig. J

A **triangle** is a three-sided figure. Figures K, L, and M are all triangles. In Figure K, each side has a different measurement. In Figure L, each side has the same measurement. In Figure M, two of the sides are the same and the third side is longer. You will learn the names for different types of triangles later.

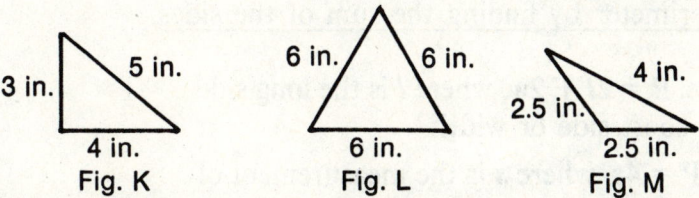

3 in.   5 in.   6 in.   6 in.   4 in.   2.5 in.

4 in.   6 in.   2.5 in.

Fig. K   Fig. L   Fig. M

The next exercise gives you a chance to test your knowledge of the geometrical terms you have learned so far.

GEOMETRY EXERCISE 1 ————————————————

Fill in each blank.

1. Lines that run in the same direction and do not cross each other are called _____ lines.

2. A four-sided figure with four right angles and four equal sides is called
a _____.

3. Lines that cross or meet each other to form right angles are called
_____ lines.

4. Two lines extending from the same point form an _____.

5. Figure X is an example of a _____.

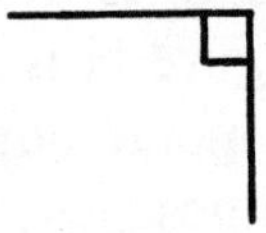

Fig. X

6. Figure Y is an example of a _____.

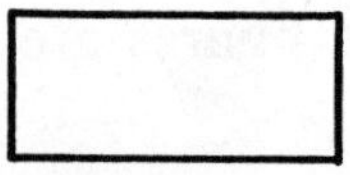

Fig. Y

7. Figure Z is an example of a _____.

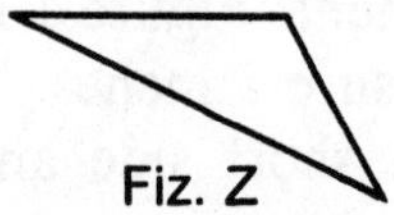

Fiz. Z

8. Lines running straight up and down are called _____
lines.

Answers start on page 291.

PERIMETER

Perimeter is the measurement of the distance around a flat figure. To find
the perimeter of a figure, add all the sides.

Following are the formulas for finding the perimeter of a rectangle, a
square, and a triangle. While it is helpful to learn these formulas, remember
that you can always find a perimeter by finding the sum of the sides.

Perimeter of a rectangle: $P = 2l + 2w$, where l is the long side
or length and w is the short side or width.

Perimeter of a square: $P = 4s$, where s is the measurement of
each side.

Perimeter of a triangle: $P = s_1 + s_2 + s_3$ where s_1, s_2, and s_3
are the measurements of the three sides.

EXAMPLE: Find the perimeter of a rectangle that is 6.5 meters long
and 3.2 meters wide.

Solution: Replace l with 6.5 meters and w with 3.2 meters in the
formula $P = 2l + 2w$.

$P = 2l + 2w$
$P = 2(6.5m) + 2(3.2m)$
$P = 13\ m + 6.4\ m$
$P = \textbf{19.4 m}$

EXAMPLE: Find the perimeter of the square shown at the right.

3 ft.

3 ft. 3 ft.

3 ft.

Solution: Replace s with 3 feet in the formula $P = 4s$.

$P = 4s$
$P = 4(3\ \text{ft.})$
$P = \textbf{12 ft.}$

EXAMPLE: Find the perimeter of the triangle shown at the right.

2.5′   4′

2.5′

Solution: Replace s_1 with 2.5′, s_2 with 2.5′, and s_3 with 4′ in the formula $P = s_1 + s_2 + s_3$.

$P = s_1 + s_2 + s_3$
$P = 2.5' + 2.5' + 4'$
$P = \textbf{9}'$

You may be given the perimeter of a figure and the length of one of the sides. You can use this information to find the length of another side.

EXAMPLE: The perimeter of a rectangle is 26 inches. The length of the rectangle is 8 inches. Find the width of the rectangle.

① $2l = 2(8) = 16$

② $P - 2l = 2w$
$26 - 16 = 2w$
$10 = 2w$

③ $10 = 2w$
$\dfrac{10}{2} = w$
$\textbf{5} = w$

Step 1. If l equals 8 inches, find $2l$.
Step 2. To find out what $2w$ (2 times the width) is equal to, subtract $2l$ (16) from the perimeter, 26.
Step 3. To find out what the width equals, find w. Divide the value of $2w$ by 2.

The width is 5 inches.

Pictures sometimes help to solve geometry problems. Some of the problems in the next exercise have no pictures. Try to draw the figures described in these problems to "see" the figures and solve the problems.

GEOMETRY EXERCISE 2 ————————————————————

Solve each problem.

1. What is the perimeter of a rectangle that is 52 feet long and 37 feet wide?

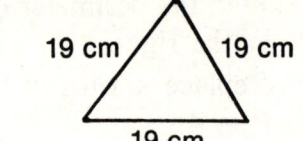

2. Find the perimeter of the triangle shown at the right.

3. Find the perimeter of a square with sides that measure $\frac{1}{2}$ inch each.

4. What is the perimeter of the figure shown at the right?

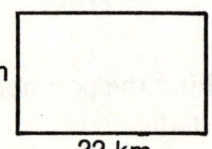

5. Find the perimeter of the figure at the right.

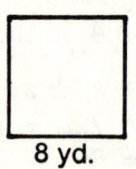

6. What is the perimeter of the triangle shown at the right?

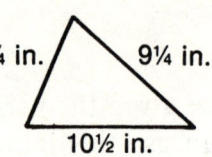

7. What is the perimeter of a rectangle that is 6.3 miles long and 5.2 miles wide?

8. Each side of a square measures 1.75 meters. Find the perimeter of the figure.

9. What is the perimeter of the triangle shown at the right?

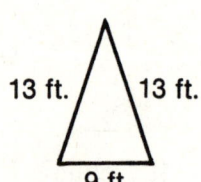

10. The perimeter of a square is 100 inches. How long is one side?

11. The perimeter of a rectangle is 50 yards. The length of the rectangle is 16 yards. Find the width.

12. One side of a triangle is 5 ft. Another side is 13 ft. The perimeter is 30 ft. Find the measurement of the third side.

Answers and solutions start on page 291.

AREA

Area is the measure of the amount of surface within the perimeter of a flat figure. Area is measured in square inches, square feet, square meters, etc.

The figure at the right represents a rectangle that is 5 inches long and 3 inches wide. You could cover the rectangle with 15 small squares. Each of these squares is one square inch. The surface, or area, of the rectangle is 15 square inches.

The formula for finding the area of a rectangle is $A = lw$ where l is the length and w is the width. Remember to multiply if there is no sign between letters.

EXAMPLE 1:   Find the area of the rectangle at the right.

Solution: Replace l with $8\frac{1}{2}$ and w with 4 in the formula $A = lw$ and multiply.

$$A = lw$$
$$A = 8\frac{1}{2} \times 4$$
$$A = \frac{17}{\cancel{2}} \times \frac{\cancel{4}^{2}}{1}$$
$$A = \textbf{34 sq. in.}$$

In a square, the length and the width are the same. Instead of multiplying length by width, we multiply side by side. The formula for finding the area of a square is $A = s^2$ where s is the measurement of one of the sides.

EXAMPLE 2:   Find the area of a square if one side measures 7 inches.

Solution: Replace s with 7 in the formula $A = s^2$.

$$A = s^2$$
$$A = 7^2$$
$$A = 7 \times 7$$
$$A = \textbf{49 sq. in.}$$

In some cases, you are given the area of a figure and are asked to find a side. To find a side of a rectangle, or a square, you divide the area by the side given.

EXAMPLE 3:   Find the length of a figure if the area is 1,400 square inches and the width is 35 inches.

Solution: Divide the area by the width.

$$\begin{array}{r} 40 \\ 35\overline{)1,400} \end{array}$$

The length is 40 inches.

Do not use the terms length and width with triangles; use the terms **base** and **height**. The **base** is the side the triangle appears to be sitting on. The **height** is a distance perpendicular to either the base or its extension. Look at the base and the height of each triangle below.

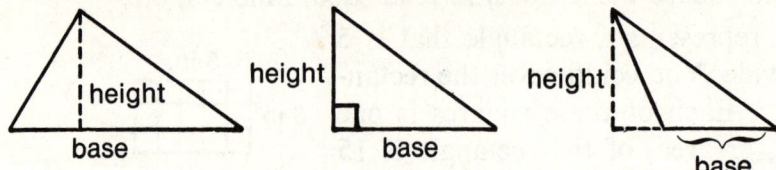

In the triangle to the left, the height is inside the triangle. In the middle triangle, the height is the left-hand side. For the triangle at the right, you can drop an imaginary line from the top of the triangle to an extension of the base to indicate the triangle's height.

You can think of the area of a triangle as one-half of the area of some four-sided figure. For example, the figure below is a rectangle with a length of 9 inches and a width of 6 inches. The shaded part of the rectangle is a ight triangle also with a base of 9 inches and a height of 6 inches.

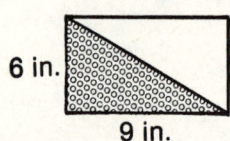

Because of this relationship, the formula for finding the area of a triangle is $A = \frac{1}{2}bh$ where b is the base and h is the height.

EXAMPLE 4: Find the area of a triangle with a base of 9 inches and a height of 6 inches.

Solution: Replace b with 9 and h with 6 in the formula $A = \frac{1}{2}bh$. Multiply.

$A = \frac{1}{2}bh$ $A = $ **27 sq. in.**

$A = \frac{1}{\cancel{2}_1} \times \frac{9}{1} \times \frac{\cancel{6}^3}{1}$

Notice that the area of the rectangle above is 54 sq. in. The area of the triangle is exactly one-half of the area of the rectangle.

EXAMPLE 5: Find the area of the triangle at the right.

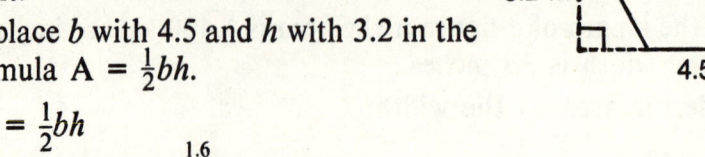

Solution: Replace b with 4.5 and h with 3.2 in the formula $A = \frac{1}{2}bh$.

$A = \frac{1}{2}bh$

$A = \frac{1}{\cancel{2}_1} \times \frac{4.5}{1} \times \frac{\cancel{3.2}^{1.6}}{1}$

$A = $ **7.2 sq. m**

While it is useful to know the formulas for the areas of the three types of figures, remember that they all involve multiplying a side (the height or width) times a perpendicular side (the base or length).

GEOMETRY EXERCISE 3 ━━━━━━━━━━━━━━━━━━━━━━━━━━━

Solve each problem.

1. Find the area of a rectangle that is 12 feet wide and 15 feet long.

2. What is the area of a square with a side that measures 14 feet?

3. Find the area of the triangle at the right.

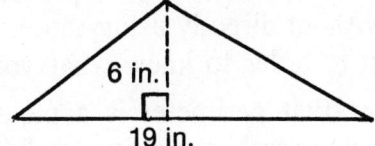

4. The side of a square measures 0.15 meter. Find the area of the square.

5. Find the area of this triangle.

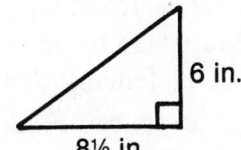

6. Find the area of this rectangle.

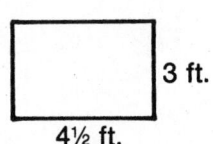

7. Find the area of the shaded part of the figure at the right.

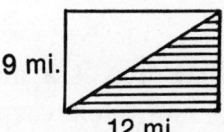

8. What is the area of a triangle with a base of 5 inches and a height of 5 inches?

9. A square measures $\frac{3}{8}$ inch on one side. Find the area of the square.

10. Find the area of the triangle shown at the right.

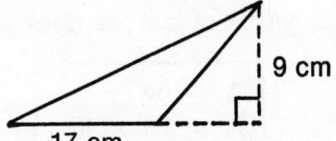

11. A rectangle has an area of 60 square inches. The width of the rectangle is 5 inches. Find the length of the rectangle.

12. A square has an area of 100 square inches. Find the measurement of one side of the square.

13. Which has a larger area: a square with sides that measure 8 inches or a rectangle with a length of 9 inches and a width of 7 inches?

Answers and solutions start on page 291.

PERIMETER AND AREA WORD PROBLEMS

On the GED test, you may have problems that require you to find area or perimeter without directly using those words. You will have to understand the problem in order to know what you are solving for.

Remember that perimeter is a measure of the distance around a flat shape. Fencing, weatherstripping, and framing are practical applications of the idea of perimeter.

EXAMPLE 1: Peter wants to put a fence around a section of his backyard. He wants the enclosed space to be 30 feet long and 10 feet wide. How many feet of fencing does he need?

Solution: Fencing goes around a space. Find the perimeter of the space. Replace l with 30′ and w with 10′ in the formula $P = 2l + 2w$.

$$P = 2l + 2w$$
$$P = 2(30) + 2(10)$$
$$P = 60' + 20'$$
$$P = 80'$$

Remember that area is a measure of the amount of surface on a flat shape. Knowing how much carpet to put on a floor, how much paint to put on walls, and how much material to buy for sewing all involve finding the area.

EXAMPLE 2: Luba wants to put carpeting on her bedroom floor. Her bedroom is 10 feet long and 9 feet wide. How many square yards of carpet does she need?

①
$$A = lw$$
$$A = 10 \times 9$$
$$A = 90 \text{ sq. ft.}$$

②
$$9\overline{)90} \quad \frac{10}{} \text{ sq. yd.}$$

Step 1. In this problem, there are two clues that tell you to find the area. "Carpet" is one clue. Carpet covers a surface. The other clue is "square yards." Only area is measured in square units. Replace l with 10′ and w with 9′ in the formula $A = lw$. Multiply to find the number of square feet needed.

Step 2. Notice that the problem asks for square yards, but the length and width are given in feet. A square yard measures 3 feet on each side. It has an area of $3^2 = 3 \times 3 = 9$ square feet.

Divide the area of the room by the number of square feet in one square yard. In other words, divide 90 by 9 to get the number of square yards of carpet needed. Luba needs 10 square yards of carpet to cover the floor of her bedroom.

Read the next problems carefully. Before beginning your work, decide whether you need to find the perimeter or the area.

GEOMETRY EXERCISE 4

Solve each problem. Be sure to put the correct label on each answer.

1. Colin is putting new wood flooring in the spare room of his house. The room is 9 feet wide and 12 feet long. How much flooring will Colin need for the room?

2. The garden in Sylvia's yard is 26 meters long and half as wide. How many meters of fencing does she need to enclose the garden?

3. Mrs. Jackson wants to put weatherstripping around the two windows in her kitchen. Each window is 36 inches wide and 42 inches tall. How many feet of weatherstripping does she need?

4. Bill wants to put 2-foot by 2-foot carpet tiles on the floor of his living room. The room is 18 feet long and 10 feet wide. Find the smallest number of carpet tiles needed to completely cover the room.

5. A gallon of floor paint will cover about 200 square feet of concrete. The floor of Harold's basement is 25 feet wide and 40 feet long. How many gallons of paint does he need to put one coat of paint on the basement floor?

6. Mark wants to put a frame around an 8″ × 10″ photograph. How many inches of framing does he need?

7. How many square inches of glass does Mark need to cover the photograph described in problem 6?

8. The figure at the right shows the dimensions of the triangular garden in the corner of Dorothy's yard. How many yards of fencing does she need to completely enclose the garden?

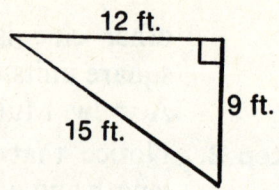

9. Dorothy wants to cover her garden (problem 8) because of a sudden chill. How many square feet of plastic does she need to buy?

10. Mary wants to make curtains for the three windows in her living room. For each pair of curtains, she needs material 6 feet wide and 8 feet long. Altogether, how many square yards of material does she need?

Answers and solutions start on page 292.

CIRCLES

A **circle** is a flat figure made of a curve; every point of the curve is the same distance from the center. The **diameter** of a circle is a line that crosses a circle through its center. It is also the measure of the distance across a circle. The **radius** of a circle is a line from the center of the circle to the curve of the circle. It is half the distance across a circle; therefore, the radius is half the diameter.

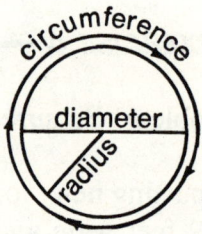

The **circumference** is the distance around a circle. It corresponds to the perimeter of figures with straight sides.

The formulas used for finding the circumference and the area of circles contain a special number that is represented by the Greek letter π (pi). π is the ratio of the circumference of a circle to its diameter.

If you carefully measure both the circumference and the diameter of any circle and then divide the circumference by the diameter, you will always get the same number (3.14), no matter how large or small the circle is. The usual value given for π is either 3.14 or its fractional form, $3\frac{1}{7}$. Often, the value of π is given as the improper fraction $\frac{22}{7}$.

The formula for finding the circumference of a circle is $C = \pi d$, where d is the diameter of the circle.

EXAMPLE 1:   Find the circumference of a circle with a diameter of 21 inches. Use $\pi = \frac{22}{7}$.

Solution: Replace π with $\frac{22}{7}$ and d with 21 inches in the formula $C = \pi d$. The circumference of the circle is 66 inches.

$$C = \pi d$$

$$C = \frac{22}{\overset{}{\underset{1}{7}}} \times \frac{\overset{3}{21}}{1} = \textbf{66 inches}$$

The formula for the area of a circle is $A = \pi r^2$ where r is the radius of the circle. Remember that area is always measured in square units.

EXAMPLE 2: Find the area of a circle with a radius of 3 inches. Use $\pi = 3.14$.

    Solution: Replace π with 3.14 and r with 3 inches in the formula $A = \pi r^2$.

$$A = \pi r^2$$
$$A = 3.14 \times 3^2$$
$$A = 3.14 \times 9 = \textbf{28.26 sq. in.}$$

Memorize the formulas for circumference and area of a circle before you try the next problems.

GEOMETRY EXERCISE 5 ────────────

Solve each problem. First, decide whether you are finding area or circumference and which formula you should use.

1. What is the circumference of a circle with a diameter of 10 yards? Use $\pi = 3.14$.

2. Find the area of the circle at the right. Use $\pi = \frac{22}{7}$.

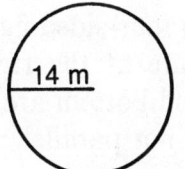
14 m

3. Find the area of a round rug that is 10 feet across. Use $\pi = 3.14$.

4. Find the circumference of the circle at the right. Use $\pi = \frac{22}{7}$.

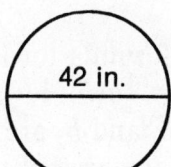
42 in.

5. What is the radius of the circle in problem 4?

6. What is the area of the circle in problem 4?

7. Find the area of a circle whose radius measures $\frac{1}{2}$ foot. Use $\pi = \frac{22}{7}$.

8. The round bottom of a lampshade has a diameter of 12.6 inches. To the nearest inch, how much trim is needed around the lampshade? Use $\pi = 3.14$.

9. Roman wants to pave a circular patio. If the distance across the patio is 14 feet, how large is the surface that he needs to cover? Use $\pi = \frac{22}{7}$.

10. To the nearest inch, how much trim will Roman need (problem 9) to surround the patio? Use $\pi = \frac{22}{7}$.

Answers and solutions start on page 293.

MORE PERIMETER AND AREA PROBLEMS

On the GED, you may have to find the perimeter or area of figures other than rectangles, squares, triangles, or circles. In such problems, you may be given the formula for the area.

Following is a description of some other geometric figures.

A **parallelogram** is a four-sided figure with two pairs of parallel sides. The sides across from each other are equal in length. The figure at the right is a parallelogram. A parallelogram is similar to a rectangle, except that the angles in a parallelogram are not right angles.

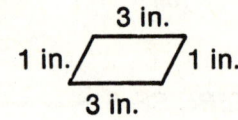

A **rhombus** is a four-sided figure with two pairs of parallel sides. All four sides of a rhombus are equal. The figure at the right is a rhombus. A rhombus is similar to a square, except that the angles in a rhombus are not right angles.

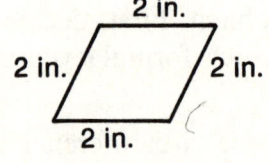

A **trapezoid** is a four-sided figure with one pair of parallel sides. The figure at the right is a trapezoid. In this picture, the top and bottom are parallel. The left side and the right side are not parallel.

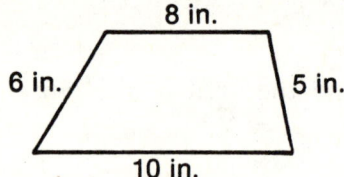

To find the perimeter of any of these figures, simply add the sides.

The formulas for the areas of these figures use the base and the height. The height is always indicated by a line perpendicular to the base.

EXAMPLE: The formula for the area of a trapezoid is $A = \frac{1}{2}h(b_1 + b_2)$ where h is the height and b_1 and b_2 are the two parallel sides. Find the area of the trapezoid shown at the right.

Solution: Replace h with 10, b_1 with 13, and b_2 with 17 in the formula $A = \frac{1}{2}h(b_1 + b_2)$. Remember to do the operation inside the parentheses first.

$A = \frac{1}{2} h(b_1 + b_2)$

$A = \frac{1}{2} \times 10(13 + 17)$

$A = \frac{1}{2} \times \frac{10}{1} \times \frac{30}{1}$

$A = \textbf{150 sq. in.}$

Notice that the area problems that follow require substitution.

GEOMETRY EXERCISE 6 ─────────────

Solve each problem.

1. The formula for the area of a parallelogram is $A = bh$ where b is the base and h is the height. Find the area of the parallelogram at the right.

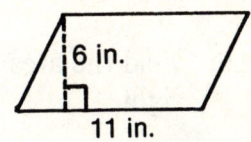

2. Find the perimeter for the parallelogram at the right.

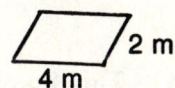

3. Find the area of the trapezoid at the right. Use the formula for the area of a trapezoid, $A = \frac{1}{2}h(b_1 + b_2)$, where h is the height and b_1 and b_2 are the two parallel sides.

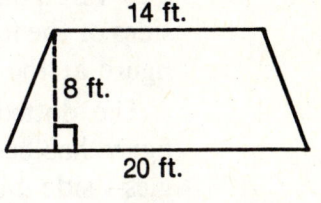

4. Find the perimeter of the trapezoid at the right.

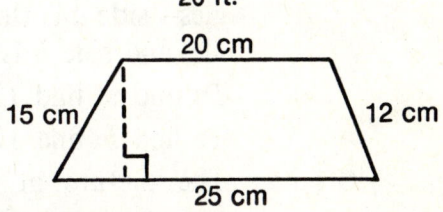

5. The formula for the surface area of a sphere is $A = 4\pi r^2$ where r is the radius. Find the surface area of a sphere with a radius of 7 inches. Use $\pi = \frac{22}{7}$.

6. The formula for the area of a rhombus is $A = bh$ where b is the base and h is the height. Find the area of a rhombus with a base of 6.4 meters and a height of 3.8 meters.

7. Find the perimeter of a rhombus with a side of $2\frac{1}{2}$ inches.

8. The formula for the area of a trapezoid can also be written $A = h \left(\frac{b_1 + b_2}{2}\right)$ where b_1 and b_2 are the parallel sides and h is the height. Find the area of a trapezoid with a height of 10 centimeters and parallel sides of 5.3 centimeters and 8.7 centimeters, respectively.

9. The formula for the area of a parallelogram is $A = bh$ where b is the base and h is the height. Find the area of the parallelogram at the right.

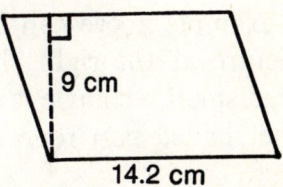

Answers and solutions start on page 293.

TWO-STEP AREA PROBLEMS

You may be asked to find the area of more complicated figures. Often, these figures are made up of two simple figures.

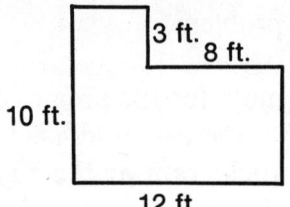

EXAMPLE: Find the area of the figure shown at the right.

Step 1.   This figure is composed of two rectangles. First, find the measurements of the sides of the figure labeled *a* and *b* on the figure at the right.

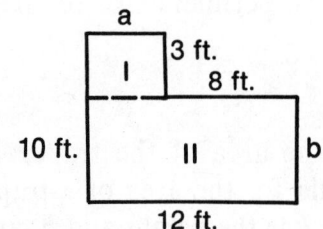

The dotted line indicates where the figure has been broken into two rectangles—side *a* is the length of one rectangle, and side *b* is the width of another.

To find *a*, find 12' − 8' = 4'.

To find *b*, find 10' − 3' = 7'.

Step 2.   The picture at the right shows the measurements of all the sides. Now you can find the area of each rectangle.

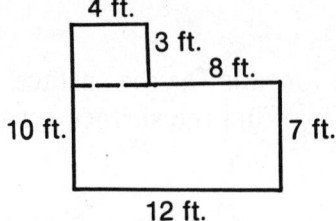

For the top rectangle, replace *l* with 4' and *w* with 3' in the formula A = *lw*. The area of the top rectangle is: A = 4' × 3' = 12 sq. ft.

For the bottom rectangle, replace *l* with 12' and *w* with 7' in the formula A = *lw*. The area of the bottom rectangle is: A = 12' × 7' = 84 sq. ft.

Step 3.   Add the area of the two rectangles. The total area is:

$$\begin{array}{r} 12 \text{ sq. ft.} \\ +\underline{84} \text{ sq. ft.} \\ \mathbf{96} \textbf{ sq. ft.} \end{array}$$

The example above can be solved more than one way. The picture at the right shows the same figure. In this picture, a small rectangle is cut away from a larger rectangle. The shaded part represents the rectangle that is cut away.

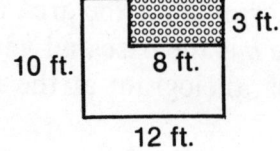

The large rectangle is 12 ft. long and 10 ft. wide. The area is A = 12 ft. × 10 ft. = 120 sq. ft.

The small rectangle is 8 ft. long and 3 ft. wide. The area is A = 8 ft. × 3 ft. = 24 sq. ft.

To find the area of the figure, subtract the area of the small rectangle from the area of the large one:

$$\begin{array}{r} 120 \text{ sq. ft.} \\ - \ 24 \text{ sq. ft.} \\ \hline \mathbf{96 \text{ sq. ft.}} \end{array}$$

GEOMETRY EXERCISE 7

Solve each problem.

1. Find the area of the figure shown at the right.

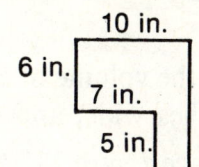

2. Find the area of the figure shown at the right.

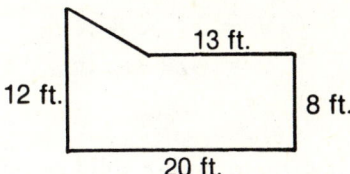

3. Find the area of the shaded part of the figure shown at the right. Use $\pi = \frac{22}{7}$.

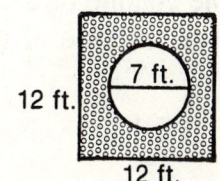

4. The diagram at the right shows the measurements of the floor of Mr. and Mrs. Porter's living room and dining room. How many square yards of carpet do they need to cover the floors of both rooms?

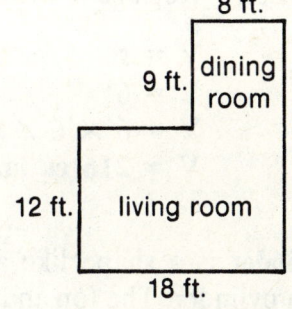

5. The shaded part of the diagram at the right shows the walk around the pool at the Centerville Community Center. Find the area of the walk.

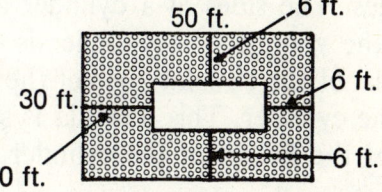

Answers and solutions start on page 294.

VOLUME

Volume is a measure of the amount of space inside a three-dimensional figure, called a **solid figure.** The rectangular solid, the cube, and the cylinder are three of the most familiar solid figures. Generally, volume is found by multiplying together length, width, and height. Volume is measured in cubic units such as cubic inches, cubic meters, etc.

A cardboard carton is in the shape of a **rectangular solid.** The corners of a rectangular solid are all right angles. The formula for the volume of a rectangular solid is V = *lwh* where *l* is the length, *w* is the width, and *h* is the height. The figure at the right is a rectangular solid.

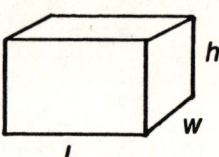

EXAMPLE 1:   Find the volume of a rectangular solid that is 11 inches long, 8 inches wide, and 6 inches high.

Solution: Replace *l* with 11″, *w* with 8″, and *h* with 6″ in the formula V = *lwh*.

V = *lwh*
V = 11 × 8 × 6
V = 528 cu. in.

A **cube** is a rectangular solid on which every side has the same measurement. The figure at the right is a cube. The formula for the volume of a cube is V = s^3 where *s* is the measurement of one side of the cube.

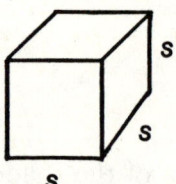

EXAMPLE 2:   Find the volume of the cube shown at the right.

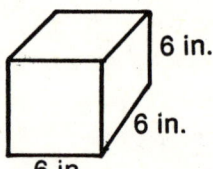

Solution: Replace *s* with 6 inches in the formula V = s^3.

V = s^3
V = 6^3
V = 6 × 6 × 6
V = 216 cu in.

A **cylinder** is a shape like a tin can. The figure at the right is a cylinder. The top and the bottom of a cylinder are circles. The sides of a cylinder are straight. The formula for the volume of a cylinder is V = $\pi r^2 h$ where *r* is the radius of the circular base of the figure and *h* is the height of the cylinder. This formula is simply the area of a circle times the height of the cylinder.

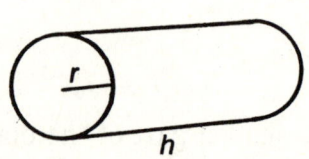

EXAMPLE 3:   Find the volume of a cylinder whose top has a radius of 7 inches and whose height measures 20 inches. Use $\pi = \frac{22}{7}$.

Solution: Replace π with $\frac{22}{7}$, *r* with 7 inches, and h with 20 inches in the formula V = $\pi r^2 h$

$$V = \pi r^2 h$$
$$V = \tfrac{22}{7} \times 7^2 \times 20$$
$$V = \tfrac{22}{7} \times \tfrac{\overset{7}{\cancel{49}}}{1} \times \tfrac{20}{1}$$
$$V = 3{,}080 \text{ cu. in.}$$

Memorize the formulas for the volume of a rectangular solid, a cube, and a cylinder before you go on.

You may not see the word volume in a volume word problem.

EXAMPLE 4: A pool is 30 feet long, 10 feet wide, and 6 feet deep. How many cubic feet of water does the pool hold?

Solution: The words cubic feet tell you to find the volume. Only volume is measured in cubic units. Replace l with 30', w with 10', and h with 6' in the formula $V = lwh$.

$$V = lwh$$
$$V = 30' \times 10' \times 6'$$
$$V = 1{,}800 \text{ cu. ft.}$$

A *sphere* is a shape like a basketball or baseball. The radius of a sphere is the distance from the center to the outside. In problems about spheres, you will be given the formula you need.

GEOMETRY EXERCISE 8

Solve each problem. Give each answer the correct label.

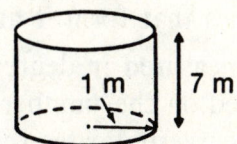

1. Find the volume of the figure shown at the right.

2. Find the volume of a cube that measures 8 inches on one edge.

3. Find the volume of the figure shown at the right. Use the formula $V = \pi r^2 h$. Let $\pi = \tfrac{22}{7}$.

4. Find the volume of a rectangular box that is $5\tfrac{1}{4}$ inches long, $3\tfrac{1}{2}$ inches wide, and $2\tfrac{2}{3}$ inches high.

5. A cube has one side that measures 1.2 centimeters. Find the volume of the cube.

6. How much concrete must be poured to make a slab that is 30 feet long, 5 feet wide, and $\frac{1}{2}$ foot high?

7. A cylindrical water tank has a radius of 9 feet and a height of 35 feet. Find its volume. The formula for the volume of a cylinder is V = $\pi r^2 h$. Let $\pi = \frac{22}{7}$.

8. The formula for the volume of a sphere is V = $\frac{4}{3}\pi r^3$, where r is the radius of the sphere. Find the volume of a sphere with a radius of 3 inches. Use $\pi = \frac{22}{7}$.

9. A rectangular box has dimensions of 6 inches by 6 inches by 10 inches. What would be the diameter of the largest ball that could fit in the box?

10. For the construction of a new building, a hole was dug measuring 30 yards long, 20 yards wide, and 4 yards deep. The trucks used to carry away the dirt can each hold 30 cubic yards. How many truckloads were needed to carry away all the dirt from the hole?

Answers and solutions start on page 295.

ANGLES

You already know that an angle is formed by two lines extending from the same point. The point is called the **vertex** of the angle. The symbol $\angle$ stands for the word "angle."

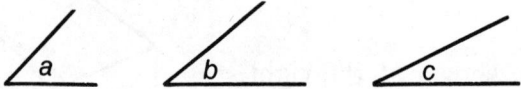

$\angle a$ and $\angle b$ above are about the same size because the lines that form them are "open" about the same amount. The length of the lines has nothing to do with the size of the angles. $\angle c$ is smaller than the other two angles because the lines that form it are "closed" more.

Angles are measured in degrees (°). Following is a list of angles whose names are based on the number of degrees in each angle. In the illustrations, a small curve indicates the opening of the angle.

An **acute angle** has less than 90°.

Examples:

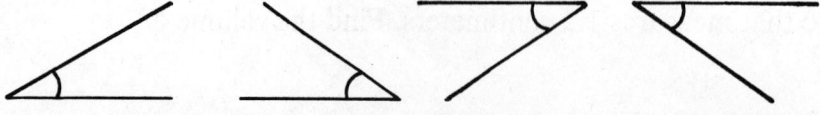

A **right angle** has exactly 90°. A right angle is often shown with a small square at the vertex.

Examples:

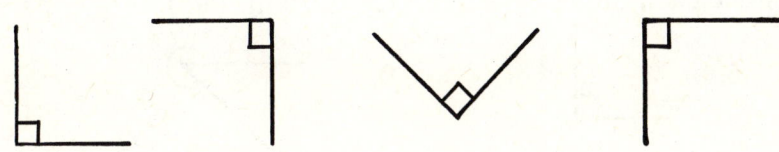

An **obtuse angle** has more than 90° but less than 180°.

Examples:

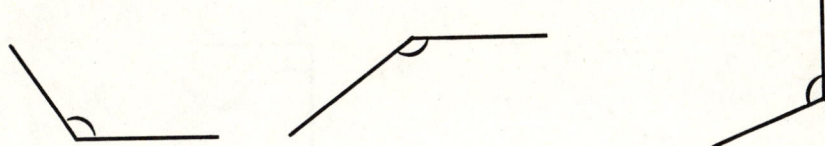

A **straight** angle has exactly 180°.

Examples:

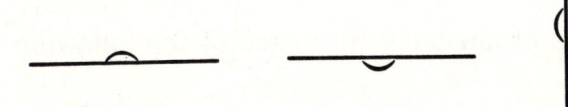

A **reflex angle** has more than 180° but less than 360°.

Examples:

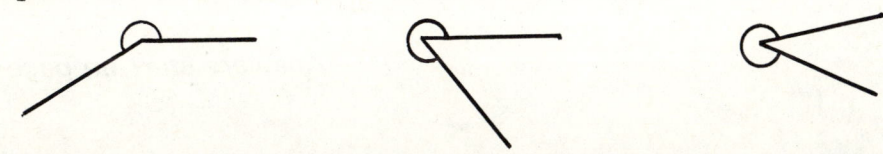

360° is the angular measurement of a complete circle.

Notice that the right angle and the straight angle contain exact numbers of degrees. For the other angles, a range of degrees is possible. For example, an acute angle could be 2° or 89° since both of these angles are more than 0° and less than 90°.

Notice also that it is important to have some indication of the opening.

For this angle   , the "inside"   is an acute angle, and

the "outside"   is a reflex angle.

Memorize the names of the angles before you try the next exercise.

GEOMETRY EXERCISE 9 ━━━━━━━━━━━━━━━━━━━━━

For items 1 to 12, identify each angle by name (acute, right, etc.).

1.

5.

9.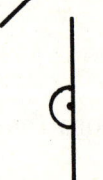

2.

6.

10.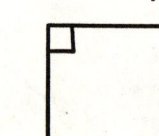

3.

7.

11.

4.

8.

12.

For items 13 to 21, tell what kind of angle contains each of the following numbers of degrees.

13. 40°_____   **16.** 220°_____   **19.** 175°_____

14. 100°_____   **17.** 90°_____   **20.** 300°_____

15. 180°_____   **18.** 83°_____   **21.** 190°_____

Answers start on page 295.

ANGLE RELATIONSHIPS

Angles are sometimes referred to with letters. For example ∠AOB refers to the straight angle shown below. The angle could also be called ∠BOA. In both cases, the vertex O is in the middle.

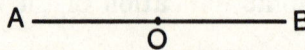

Another line extending from point O forms two new angles. ∠BOC is acute, and ∠AOC is obtuse.

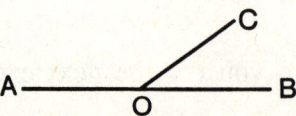

Suppose ∠BOC measures 50°. To find the measurement of ∠AOC take 50° from 180° since the straight angle ∠AOB measures 180°.

$$∠AOC = 180° - 50° = 130°$$

∠AOC and ∠BOC are called supplementary angles.

Supplementary angles are two angles whose sum is 180°.

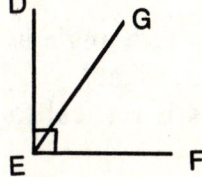

In the picture at the right, ∠DEF is a right angle. Suppose ∠GEF = 35°. To find ∠DEG, take 35° from 90°.

$$∠DEG = 90° - 35° = 55°$$

∠DEG and ∠GEF are called complementary angles.

Complementary angles are two angles whose sum is 90°.

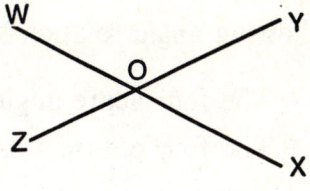

In the picture at the right, two lines **intersect** (cross) at point O. Suppose ∠WOZ measures 70°. ∠WOY must measure 180° − 70° = 110° because ∠WOY and ∠WOZ are supplementary.

∠YOX measures 70° because ∠WOY and ∠YOX are supplementary.

∠XOZ measures 110° because ∠YOX and ∠XOZ are supplementary.

We have the following angle measurements.

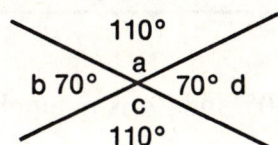

In the picture above, ∠a and ∠c, which both measure 110°, are called vertical angles. ∠b and ∠d which both measure 70° are also called vertical angles.

Vertical angles are the two angles opposite each other when two straight lines intersect. Vertical angles are equal.

The following examples illustrate problems using supplementary, complementary, and vertical angles.

EXAMPLE 1: Find the supplement of an angle measuring 65°.

Solution: Take 65° from 180°. The supplement of a 65° angle measures **115°**.

$$\begin{array}{r} 180° \\ - 65° \\ \hline \mathbf{115°} \end{array}$$

EXAMPLE 2:   Find ∠x in the figure at the right. ∠AOB = 90°.
　　Solution: Take 30° from 90°. ∠x measures **60°**.

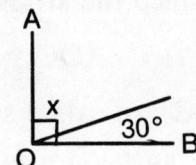

$$\begin{array}{r} 90° \\ -30° \\ \hline 60° \end{array}$$

EXAMPLE 3:   Which angle is vertical to ∠u in the picture at
　　　　　　　the right?
　　Solution: ∠s is vertical to ∠u because ∠s is opposite ∠u.

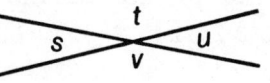

As you saw earlier, parallel lines are lines that run side by side and never cross. Sometimes, parallel lines are crossed by a third line, called a **transversal.** When two paralled lines are crossed by a transversal, a set of four angles is created at the points of intersection.

The following angle relationships are created:

- The four acute angles are equal.
- The four obtuse angles are equal.
- Each acute angle is supplementary to each obtuse angle.

EXAMPLE 4:   What is the value of ∠a?

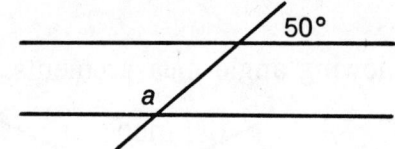

　　Solution: The acute angle is equal to 50° and ∠a is a supplementary
　　　　　　　obtuse angle.
　　　　　　　∠a = 180° − 50° = **130°**

GEOMETRY EXERCISE 10

Solve each problem.

1. If an angle measures 48°, its complement measures how many degrees?

2. How many degrees are in the supplement of a 48° angle?

3. Find the complement of an angle measuring 25°.

4. Find the supplement of an angle measuring 63°.

Problems 5 through 8 refer to the following illustration.

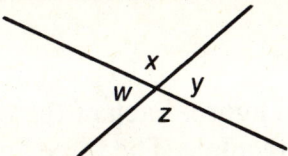

5. Which angle is vertical to ∠y?

6. What is the sum of ∠w, ∠x, ∠y, and ∠z?

7. If ∠w measures 75°, how many degrees are in ∠x?

8. If ∠x measures 119°, how many degrees are in ∠w?

9. In the picture at the right ∠a = ∠c. If ∠b = 80°, how many degrees are in ∠a?

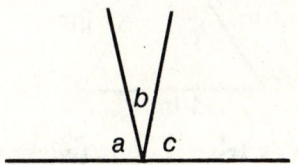

10. The supplement of an acute angle is always what kind of angle?

11. What is ∠m in the picture at the right?

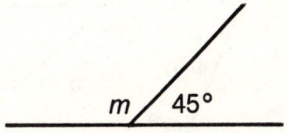

12. If an angle measures 37°, an angle vertical to it must measure how many degrees?

13. ∠XOZ in the picture at the right measures 110.5°. Find the number of degrees in ∠YOZ.

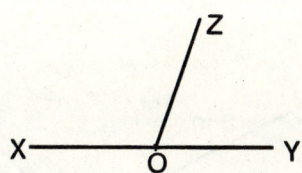

14. What is the value of ∠z at right?

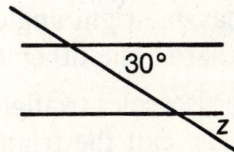

15. Find the value of ∠b.

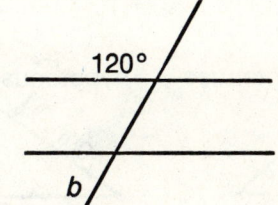

Answers and solutions start on page 295.

TRIANGLES

You already learned that a triangle is a three-sided figure. Each of the three points where the sides meet is called the vertex of an angle. The three angles inside any triangle add up to 180°.

The following are the most important types of triangles.

An **equilateral triangle** has three equal sides. The name equilateral means equal sides. It also has three equal angles. Each angle measures 180° ÷ 3 or 60°. An equilateral triangle can also be called **equiangular,** which means equal angles.

Examples:

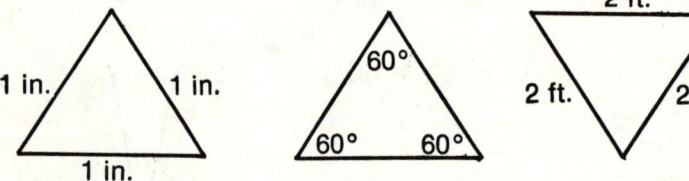

An **isosceles triangle** has two equal sides. The third side may be longer or shorter than the other two sides. An isosceles triangle also has two equal angles. The two equal angles are called the **base angles.** The angle with a different measurement is called the **vertex angle.**

Examples:

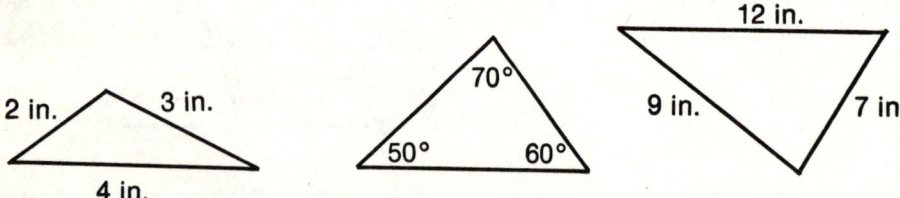

A **scalene triangle** has three sides with different measurements. It also has three angles of different sizes.

Examples:

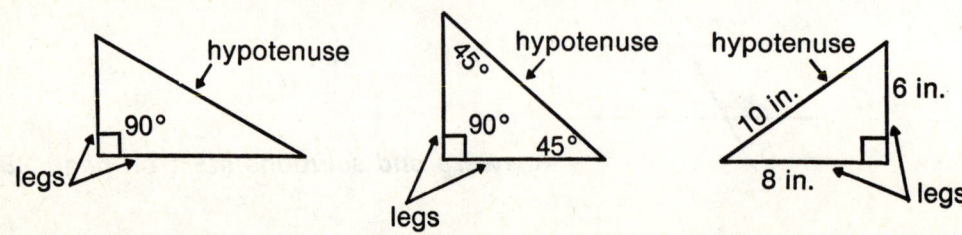

A **right triangle** has one right angle. The side across from the right angle is called the **hypotenuse.** The other two sides are called the **legs.**

Either a scalene or isosceles triangle can be a right triangle. If either kind contains a right angle, call the triangle a right triangle.

The symbol △ stands for the word "triangle." A side of a triangle is referred to by the two letters that form the end points of each side. For △ABC, side BC is the hypotenuse. Sides AB and AC are the legs of this right triangle.

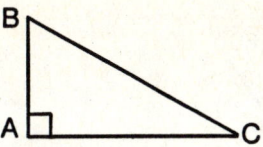

The angles in a triangle can be referred to by the letter that forms the vertex of the angle. For △ABC to the right, ∠A is the right angle. Sometimes, three letters are used to refer to an angle of a triangle. The vertex letter must be in the middle. For this triangle, ∠A is the same as ∠BAC and ∠CAB.

EXAMPLE 1: In triangle PQR, ∠P = 35° and ∠Q = 45°. Find the measurement of ∠R.

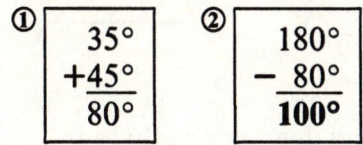

Step 1. Add the measurements of ∠P and ∠Q.

Step 2. Subtract the sum from 180°. The measurement of ∠R is **100°.**

EXAMPLE 2: In △STU, ∠S = 60° and ∠T = 30°. What kind of triangle is △STU?

①
$$\begin{array}{r} 60° \\ +30° \\ \hline 90° \end{array}$$
②
$$\begin{array}{r} 180° \\ -90° \\ \hline 90° \end{array}$$

Step 1. Add the measurements of ∠S and ∠T.

Step 2. Subtract the sum from 180°. The measurement of ∠U is 90°. A triangle with one right angle is a **right triangle.**

Memorize the triangle definitions in this section before you try the next exercise.

GEOMETRY EXERCISE 11 ─────────────────────

Solve each problem.

1. Identify each of the following triangles (right, isosceles, scalene or equilateral).

   a. 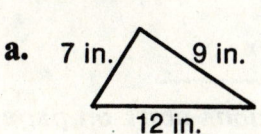   **b.** 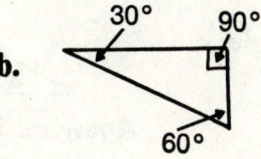   **c.**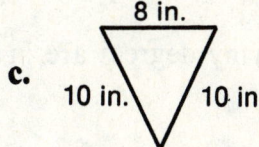

d.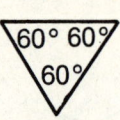

f.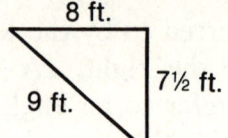

h.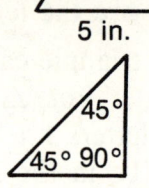

e.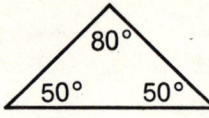

g.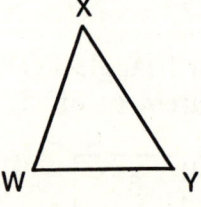

i.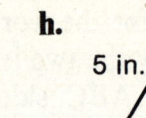

2. In the triangle at the right, ∠W = 65° and ∠X = 50°. Find the number of degrees in ∠Y.

3. What kind of triangle is △ WXY in problem 2?

4. In △ ABC, ∠A = 30° and ∠B = 60°. How many degrees are there in ∠C?

5. What kind of triangle is △ ABC in problem 4?

6. The vertex angle of an isosceles triangle measures 82°. How many degrees are there in each base angle?

7. Each base angle of an isosceles triangle measures 63°. How many degrees are there in the vertex angle?

8. In the triangle at the right, side DE measures 8 inches and side DF measures 5 inches. The perimeter of the triangle is 21 inches. What kind of triangle is △ DEF?

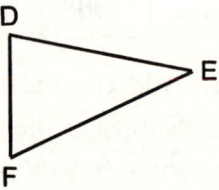

9. In triangle PQR, ∠P = 25° and ∠Q = 35°. What kind of triangle is △ PQR?

10. In triangle XYZ at the right, side XY is 4 inches long and side YZ is 5 inches long. The perimeter of the triangle is 16 inches. What kind of triangle is △ XYZ?

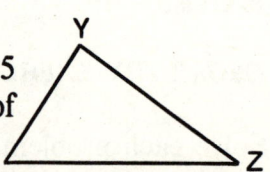

11. What is the sum of the two acute angles in a right triangle?

12. Find how many degrees are in angle *x*.

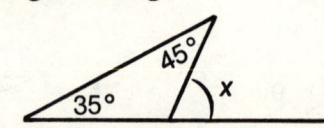

Answers and solutions start on page 296.

SIMILARITY

Similar figures have the same shape but different sizes. Corresponding sides of similar figures can be written as a proportion. The drawing below shows two similar rectangles.

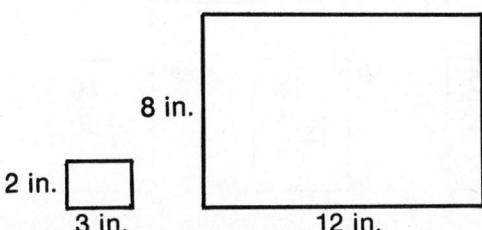

For the small rectangle above, the ratio of the width to the length is 2:3 and can be written $\frac{2}{3}$. The ratio of the width to the length of the larger rectangle is the same. These two figures are similar because the ratios of the widths to the lengths are proportional. $\frac{2}{3} = \frac{8}{12}$.

Two triangles are similar if the corresponding angles are equal and the sides are in proportion. Triangle ABC below is similar to triangle XYZ because each triangle has angles measuring 30°, 60°, and 90°.

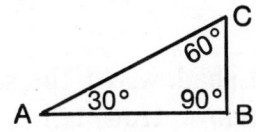

EXAMPLE 1: In triangle DEF, ∠D = 50° and ∠E = 75°. In triangle GHI, ∠G = 55° and ∠I = 50°. Are these two triangles similar?

①

$$\begin{array}{cc} 50° & 180° \\ +75° & -125° \\ \hline 125° & 55° = \angle F \end{array}$$

②

$$\begin{array}{cc} 55° & 180° \\ +50° & -105° \\ \hline 105° & 75° = \angle H \end{array}$$

Step 1. Find ∠F. Take the sum of ∠D and ∠E from 180°.

Step 2. Find ∠H. Take the sum of ∠G and ∠I from 180°. Each triangle has angles measuring 50°, 55°, and 75°. In △ DEF, ∠D = 50°, ∠E = 75°, and ∠F = 55°; in △GHI, ∠G = 55°, ∠H = 75°, and ∠I = 50. **Both triangles have the same angles and therefore are similar.**

Remember that corresponding sides of similar figures are in proportion.

EXAMPLE 2:   In triangle ABC below, side BC measures 9 feet and side AC measures 18 feet. In triangle GHI, side HI measures 5 feet. Find the length of side GH.

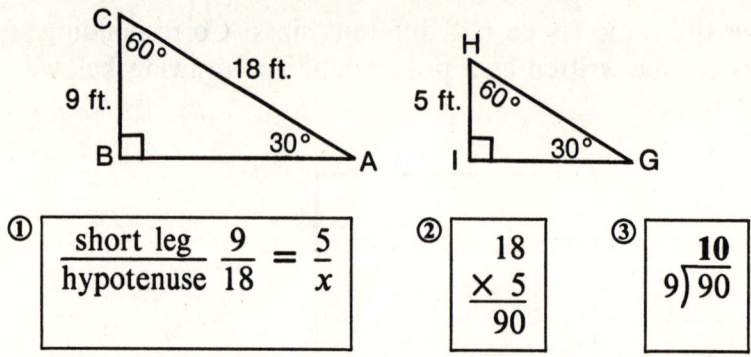

① $\dfrac{\text{short leg}}{\text{hypotenuse}}\ \dfrac{9}{18} = \dfrac{5}{x}$

② $\begin{array}{r} 18 \\ \times\ 5 \\ \hline 90 \end{array}$

③ $9\overline{)90}$ = 10

Step 1.   First notice that these triangles are similar. Each has angles measuring 30°, 60°, and 90°.

Set up a proportion with the short leg and the hypotenuse for each triangle. The short leg of triangle ABC is 9 feet, and the hypotenuse is 18 feet. The short leg of triangle GHI is 5 feet, and the hypotenuse is unknown, or x.

Step 2.   Cross multiply the numbers that are diagonal to each other (5 and 18).

Step 3.   Then divide by the other number in the proportion, 9. The length of side GH is **10 feet.**

EXAMPLE 3:   A vertical yardstick casts a two-foot shadow. At the same time, a building casts a 48-foot shadow. How tall is the building?

① $\dfrac{\text{height}}{\text{shadow}}\ \dfrac{3}{2} = \dfrac{x}{48}$

② $\begin{array}{r} 48 \\ \times\ 3 \\ \hline 144 \end{array}$

③ $2\overline{)144}$ = 72

Step 1.   First notice that this problem has two similar triangles. Drawing a picture as you read the problem may help you to see this. One leg of each triangle is the height of the object. The other leg is the length of the shadow. The hypotenuse of each triangle is the line from the top of the object to the end of the shadow. The picture below shows the triangles. Set up a proportion. Let x = the unknown, the height of the building. Use 3 as the height. There are 3 feet in a yard. Since the shadow is given in feet, you should use the same unit of measurement.

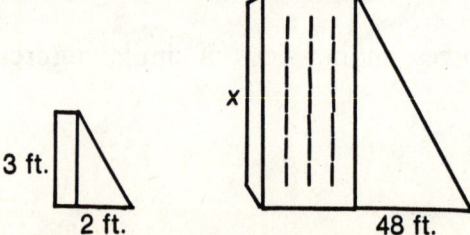

3 ft.   2 ft. x 48 ft.

Step 2. Cross multiply the numbers that are diagonal to each other, 48 and 3.

Step 3. Divide by the remaining number, 2. The building is **72 feet tall.**

GEOMETRY EXERCISE 12 ────────────

Solve each problem.

1. In triangle MNO, ∠M = 45° and ∠N = 85°. In triangle PQR, ∠P = 50° and ∠Q = 45°. Are these two triangles similar?

2. In triangle ABC, ∠A = 60° and ∠B = 50°. In triangle DEF, ∠D = 50° and ∠E = 80°. Are these two triangles similar?

3. Are the triangles shown at the right similar?

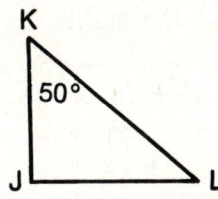 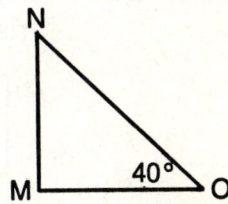

4. In triangle JKL above, side JK measures 8 inches and side JL measures 12 inches. In triangle MNO, side MN measures 14 inches. Find the length of side MO.

5. A six-foot-tall vertical post casts a five-foot shadow. At the same time, a tree casts a 65-foot shadow. How tall is the tree?

6. In the picture at the right ∠S and ∠W are both right angles. Are the two triangles similar?

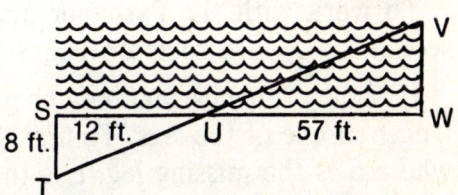

7. In the picture with problem 6, VW is the distance across a river. Find the measurement of VW.

8. In the figure at the right, ∠CDE and ∠CAB are both right angles. Are triangles CDE and CAB similar?

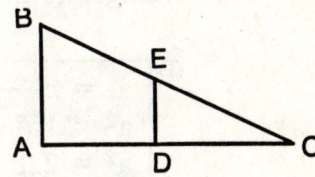

9. In problem 8, CD = 15 inches, AD = 10 inches, and DE = 6 inches. Find AB.

Answers and solutions start on page 296.

PYTHAGOREAN THEOREM

A Greek mathematician named Pythagoras discovered a special relationship among the sides of a right triangle. We call his discovery the **Pythagorean Theorem.**

In words, the Pythagorean Theorem says: The square of the hypotenuse of a right triangle is equal to the sum of the squares of the other two sides. This theorem gives the basis for finding either the leg of a right triangle or the hypotenuse if the other two sides are given.

The formula for the theorem is $c^2 = a^2 + b^2$ where c is the hypotenuse and *a* and *b* are the legs of a right triangle.

EXAMPLE 1:   Find the length of the hypotenuse in the triangle at the right.

①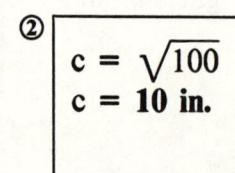

②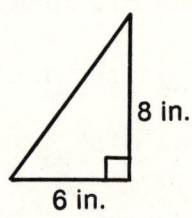

Step 1.   Replace *a* with 6 and *b* with 8 in the formula $c^2 = a^2 + b^2$. Find the value of c^2.

Step 2.   The formula gives the value of the hypotenuse squared. To get the length of the hypotenuse, find the square root of 100.

To work with the Pythagorean Theorem, you may need to review the section on square roots on pages 236 to 239.

In some problems, you may be given the length of the hypotenuse and the length of one of the legs. To find the other leg use the formula $a^2 = c^2 - b^2$ where a is the missing leg, c is the length of the hypotenuse, and b is the length of the leg given in the problem.

EXAMPLE 2:   Find the length of the missing leg in the triangle at the right.

①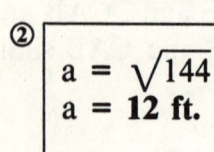

②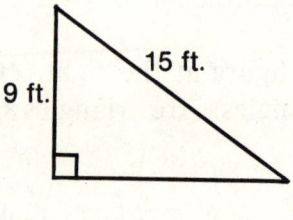

Step 1.   Replace c with 15 and b with 9 in the formula $a^2 = c^2 - b^2$.

Step 2.   Find the square root of 144.

In some problems, you will have to recognize that a figure, pictured or described, is a right triangle and that you are being asked to solve for a hypotenuse or a leg.

EXAMPLE 3: A boat sails 15 miles east of port and then 20 miles south. How far is the boat from the port?

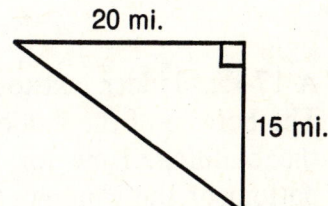
20 mi.
15 mi.

Step 1: See that the distance of the boat from the port can be found by finding the length of the hypotenuse of a right triangle. Making a drawing may help you to see how a problem can be solved.

Step 2: Substitute the values given for the legs of the triangle in the formula $c^2 = a^2 + b^2$.

$$c^2 = 20^2 + 15^2$$
$$c^2 = 400 + 225$$
$$c^2 = 625$$
$$c = \sqrt{625}$$
$$c = 25$$

The boat is **25 miles** from the port.

GEOMETRY EXERCISE 13

Solve each problem.

1. Find the hypotenuse of the triangle at the right.

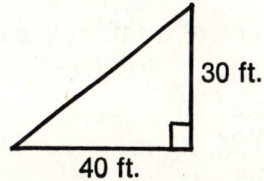
30 ft.
40 ft.

2. One leg of a right triangle measures 10 inches. The other measures 24 inches. Find the length of the hypotenuse.

3. In triangle XYZ at the right, side XY = 12 inches and YZ = 5 inches. Find XZ.

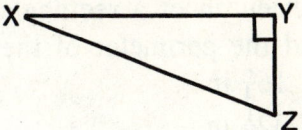

4. What is the length of the hypotenuse of a right triangle whose legs measure 12 yards and 16 yards?

5. Ralph drove 20 miles west from Eastport to Middletown and then 15 miles south to Southport. He returned to Eastport on Route 3. How far is it from Southport to Eastport on Route 3?

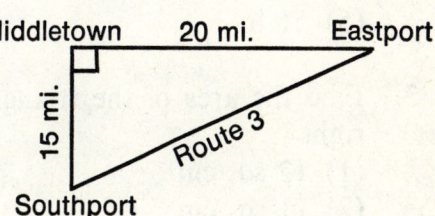
Middletown   20 mi.   Eastport
15 mi.
Route 3
Southport

6. One leg of a right triangle measures 18 yards. The hypotenuse measures 30 yards. Find the length of the other leg.

7. In triangle ABC at the right, AB = 16 ft. and AC = 34 ft. Find the length of BC.

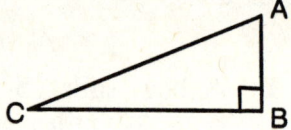

8. A 17-foot ladder just touches the bottom of a window. The bottom of the ladder is 8 feet from the bottom of the building. How far is it from the ground to the bottom of the window?

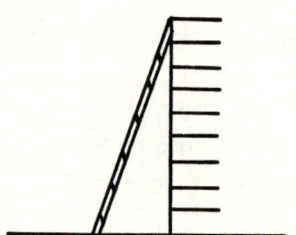

9. The length of the rectangle at the right is 2.4 meters and the width is 1.8 meters. Find the measurement of the diagonal distance SU.

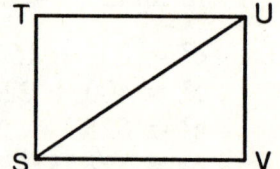

Answers and solutions start on page 297.

GEOMETRY REVIEW

For each problem fill in the circle that corresponds to the best answer.

1. Each side of a square measures 3.2 meters. Find the perimeter of the square.
   (1) 10.24 m
   (2) 10.24 sq. m
   (3) 12.8 m
   (4) 3.2 sq. m
   (5) 9.6 m

   1 ① ② ③ ④ ⑤

2. The length of a rectangle is 12 inches. The width is $4\frac{1}{4}$ inches. Find the perimeter of the rectangle.
   (1) $16\frac{1}{4}$ in.
   (2) $32\frac{1}{2}$ in.
   (3) $32\frac{1}{4}$ in.
   (4) 50 in.
   (5) 51 in.

   2 ① ② ③ ④ ⑤

3. Find the area of the triangle shown at the right.
   (1) 12 sq. cm
   (2) 11 sq. cm
   (3) 9.5 sq. cm
   (4) 6 sq. cm
   (5) 3 sq. cm

   3 ① ② ③ ④ ⑤

   1.5 cm ⌐‾‾‾‾‾‾‾‾
   8 cm

4. The length of a rectangle is 120 yards. The width is 60 yards. Find the area of the rectangle.
   (1) 72,000 sq. yds.
   (2) 7,200 sq. yds.
   (3) 720 sq. yds.
   (4) 360 sq. yds.
   (5) 180 sq. yds.

4 ① ② ③ ④ ⑤

5. The hall in Phoebe's apartment is 3 feet wide and 18 feet long. She wants to put carpet on the hall floor. At $16 a square yard, how much will the carpet for the hall cost?
   (1) $864
   (2) $96
   (3) $108
   (4) $54
   (5) $72

5 ① ② ③ ④ ⑤

6. Sharon wants to put fencing around the vegetable garden in her yards. The garden is 20 feet long and 8 feet wide. She will leave a four-foot opening for a walkway into the garden. How many feet of fencing does she need?
   (1) 160 ft.
   (2) 28 ft.
   (3) 52 ft.
   (4) 56 ft.
   (5) 60 ft.

6 ① ② ③ ④ ⑤

7. Find the circumference of a circle with a diameter of 30 feet. Use $\pi = 3.14$.
   (1) 706.5 ft.
   (2) 120 ft.
   (3) 94.2 ft.
   (4) 47.1 ft.
   (5) 30 ft.

7 ① ② ③ ④ ⑤

8. What is the area of a circle with a radius of 35 inches? Use $\pi = \frac{22}{7}$.
   (1) 3,850 sq. in.
   (2) 3,850 in.
   (3) 220 sq. in.
   (4) 220 in.
   (5) 1,540 sq. in.

8 ① ② ③ ④ ⑤

9. Find the area of the figure shown at the right.
   (1) 130 sq. in.
   (2) 120 sq. in.
   (3) 100 sq. in.
   (4) 98 sq. in.
   (5) 82 sq. in.

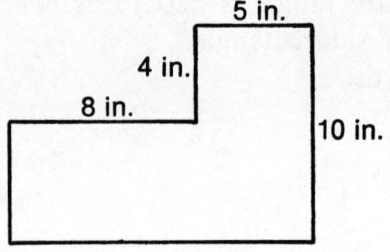

9 ① ② ③ ④ ⑤

10. Mr. Porter is building a garage. The garage floor will be made of poured concrete. It will be 22 feet long, 18 feet wide, and $\frac{1}{2}$ foot deep. How many cubic feet of concrete does Mr. Porter need to make the floor?
    (1) 396 cu. ft.
    (2) 198 cu. ft.
    (3) 99 cu. ft.
    (4) 149 cu. ft.
    (5) 792 cu. ft.

10 ① ② ③ ④ ⑤

11. What kind of angle is ∠AOB in the picture at the right?
    (1) acute
    (2) right
    (3) obtuse
    (4) straight
    (5) reflex

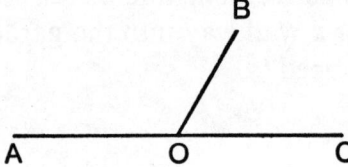

11 ① ② ③ ④ ⑤

12. What kind of angle is ∠BOC in the picture above?
    (1) acute
    (2) right
    (3) obtuse
    (4) straight
    (5) reflex

12 ① ② ③ ④ ⑤

13. How many degrees are in the complement of a 73° angle?
    (1) 117°
    (2) 107°
    (3) 27°
    (4) 17°
    (5) 7°

13 ① ② ③ ④ ⑤

14. In the picture at the right ∠a = 65°. How many degrees are there in ∠d?
    (1) 15°
    (2) 25°
    (3) 35°
    (4) 105°
    (5) 115°

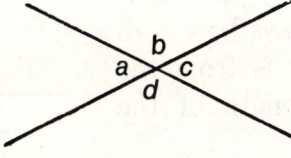

15. In triangle XYZ, ∠X = 33° and ∠Y = 57°. What kind of triangle is △XYZ?
    (1) equilateral
    (2) isosceles
    (3) obtuse
    (4) right
    (5) none of these

16. The vertex angle of an isosceles triangle measures 102°. How many degrees are there in each base angle?
    (1) 78
    (2) 51
    (3) 39
    (4) 60
    (5) 64

17. Find the length of CD in the figure at the right.
    (1) 7.2 ft.
    (2) 3.6 ft.
    (3) 2.5 ft.
    (4) 1.8 ft.
    (5) .9 ft.

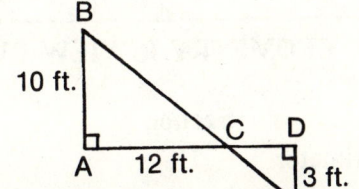

18. Two trees stand side by side. At a certain time, the 30-foot high tree casts a shadow 40-feet long. How long is the shadow of the other tree that stands 45 feet tall?
    (1) 30 ft.
    (2) 40 ft.
    (3) 45 ft.
    (4) 60 ft.
    (5) 1,800 ft.

19. A ladder is placed against a wall. The picture at the right shows the distance from the foot of the ladder to the wall. It also shows how high the top of the ladder is from the floor. What is the length of the ladder?

    (1) 10 ft.
    (2) 14 ft.
    (3) 18 ft.
    (4) 100 ft.
    (5) 110 ft.

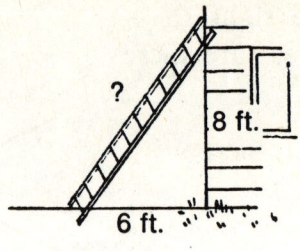

19 ① ② ③ ④ ⑤

20. In the rectangle at the right, side BC measures 36 inches and diagonal line AC measures 39 inches. Find side AB.

    (1) 15 in.
    (2) 18 in.
    (3) 30 in.
    (4) 33 in.
    (5) 42 in.

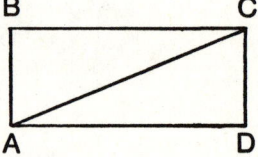

20 ① ② ③ ④ ⑤

Answers and solutions start on page 298.

GEOMETRY REVIEW EVALUATION

Problem	Section	Starting Page
1, 2	Perimeter	256
3, 4	Area	259
5, 6	Perimeter and Area Word Problems	262
7, 8	Circles	264
9	More Area Problems	266
10	Volume	269
11, 12	Angles	272
13, 14	Angle Relationships	274
15, 16	Triangles	278
17, 18	Similarity	281
19, 20	Pythagorean Theorem	284

Passing score: __15__ right out of 20 problems.
Your score: _____ right out of 20 problems.

If you had less than a passing score, review the sections for the problems you got wrong. Then repeat this test before you go on to the next section.

If you had a passing score, correct any problem you got wrong. Then go on to algebra.

ANSWERS AND SOLUTIONS

Geometry Exercise 1

1. parallel
2. square
3. perpendicular
4. angle
5. right angle
6. rectangle
7. triangle
8. vertical

Geometry Exercise 2

1. 178 ft. $P = 2l + 2w$
$P = 2(52) + 2(37)$
$P = 104 + 74 = 178$

2. 57 cm $P = s_1 + s_2 + s_3$
$P = 19 + 19 + 19 = 57$

3. 2 in. $P = 4s$
$P = 4(\frac{1}{2}) = 2$

4. 72 km $P = 2l + 2w$
$P = 2(22) + 2(14)$
$P = 44 + 28$
$P = 72$

5. 32 yd. $P = 4s$
$P = 4(8) = 32$

6. $27\frac{1}{2}$ in. $P = s_1 + s_2 + s_3$
$7\frac{3}{4} = 7\frac{3}{4}$ in.
$9\frac{1}{4} = 9\frac{1}{4}$
$+10\frac{1}{2} = 10\frac{2}{4}$
$26\frac{6}{4} = 27\frac{2}{4} = 27\frac{1}{2}$

7. 23 mi. $P = 2l + 2w$
$P = 2(6.3) + 2(5.2)$
$P = 12.6 + 10.4 = 23$

8. 7 m $P = 4s$
$P = 4(1.75) = 7$

9. 35 ft. $P = s_1 + s_2 + s_3$
$P = 13 + 13 + 9 = 35$

10. 25 in.
You have the perimeter. You want one side. Divide the perimeter by the number of sides, 4.
$$4\overline{)100}\quad 25$$

11. 9 yd.
You have the perimeter and the length. You want the width.
Step 1. Multiply the length by 2.
$2l = 2(16) = 32$
Step 2. Subtract 32 yd. from the perimeter, 50 yd.
$50 - 32 = 18$
Step 3. Divide 18 yd. by 2.
$18 \div 2 = 9$

12. 12 ft.
You have the perimeter and two sides. You want to find the third side.
Step 1. Add the sides you have been given.
$5 + 13 = 18$
Step 2. Subtract the total from the perimeter, 30 ft.
$30 - 18 = 12$

Geometry Exercise 3

1. 180 sq. ft. $A = lw$
$A = 15 \times 12 = 180$

2. 196 sq. ft. $A = s^2$
$A = 14^2 = 14 \times 14 = 196$

3. 57 sq. in. $A = \frac{1}{2}bh$
$A = \frac{1}{2} \times \frac{19}{1} \times \frac{6}{1} = 57$

4. .0225 sq. m $A = s^2$
$A = (0.15)^2 = .15 \times .15 = .0225$
Notice that the answer has four decimal places.

5. $25\frac{1}{2}$ sq. in. $A = \frac{1}{2}bh$
$A = \frac{1}{2} \times 8\frac{1}{2} \times 6$
$A = \frac{1}{2} \times \frac{17}{2} \times \frac{6}{1} = \frac{51}{2} = 25\frac{1}{2}$

6. $13\frac{1}{2}$ sq. ft. $A = lw$
$A = 4\frac{1}{2} \times 3 = \frac{9}{2} \times \frac{3}{1} = \frac{27}{2}$
$= 13\frac{1}{2}$

Exercise 3 cont'd.

7. 54 sq. mi.

The shaded area is a triangle, $\frac{1}{2}$ the area of the rectangle.

$A = \frac{1}{2}bh$

$A = \frac{1}{\overset{}{\underset{1}{2}}} \times \frac{\overset{6}{\cancel{12}}}{1} \times \frac{9}{1} = 54$

8. $12\frac{1}{2}$ sq. in.　$A = \frac{1}{2}bh$

$A = \frac{1}{2} \times \frac{5}{1} \times \frac{5}{1} = \frac{25}{2} = 12\frac{1}{2}$

9. $\frac{9}{64}$ sq. in.　$A = s^2$

$A = (\frac{3}{8})^2 = \frac{3}{8} \times \frac{3}{8} = \frac{9}{64}$

10. $76\frac{1}{2}$ sq. cm　$A = \frac{1}{2}bh$

$A = \frac{1}{2} \times \frac{17}{1} \times \frac{9}{1} = \frac{153}{2} = 76\frac{1}{2}$

11. 12 in.

You have the area and the width. You want the length. Divide the area by the width.

$$5\overline{)60}\;^{12}$$

12. 10 in.

You have the area. You want one side. Find the square root of 100.

$\sqrt{100} = 10$

13. The square　$A = s^2$

$A = 8^2 = 64\, sq.\, in.$

$A = lw$
$A = 9 \times 7 = 63\, sq.\, in.$

Geometry Exercise 4

1. 108 sq. ft.

Find the area of the floor

$A = lw$
$A = 12 \times 9 = 108$

2. 78 m

Step 1.　Find the width of the garden.

$w = \frac{1}{2}l$

$\frac{1}{2} \times 26 = 13$

Step 2.　Find the perimeter.

$P = 2l + 2w$
$P = 2(26) + 2(13)$
$= 52 + 26 = 78\ m$

3. 26 ft.

Step 1.　Find the perimeter of each window.

$P = 2l + 2w$
$P = 2(42) + 2(36) = 84 + 72 = 156\ in.\ per\ window$

Step 2.　Multiply the perimeter by 2 for the two windows.

$2 \times 156 = 312\ in.\ for\ two\ windows$

Step 3.　Change the inches to feet. Divide by 12 or use the proportional method.

$12\overline{)312}\;^{26}$　or　$\dfrac{12\ in.}{1\ ft.} = \dfrac{312\ in.}{x}$

$12x = 312$
$x = 26$

4. 45 carpet tiles

Step 1.　Find the area of the room.

$A = lw$
$A = 18 \times 10 = 180$

Step 2.　Find the area of one carpet tile.

$A = s^2$
$A = 2 \times 2 = 4$

Step 3.　Divide the area of the room by the area of one carpet tile to find the smallest number of tiles needed.

$$4\overline{)180}\;^{45}$$

5. 5 gallons

Step 1.　Find the area of the floor.

$A = lw$
$A = 40 \times 25 = 1,000$

Step 2.　Divide the area of the floor by the number of square feet a gallon will cover.

$$200\overline{)1,000}\;^{5}$$

6. 36 in.

Find the perimeter of the picture.

$P = 2l + 2w$
$P = 2(10) + 2(8) = 20 + 16 = 36$

7. 80 sq. in.

Find the area of the picture.

$A = lw$
$A = 10 \times 8 = 80\ sq.\ in.$

8. 12 yd.
    Step 1. Find the perimeter of the yard.
$$P = s_1 + s_2 + s_3$$
$$P = 15 + 12 + 9 = 36 \text{ feet}$$
    Step 2. Change the perimeter to yards. Divide by 3 or use the proportional method.

$$3\overline{)36} \xrightarrow{} 12 \text{ yd.} \quad \text{or} \quad \frac{3 \text{ ft.}}{1 \text{ yd.}} = \frac{36 \text{ ft.}}{x}$$
$$3x = 36$$
$$x = 12$$

9. 54 sq. ft. $A = \frac{1}{2}bh$
$$A = \frac{1}{2} \times \frac{\overset{6}{\cancel{12}}}{1} \times \frac{9}{1} = 54$$

10. 16 sq. yd.
    Step 1. Find the area of each pair of curtains.
$$A = lw$$
$$A = 8 \times 6 = 48 \text{ sq. ft.}$$
    Step 2. Multiply the area of one pair of curtains by 3 for the three windows.
$$3 \times 48 = 144 \text{ sq. ft.}$$
    Step 3. Change square feet to square yards.
$$1 \text{ sq. yd.} = 3' \times 3' = 9 \text{ sq. ft.}$$
Divide 144 sq. ft. by 9 or use the proportional method.

$$9\overline{)144} \xrightarrow{} 16 \quad \text{or} \quad \frac{9 \text{ sq. ft.}}{1 \text{ sq. yd.}} = \frac{144 \text{ sq. ft.}}{x}$$
$$9x = 144$$
$$x = 16$$

Geometry Exercise 5

1. 31.4 yd. $C = \pi d$
$$C = 3.14 \times 10 = 31.40$$

2. 616 sq. m $A = \pi r^2$
$$A = \frac{22}{7} \times \frac{\overset{2}{\cancel{14}}}{1} \times \frac{14}{1} = 616$$

3. 78.5 sq. ft. $A = \pi r^2$
$$A = 3.14 \times 5 \times 5$$
$$A = 3.14 \times 25 = 78.5$$

4. 132 in. $C = \pi d$
$$C = \frac{22}{7} \times \frac{\overset{6}{\cancel{42}}}{1} = 132$$

5. 21 in.
The radius is $\frac{1}{2}$ the diameter.
$$\frac{1}{2} \times \frac{\overset{21}{\cancel{42}}}{1} = 21$$

6. 1,386 sq. in. $A = \pi r^2$
$$A = \frac{22}{7} \times \frac{\overset{3}{\cancel{21}}}{1} \times \frac{21}{1} = 1,386$$

7. $\frac{11}{14}$ sq. ft. $A = \pi r^2$
$$A = \frac{\overset{11}{\cancel{22}}}{7} \times \frac{1}{2} \times \frac{1}{2} = \frac{11}{14}$$

8. 40 in. $C = \pi d$
$$C = 3.14 \times 12.6 = 39.564 \text{ to the}$$
nearest inch = 40 in.

9. 154 sq. ft.
The radius, $\frac{1}{2}$ the diameter, is
$\frac{1}{2} \times 14 = 7$ ft.
$$A = \pi r^2$$
$$A = \frac{22}{7} \times \frac{1}{\cancel{7}} \times 7 = 154$$

10. 528 in. $C = \pi d$
$$C = \frac{22}{7} \times \overset{2}{\cancel{14}} = 44 \text{ feet}$$

Use the proportional method.
$$\frac{1 \text{ ft.}}{12 \text{ in.}} = \frac{44 \text{ ft.}}{x}$$
$$x = 44 \times 12$$
$$x = 528$$

Geometry Exercise 6

1. 66 sq. in. $A = bh$
$$A = 11'' \times 6'' = 66$$

2. 12 m $P = 2 + 2 + 4 + 4 = 12$

3. 136 sq. ft. $A = \frac{1}{2}h(b_1 + b_2)$
$$A = \frac{1}{2} \times 8 \ (14 + 20)$$
$$A = \frac{1}{2} \times \frac{\overset{4}{\cancel{8}}}{1} \times \frac{34}{1} = 136$$

4. 72 cm $P = 15 + 20 + 12 + 25 = 72$

5. 616 sq. in. $A = 4\pi r^2$
$$A = 4 \times \frac{22}{7} \times 7^2$$
$$A = \frac{4}{1} \times \frac{22}{\cancel{7}} \times \frac{\overset{7}{\cancel{49}}}{1} = 616$$

6. 24.32 sq. m $A = bh$
$$A = 6.4 \times 3.8 = 24.32$$

Exercise 6 cont'd.

7. 10 in. $P = 2\frac{1}{2} + 2\frac{1}{2} + 2\frac{1}{2} + 2\frac{1}{2} = 10$

8. 70 sq. cm $A = h\left(\frac{b_1 + b_2}{2}\right)$
$A = 10\left(\frac{5.3 + 8.7}{2}\right)$
$A = 10\left(\frac{14}{2}\right) = 10 \times 7 = 70$

9. 127.8 sq. cm $A = bh$
$A = 14.2 \times 9 = 127.8$

Geometry Exercise 7

1. 75 sq. in.
Step 1. Find the measurement of the horizontal side of the bottom part of the figure.

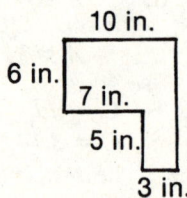

Step 2. Find the area of each part.
top: $A = 6 \times 10 = 60$ sq. in.
bottom: $A = 5 \times 3 = 15$ sq. in.
Step 3. Add the two areas.
60 sq. in. + 15 sq. in. = 75 sq. in.

2. 174 sq. ft.
Step 1. Find the measurements of the base and height of the triangular part of the figure.

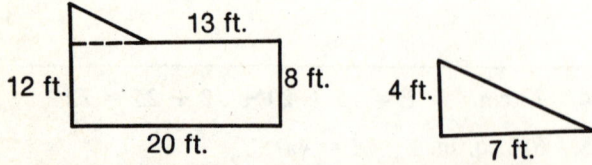

Step 2. Find the area of each part.
triangle: $A = \frac{1}{2}bh = \frac{1}{2} \times \frac{7}{1} \times \frac{4}{1}$
$= 14$ sq. ft.
rectangle: $A = lw = 20 \times 8$
$= 160$ sq. ft.

Step 3. Add the areas.
160 sq. ft. + 14 sq. ft. = 174 sq. ft.

3. $105\frac{1}{2}$ sq. ft.
Step 1. Find the area of the square.
$A = 12 \times 12 = 144$ sq. ft.
Step 2. Find the area of the circle. (The radius is $\frac{1}{2}$ of the diameter or $3\frac{1}{2}$.)
$A = \frac{22}{7} \times \frac{7}{2} \times \frac{7}{2} = \frac{77}{2} = 38\frac{1}{2}$ sq. ft.
Step 3. Subtract the area of the circle from the area of the square.

$$144 = 143\frac{2}{2}$$
$$-38\frac{1}{2} = 38\frac{1}{2}$$
$$\overline{\quad\quad 105\frac{1}{2}}$$

4. 32 sq. yd.
Step 1. Find the area of the dining room.
$A = 9 \times 8 = 72$ sq. ft.
Step 2. Find the area of the living room.
$A = 18 \times 12 = 216$ sq. ft.
Step 3. Add the areas.
72 sq. ft. + 216 sq. ft. = 288 sq. ft.
Step 4. Change the square feet to square yards.
Divide 288 by 9 or use the proportional method.

$$9\overline{)288}\;^{32} \quad \text{or} \quad \frac{9 \text{ sq. ft.}}{1 \text{ sq. yd.}} = \frac{288 \text{ sq. ft.}}{x}$$
$$9x = 288$$
$$x = 32$$

5. 888 sq. ft.
Step 1. Find the dimensions of the pool.
The length is $50 - 10 - 6 = 34'$
The width is $30 - 6 - 6 = 18'$
Step 2. Find the area of the large rectangle.
$A = 50 \times 30 = 1,500$ sq. ft.
Step 3. Find the area of the pool.
$A = 34 \times 18 = 612$ sq. ft.
Step 4. Subtract the two areas.
1,500 sq. ft. − 612 sq. ft. = 888 sq. ft.

Geometry Exercise 8

1. 1,620 cu. in. $V = lwh$

$$
\begin{array}{r}
15 \\
\times 12 \\
\hline
30 \\
15 \\
\hline
180 \\
\times 9 \\
\hline
1{,}620
\end{array}
$$

2. 512 cu. in. $V = s^3$

$$
\begin{array}{r}
8 \\
\times 8 \\
\hline
64 \\
\times 8 \\
\hline
512
\end{array}
$$

3. 22 cu. m $V = \pi r^2 h$

$$V = \frac{22}{7} \times 1^2 \times 7$$

$$V = \frac{22}{\cancel{7}_1} \times \frac{1}{1} \times \frac{\cancel{7}^1}{1} = 22$$

4. 49 cu. in. $V = lwh$

$$V = 5\frac{1}{4} \times 3\frac{1}{2} \times 2\frac{2}{3}$$

$$V = \frac{\cancel{21}^7}{\cancel{4}_1} \times \frac{7}{\cancel{2}_1} \times \frac{\cancel{8}^{\cancel{4}\,1}}{\cancel{3}_1} = 49$$

5. 1.728 cu. cm $V = s^3$

$$V = 1.2 \times 1.2 \times 1.2 = 1.728$$

6. 75 cu. ft. $V = lwh$

$$V = \frac{\cancel{30}^{15}}{1} \times \frac{5}{1} \times \frac{1}{\cancel{2}_1} = 75$$

7. 8,910 cu. ft. $V = \pi r^2 h$

$$V = \frac{22}{7} \times 9^2 \times 35$$

$$V = \frac{22}{7} \times \frac{81}{1} \times \frac{\cancel{35}^5}{1} = 8{,}910$$

8. $113\frac{1}{7}$ cu. in. $V = \frac{4}{3}\,\pi\,r^3$

$$V = \frac{4}{3} \times \frac{22}{7} \times 3^3$$

$$V = \frac{4}{\cancel{3}_1} \times \frac{22}{7} \times \frac{\cancel{27}^9}{1} = 113\frac{1}{7}$$

9. 6 inches.

The diameter is the distance across a circle. The box could only accommodate a sphere with a diameter of 6 inches.

10. 80 truckloads

Step 1.   Find the volume of the hole.
$V = lwh$
$V = 30 \times 20 \times 4 = 2{,}400$ cu. yd.

Step 2.   Divide the volume of the hole by the volume of dirt a truck will hold.

$$
\begin{array}{r}
80 \\
30\overline{)2{,}400}
\end{array}
$$

Geometry Exercise 9

1.	acute	8.	obtuse	15.	straight
2.	right	9.	reflex	16.	reflex
3.	obtuse	10.	straight	17.	right
4.	reflex	11.	right	18.	acute
5.	right	12.	acute	19.	obtuse
6.	straight	13.	acute	20.	reflex
7.	acute	14.	obtuse	21.	reflex

Geometry Exercise 10

1. 42°
$$
\begin{array}{r}
90° \\
- 48° \\
\hline
42°
\end{array}
$$

2. 132°
$$
\begin{array}{r}
180° \\
- 48° \\
\hline
132°
\end{array}
$$

3. 65°
$$
\begin{array}{r}
90° \\
- 25° \\
\hline
65°
\end{array}
$$

4. 117°
$$
\begin{array}{r}
180° \\
- 63° \\
\hline
117°
\end{array}
$$

5. ∠w

6. 360°

7. 105°
$$
\begin{array}{r}
180° \\
- 75° \\
\hline
105°
\end{array}
$$

8. 61°
$$
\begin{array}{r}
180° \\
-119° \\
\hline
61°
\end{array}
$$

9. 50°
$$
\begin{array}{r}
180° \\
-80° \\
\hline
100°
\end{array}
\qquad
\begin{array}{r}
50° \\
2\overline{)100°}
\end{array}
$$

10. obtuse

The supplement to an angle that measures less than 90° must be more than 90° and the two angles must total less than 180°.

Exercise 10 cont'd.

11. 135°   180°
          $\frac{-\ 45°}{135°}$

12. 37°
     Vertical angles
     are equal.

13. 69.5°   180.0°
           $\frac{-\ 110.5°}{69.5°}$

14. 30°
     All the acute
     angles are
     equal.

15. 60°
     The angles are
     supplementary.
         180°
         $\frac{-120°}{60°}$

Geometry Exercise 11

1. a. scalene
   b. right
   c. isosceles
   d. equiangular
       or equilateral
   e. isosceles

   f. scalene
   g. scalene
   h. equilateral or equi-
       angular
   i. isosceles and right

2. 65°   65°   180°
        $\frac{+50°}{115°}$ $\frac{-115°}{65°}$

3. isosceles
     Two angles in triangle WXY are equal.

4. 90°   30°   180°
        $\frac{+60°}{90°}$ $\frac{-90°}{90°}$

5. right
     One angle measures 90°.

6. 49°   180°   49°
        $\frac{-82°}{98°}$ $2\overline{)98°}$

7. 54°   63°   180°
        $\frac{\times 2}{126°}$ $\frac{-126°}{54°}$

8. isosceles   8″   21″
                 $\frac{+5″}{13″}$ $\frac{-13″}{8″}$
     Side EF measures 8″. Two sides have
     the same measurement.

9. scalene   25°   180°
               $\frac{+35°}{60°}$ $\frac{-60°}{120°}$
     The three angles are different.

10. scalene   4″   16″
                $\frac{+5″}{9″}$ $\frac{-9″}{7″}$
     All three sides have different measure-
     ments.

11. 90°
     The right angle measures 90°. The other
     two angles must add up to 90° for the
     triangle to total 180°.

12. 80°
     The two angles given measure 45° and 35°
     for a total of 80°. The remaining angle in
     the triangle must measure 180° − 80° =
     100°. The angle to be found, angle x, is the
     supplement of the 100° angle. 180° − 100°
     = 80°.

Geometry Exercise 12

1. yes   45°   180°
           $\frac{+85°}{130°}$ $\frac{-130°}{50°}$ = ∠O

           50°   180°
           $\frac{+45°}{95°}$ $\frac{-95°}{85°}$ = ∠R

     Each triangle has angles of 45°, 50°, and
     85°.

2. no   60°   180°
           $\frac{+50°}{110°}$ $\frac{-110°}{70°}$ = ∠C

           50°   180°
           $\frac{+80°}{130°}$ $\frac{-130°}{50°}$ = ∠F

     The angles in these triangles are not the
     same.

3. yes

$$
\begin{array}{r} 90° \\ +50° \\ \hline 140° \end{array} \qquad \begin{array}{r} 180° \\ -140° \\ \hline 40° = \angle L \end{array}
$$

$$
\begin{array}{r} 90° \\ +40° \\ \hline 130° \end{array} \qquad \begin{array}{r} 180° \\ -130° \\ \hline 50° = \angle N \end{array}
$$

Each triangle has angles of 40°, 50° and 90°.

4. 21 inches

$$\frac{\text{short side}}{\text{long side}} \quad \frac{8}{12} = \frac{14}{x}$$

$$
\begin{array}{r} 14 \\ \times 12 \\ \hline 28 \\ 14 \\ \hline 168 \end{array} \qquad \begin{array}{r} 21 \\ 8\overline{)168} \end{array}
$$

5. 78 feet

$$\frac{\text{height}}{\text{shadow}} \quad \frac{6}{5} = \frac{x}{65}$$

$$
\begin{array}{r} 65 \\ \times 6 \\ \hline 390 \end{array} \qquad \begin{array}{r} 78 \\ 5\overline{)390} \end{array}
$$

6. yes

$\angle SUT = \angle VUW$ because they are vertical. $\angle S = \angle W$ because they are both right angles.

If we subtract the sum of one right angle and one of the vertical angles from 180°, we get the same third angle for each triangle.

7. 38 feet

$$\frac{\text{short side}}{\text{long side}} \quad \frac{8}{12} = \frac{x}{57}$$

$$
\begin{array}{r} 57 \\ \times 8 \\ \hline 456 \end{array} \qquad \begin{array}{r} 38 \\ 12\overline{)456} \\ \underline{36} \\ 96 \\ \underline{96} \end{array}
$$

8. yes

Both triangles have a right angle. The triangles also share $\angle C$. If you subtract the sum of a right angle and $\angle C$ from 180°, you get the same third angle for each triangle.

9. 10 inches

The entire length of side AC is 15" + 10" = 25".

$$\frac{\text{side}}{\text{bottom}} \quad \frac{6}{15} = \frac{x}{25}$$

$$
\begin{array}{r} 25 \\ \times 6 \\ \hline 150 \end{array} \qquad \begin{array}{r} 10 \\ 15\overline{)150} \end{array}
$$

Geometry Exercise 13

1. 50 feet
$$c^2 = a^2 + b^2$$
$$c^2 = 30^2 + 40^2$$
$$c^2 = 900 + 1600$$
$$c^2 = 2500$$
$$c = \sqrt{2500}$$
$$c = 50$$

2. 26 inches
$$c^2 = a^2 + b^2$$
$$c^2 = 10^2 + 24^2$$
$$c^2 = 100 + 576$$
$$c^2 = 676$$
$$c = \sqrt{676}$$
$$c = 26$$

3. 13 inches
$$c^2 = a^2 + b^2$$
$$c^2 = 12^2 + 5^2$$
$$c^2 = 144 + 25$$
$$c^2 = 169$$
$$c = \sqrt{169}$$
$$c = 13$$

4. 20 yards
$$c^2 = a^2 + b^2$$
$$c^2 = 12^2 + 16^2$$
$$c^2 = 144 + 256$$
$$c^2 = 400$$
$$c = \sqrt{400}$$
$$c = 20$$

5. 25 miles
$$c^2 = a^2 + b^2$$
$$c^2 = 15^2 + 20^2$$
$$c^2 = 225 + 400$$
$$c^2 = 625$$
$$c = \sqrt{625}$$
$$c = 25$$

6. 24 yards
$$a^2 = c^2 - b^2$$
$$a^2 = 30^2 - 18^2$$
$$a^2 = 900 - 324$$
$$a^2 = 576$$
$$a = \sqrt{576}$$
$$a = 24$$

7. 30 feet
$$a^2 = c^2 - b^2$$
$$a^2 = 34^2 - 16^2$$
$$a^2 = 1156 - 256$$
$$a^2 = 900$$
$$a = \sqrt{900}$$
$$a = 30$$

8. 15 feet
$$a^2 = c^2 - b^2$$
$$a^2 = 17^2 - 8^2$$
$$a^2 = 289 - 64$$
$$a^2 = 225$$
$$a = \sqrt{225}$$
$$a = 15$$

9. 3 meters

The diagonal distance is the length of the hypotenuse of a right triangle.
$$c^2 = a^2 + b^2$$
$$c^2 = (2.4)^2 + (1.8)^2$$
$$c^2 = 5.76 + 3.24$$
$$c^2 = 9$$
$$c = \sqrt{9}$$
$$c = 3$$

Geometry Review

1. (3) 12.8 m
$$P = 4s$$
$$= 4 \times 3.2\ m$$
$$= 12.8\ m$$

2. (1) $32\frac{1}{2}$ in.
$$P = 2l + 2w$$
$$= 2(12) + 2(4\frac{1}{4})$$
$$= 24 + 8\frac{2}{4}$$
$$= 32\frac{1}{2}\ in.$$

3. (4) 6 sq. cm
$$A = \frac{1}{2}bh$$
$$= \frac{1}{\cancel{2}} \times \frac{\cancel{8}^{4}}{1} \times \frac{1.5}{1}$$
$$= 6\ sq.\ cm$$

4. (2) 7,200 sq. yd.
$$A = lw$$
$$= 120 \times 60$$
$$= 7,200\ sq.\ yd.$$

5. (2) $96
Find the area.
$$A = lw$$
$$= 18 \times 3$$
$$= 54\ sq.\ ft.$$

Change sq. ft. to sq. yd.
1 sq. yd. = 9 sq. ft.

$$\begin{array}{c} 6\ sq.\ yd. \\ 9\overline{)54} \end{array} \qquad \begin{array}{r} \$16 \\ \times 6 \\ \hline \$96 \end{array}$$

6. (3) 52 ft.
Find the perimeter.
$$P = 2l + 2w$$
$$= 2(20) + 2(8)$$
$$= 40 + 16$$
$$= 56\ ft.$$
Subtract four feet for the walkway.
56 − 4 = 52 ft.

7. (3) 94.2 ft.
$$C = \pi d$$
$$= 3.14 \times 30 = 94.2\ ft.$$

8. (1) 3,850 sq. in.
$$A = \pi r^2$$
$$= \frac{22}{\cancel{7}} \times \frac{\cancel{35}^{5}}{1} \times \frac{35}{1} = 3,850\ sq.\ in.$$

9. (4) 98 sq. in.
Find the measurements of the missing sides of the figure.

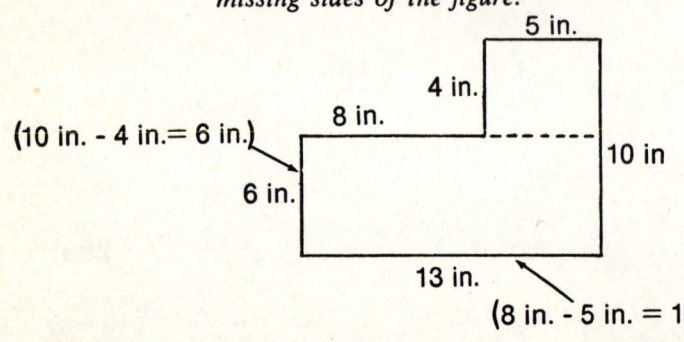

(10 in. − 4 in.= 6 in.)

(8 in. − 5 in. = 13 in.)

Find the area of each rectangle and add.

Top: $A = 5 \times 4 = \quad 20$ sq. in.
Bottom: $A = 13 \times 6 = \underline{+78}$ sq. in.
$\qquad\qquad\qquad\qquad\quad 98$ sq. in.

10. (2) 198 cu. ft.
Find the volume.
$$V = lwh$$
$$= \frac{22}{1} \times \frac{\cancel{18}^{9}}{1} \times \frac{1}{\cancel{2}} = 198\ cu.\ ft.$$

11. (3) obtuse

12. (1) acute

13. (4) 17°
$$\begin{array}{r} 90° \\ -73° \\ \hline 17° \end{array}$$

14. (5) 115°
$$\begin{array}{r} 180° \\ -\ 65° \\ \hline 115° \end{array}$$

15. (4) right
$$\begin{array}{r} 33° \\ +57° \\ \hline 90° \end{array} \qquad \begin{array}{r} 180° \\ -\ 90° \\ \hline 90° \end{array} = \angle Z$$

16. (3) 39°
$$\begin{array}{r} 180° \\ -102° \\ \hline 78° \end{array} \qquad \begin{array}{r} 39° \\ 2\overline{)78°} \end{array}$$

17. (2) 3.6 ft.
$$\frac{short\ side}{long\ side}\ \frac{10}{12} = \frac{3}{x}$$

$$\begin{array}{r} 12 \\ \times\ 3 \\ \hline 36 \end{array} \qquad \begin{array}{r} 3.6 \\ 10\overline{)36.0} \end{array}$$

18. (4) 60 ft.
$$\frac{tree}{shadow}\ \frac{30}{40} = \frac{45}{x}$$

$$\begin{array}{r} 45 \\ \times 40 \\ \hline 1,800 \end{array} \qquad \begin{array}{r} 60 \\ 30\overline{)1800} \end{array}$$

19. (1) 10 ft.
$$c^2 = a^2 + b^2$$
$$c^2 = 6^2 + 8^2$$
$$c^2 = 36 + 64$$
$$c^2 = 100$$
$$c = \sqrt{100} = 10$$

20. (1) 15 in.
$$a^2 = c^2 - b^2$$
$$a^2 = 39^2 - 36^2$$
$$a^2 = 1521 - 1296$$
$$a^2 = 225$$
$$a = \sqrt{225}$$
$$a = 15$$

Algebra

Algebra is an extension of the skills you learned in arithmetic. You have already learned how to find powers and roots, to use formulas, and to substitute numbers. These are common algebraic skills. In this section, you will learn additional skills that you will need for the algebra questions on the GED.

THE NUMBER LINE AND SIGNED NUMBERS

In arithmetic, the numbers we use can be represented on a line like this:

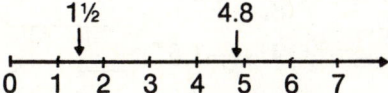

Fractions, decimals, and mixed numbers are indicated in the spaces between the whole numbers. Except for zero, all the numbers on this line are said to be **positive,** which means more than zero (0).

However, we can also work with numbers that are less than zero (0), called **negative numbers.**

Temperatures on a thermometer are the most familiar use of positive and negative numbers. When we say it is 70° outside, we mean that it is 70° above zero (0), a positive number. However, when it is −20°, that is another way of saying it is 20° below zero (0).

We can extend the line shown above to include both positive and negative numbers. This is similar to turning a thermometer on its side.

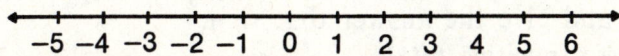

This is called the **number line.** The arrows at both ends indicate that both the positive and negative values can be extended.

The positive numbers (shown with a + sign) represent amounts above zero; moving to the right on the line indicates increasing amounts, such as a gain in weight or a rise in temperature. The negative numbers (shown with a − sign) represent amounts below zero; moving to the left on the line indicates decreasing amounts, such as a loss of weight or a drop in temperature. Zero (0) has no sign. It is neither positive nor negative.

Positive numbers do not have to be written with a plus (+) sign in front of them. The number 8 is understood to be a +8. Negative numbers, however, must always have a minus (−) sign in front of them.

ALGEBRA EXERCISE 1

Use the number line below to tell which letter corresponds to each of the following numbers. Problem 1 is already done. The letter J corresponds to +7.

```
      A     B          C D          E  F G     H          I        J
  ←───┼──┼──┼──┼──┼──┼──┼──┼──┼──┼──┼──┼──┼──┼──┼──┼──┼──┼──→
     −8 −7 −6 −5 −4 −3 −2 −1   0 +1 +2 +3 +4 +5 +6 +7 +8
```

1. +7 _____J_____	**6.** $\frac{5}{4}$ _____
2. −3 _____	**7.** $+\frac{16}{3}$ _____
3. $-6\frac{1}{2}$ _____	**8.** 2.75 _____
4. −8 _____	**9.** −3.5 _____
5. $+\frac{2}{3}$ _____	**10.** −0.2 _____

Answers and solutions start on page 341.

ADDITION OF SIGNED NUMBERS

In the next four sections, you will learn the rules for adding, subtracting, multiplying, and dividing signed numbers. These are tools you will need throughout algebra.

You can add signed numbers by following the rules below.

Rules for Adding Signed Numbers

1. To add two or more numbers that have the same sign, combine the numbers and give the answer that sign:
- Positive numbers give a positive total.
- Negative numbers give a negative total.

2. To add a positive number and a negative number, find the difference between the two numbers and give the answer the sign of the larger number.

Problems containing negative numbers may not look like addition problems. However, if there is no indication of subtraction (a minus sign between parentheses, the words subtract or difference, etc.), assume that you should follow the rules for adding signed numbers.

EXAMPLE 1: Find $-16 -8$.

       ① & ②
$$\begin{array}{r} -16 \\ -\ 8 \\ \hline -24 \end{array}$$

Step 1. Since they have the same sign, combine the numbers.

Step 2. Since they are both negative numbers, the answer is negative.

This can be seen as similar to a drop in a negative temperature. If it is already $-16°$ and it drops 8 degrees more $(-8°)$, it will be $-24°$.

EXAMPLE 2: Add $+73$ and -104.

  ①
$$\begin{array}{r} 104 \\ 73 \\ \hline 31 \end{array}$$
  ②
$$\begin{array}{r} -104 \\ 73 \\ \hline -\ 31 \end{array}$$

Step 1. Since the numbers have different signs, you must find the difference between them. The difference between 104 and 73 is 31.

Step 2. -104 was the larger number of the two. Your final answer is a negative number.

This is similar to adding a win and loss record for a team. If the team had won 73 games but had lost 104, its overall standing would be represented by -31.

You can use the rules for adding signed numbers to find the total of a group of positive and negative numbers.

Imagine that at the end of the month you have $20. You get a gas bill for $8 and a telephone bill for $14. Then your brother pays you $5 that he borrowed and a cousin calls to remind you that you owe him $10. To find out how much money you have (or you owe), you can add a series of signed numbers.

Think of the $20 you started with and the $5 your brother paid you as positive numbers. Altogether this is $25 or +25. In algebra, this is written as $(+20) + (+5) = +25$.

Think of the bills for $8 and $14 and the $10 you owe your cousin as negative numbers. Their total is $32, but you think of these as -32 since this is money you owe. In algebra, this is written as $(-8) + (-14) + (-10) = -32$.

The positive total was +25, and the negative total was −32. The difference between these two is $7 or −7. In algebra, this is written as (+25) + (−32) = −7. In other words, you end up owing $7.

More Rules for Adding Signed Numbers

1. Combine all the positive numbers for a positive total.
2. Combine all the negative numbers for a negative total.
3. Find the difference between the two totals and give the answer the sign of the larger total.

Addition of signed number problems can be written several ways. Look at the following examples carefully to see why they are all addition problems.

EXAMPLE 3:   (−9) + (10) + (−8) + (+4) =?

① $\begin{array}{r} +10 \\ +\ 4 \\ \hline +14 \end{array}$ ② $\begin{array}{r} -\ 9 \\ -\ 8 \\ \hline -17 \end{array}$ ③ $\begin{array}{r} -17 \\ +14 \\ \hline -\ 3 \end{array}$

Step 1. Combine the positive numbers 10 and +4 and make the answer positive.

Step 2. Combine the negative numbers −9 and −8 and make the answer negative.

Step 3. Subtract 14 from 17, and give the answer the sign of the larger number.

Notice that in this example, the numbers within parentheses are either positive or negative. Also, there are plus signs between the parentheses to indicate addition.

EXAMPLE 4:   Simplify: − 6 − 3 − 4.

$$\begin{array}{r} -\ 6 \\ -\ 3 \\ -\ 4 \\ \hline -13 \end{array}$$

Solution: In this example, there are no positive numbers. Simply combine the negative numbers and make the answer negative. The total is −13.

EXAMPLE 5:   Add: − 5 + 4 + 7 − 6.

① $\begin{array}{r} +\ 4 \\ +\ 7 \\ \hline +11 \end{array}$ ② $\begin{array}{r} -\ 5 \\ -\ 6 \\ \hline -11 \end{array}$ ③ $\begin{array}{r} +11 \\ -11 \\ \hline 0 \end{array}$

Step 1. Combine the positive numbers +4 and +7 for a positive total. (+11)

Step 2. Combine the negative numbers −5 and −6 for a negative total. (−11)

Step 3. Subtract 11 from 11. The difference is 0. (Remember that 0 has no sign).

ALGEBRA EXERCISE 2 ───────────────

Solve each problem.

1. Simplify: − 9 − 3.

2. 6 − 8 = ?

3. Find the sum of +12 and −20$\frac{1}{2}$.

4. Find the sum of +12$\frac{1}{2}$ and −20.

5. (+5) + (−8) + (+9) = ?

6. Find the sum of +8, −6, and −5.

7. −18 −2 + 6 = ?

8. Add −6, 8, −4, and +11.

9. 40 − 7 − 5 − 3 − 10 − 2 = ?

10. Find the sum of +3, −10, and +7.

11. (−10) + (5) + (−3) + (−9) = ?

12. (−9) + (−4) + (8) + (7) + (−2) = ?

13. At 5 a.m., it was −10°. By 8 a.m., the temperature had dropped another four degrees. What was the 8 a.m. temperature?

14. Although the quarterback had rushed to gain 15 yards, he had been tackled for a loss of 37 yards. What was his net yardage for the day?

15. Mercedes had $235 in her checking account. She deposited $55 and wrote checks for $30, $150, and $120. Did she have enough money in her account to cover the checks?

Answers and solutions start on page 341.

SUBTRACTION OF SIGNED NUMBERS

With negative numbers it is possible to do subtraction problems that you could not do before. For example, you can take 9 from 4. Begin at +4 on the number line, and move 9 spaces to the left.

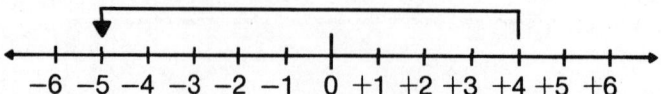

You should be at −5. In algebra, this problem is written as: (+4) − (+9) = −5.

As the previous section indicated, in a problem containing negative numbers, there will be a clear indication that subtraction is required. In the problem (+4) − (+9), the minus sign between parentheses indicates subtraction.

The only difference between adding and subtracting signed numbers is that a subtraction sign changes the sign of the number following it. Once that change takes place, you no longer consider the subtraction sign there. You proceed as you did in adding signed numbers: combine numbers with like signs and give the answer that sign, or you find the difference between numbers with different signs and give the answer the sign of the larger number.

Rules for Subtracting Signed Numbers

1. Change the sign of the number being subtracted to the opposite sign and drop the subtraction sign.

2. Follow the rules for adding signed numbers.

EXAMPLE 1: What is (−8) − (+3)?

① & ② $(-8) - (+3) = -8 - 3 = \mathbf{-11}$

Step 1. The − sign outside the parentheses means the (+3) that follows is being subtracted. Change +3 to −3 and drop the subtraction sign. You can also drop the parentheses since there is now only one sign between numbers.

Step 2. Since both signs are now the same, combine the numbers and use that sign.

EXAMPLE 2: Find (−10) − (−2).

① & ② $(-10) - (-2) = -10 + 2 = \mathbf{-8}$

Step 1. The − sign outside parentheses means the −2 is being subtracted. Change −2 to +2 and drop the parentheses and the subtraction sign.

Step 2. Since the signs are different, find the difference and use the sign of the larger number.

EXAMPLE 3: Simplify $(+6) − (−4) + (−2) − (+5)$.

①
$$(+6) − (−4) + (−2) − (+5) =$$
$$+6 +4 − 2 − 5$$

②
$$\begin{array}{r} + 6 \\ + 4 \\ \hline +10 \end{array}$$

③
$$\begin{array}{r} −2 \\ −5 \\ \hline −7 \end{array}$$

④
$$\begin{array}{r} +10 \\ − 7 \\ \hline + 3 \text{ or } 3 \end{array}$$

Step 1. The two − signs outside parentheses mean that −4 and +5 are being subtracted. Change −4 to +4 and +5 to −5. Keep any numbers following an addition sign the same but drop the addition sign.

Step 2. Combine the positive numbers and make the answer positive.

Step 3. Combine the negative numbers and make the answer negative.

Step 4. Find the difference between 7 and 10 and give the answer the sign of the larger number.

ALGEBRA EXERCISE 3 ———

Solve each problem.

1. $(+6) − (+4) =$

2. $(−8) − (+3) =$

3. $(−9) − (−8) =$

4. $(+10) − (−9) =$

5. Subtract: $(+8) − (7) =$

6. Find $(−9) − (−9) =$

7. Take 12 away from −10.

8. $(+6) − (−3) + (−2) =$

9. $(−9) − (+4) − (+10) =$

10. $(−15) − (20) + (+6) =$

11. $(−8) + (−13) − (+6) =$

12. $(−3) + (−4) − (−5) − (−6) =$

13. The coastline is 22 feet above sea level and Rainbow Valley is 43 feet below sea level. How many feet lower than the coastline is Rainbow Valley?

14. By daybreak, it was 7° above zero, but at dusk the temperature dropped to −17°. By how much had the temperature dropped?

Answers and solutions start on page 341.

MULTIPLICATION OF SIGNED NUMBERS

Following are some practical examples that illustrate multiplication of signed numbers.

If you gain two pounds a week for five weeks, you will weigh ten pounds more than you weigh now. In algebra, this is written as $(+2)(+5) = +10$. If there is no sign between parentheses, multiply.

If you lose two pounds a week for five weeks, you will weigh ten pounds <u>less</u> than you weigh now. In algebra, this is written as $(-2)(+5) = -10$.

If you have been gaining two pounds a week for five weeks, you weighed ten pounds <u>less</u> five weeks ago. In algebra, this is written as $(+2)(-5) = -10$. The $-$ sign before the 5 suggests weeks in the past.

If you had been losing two pounds a week for the past five weeks, you weighed ten pounds <u>more</u> five weeks ago. In algebra, this is written as $(-2)(-5)=+10$. The $-$ sign before the 5 again suggests weeks in the past.

There is a simple pattern to these examples. When the signs of two numbers being multiplied are alike, the answer is positive. When the signs of two numbers being multiplied are not alike, the answer is negative.

Rules for Multiplying Two Signed Numbers

1. If the signs of the numbers are alike, the answer is positive.
2. If the signs of the numbers are different, the answer is negative.

EXAMPLE 1: What is the product of (-8) and (-7)?
    Solution: Since the signs are alike, the answer is positive.
        $(-8)(-7) = \mathbf{+56}$

EXAMPLE 2: What is $(12)(-3)$?
    Solution: Since the signs are different, the answer is negative.
        $(12)(-3) = \mathbf{-36}$

EXAMPLE 3: Find $-3 \cdot 10$.
    Solution: The dot between the numbers is another symbol for multiplication. Since the signs are different, the answer is negative.
        $-3 \cdot 10 = \mathbf{-30}$

EXAMPLE 4: What is $(-6)(+2)(-3)(+4)$?
    Solution: The product of these four numbers is $+144$. The rules for multiplication involve only two numbers at a time. -6 times $+2$ is negative because the signs are different. The product, -12, times -3 is positive because the signs are alike. The

product +36 times +4 is also positive because the signs are alike.

$$(-6)(+2) = -12$$
$$(-12)(-3) = +36$$
$$(+36)(+4) = +144$$

There are shortcuts for multiplying more than two signed numbers.

Rules for Multiplying More Than Two Signed Numbers

1. An even number of − signs gives a positive answer.
2. An odd number of − signs gives a negative answer.

EXAMPLE 5: Find $(-2)(-1)(+3)(-5)$.

Solution: This problem has three − signs. Three is an odd number, so the answer has a negative sign. The product of these four numbers is −30.

$$(-2)(-1)(+3)(-5) = -30$$

Look at Example 4. In that example, there were two negative numbers. Two is an even number, so the answer was positive.

ALGEBRA EXERCISE 4

Solve the problems below.

1. $(-2)(+9) =$

2. $(-6)(-6) =$

3. $(+5)(-9) =$

4. $(+8)(3) =$

5. $(-\frac{3}{4})(12) =$

6. $(+\frac{2}{3})(-\frac{3}{4}) =$

7. $-8 \cdot 6 \cdot -2 =$

8. $-7 \cdot -103 =$

9. $+5 \cdot +4 \cdot -2 =$

10. $-8 \cdot -\frac{5}{8} =$

11. $(4)(-2)(-1)(-6)(4) =$

12. Find the product of −14 and +2.

13. Find the product of +14 and −2.

14. Maureen had been servicing 17 accounts. Then she lost 2 accounts every week for a month. How many accounts were left?

15. Sunshine Day Care Center has gained 3 new children every month. How many fewer children were there in the school six months ago?

Answers and solutions start on page 342.

DIVISION OF SIGNED NUMBERS

The rules for dividing signed numbers are similar to those for multiplying signed numbers.

Rules for Dividing Signed Numbers

1. If the signs are alike, the answer is positive.

2. If the signs are different, the answer is negative.

EXAMPLE 1: What is $(+30) \div (-6)$?

  Solution: Since the signs are different, the answer is negative.

$$\frac{+30}{-6} = -5$$

EXAMPLE 2: What is $\frac{-28}{-12}$?

  Solution: Remember that the fraction bar indicates that you can divide the denominator into the numerator. Since the signs are alike, the answer is positive. Improper fractions as answers are common in algebra. Either $2\frac{1}{3}$ or $\frac{7}{3}$ would be a correct answer. On a multiple choice test, such as the GED, you should check the form of the answers to see which is appropriate.

$$\frac{-28}{-12} = +2\frac{4}{12} = +2\frac{1}{3} \text{ or } +\frac{7}{3}$$

EXAMPLE 3: Simplify $\frac{+9}{-15}$

  Solution: In this example, you cannot divide; you can only reduce. Since the signs are different, the answer is negative.

$$\frac{+9}{-15} = -\frac{3}{5}$$

ALGEBRA EXERCISE 5 ━━━━━━━━━━━━━━━━━

Solve each problem.

1. $\dfrac{-40}{-20} =$

2. Find: $(-12) \div (+6) =$

3. Simplify $\dfrac{72}{-9} =$

10. $(-108) \div (-9) =$

4. $\dfrac{+16}{-24} =$

11. $\dfrac{-48}{-60} =$

5. $(-15) \div (+5) =$

12. $\dfrac{+65}{-5} =$

6. $\dfrac{30}{-36} =$

13. $(-36) \div (-24) =$

7. Find the value of $\dfrac{-8}{-1} =$

14. $\dfrac{-63}{+35} =$

8. $\dfrac{-13}{26} =$

15. $\dfrac{75}{-100} =$

9. $(144) \div (-24) =$

Answers and solutions start on page 342.

Before you go on to equations, review all of the signed number skills you have learned so far.

ALGEBRA EXERCISE 6 ━━━━━━━━━━━━━

Solve each problem.

1. $(15) - (-9) =$

7. $(-29) + (-14) =$

2. $(-6)(+20) =$

8. $(-20) - (-21) =$

3. $-4 + 3 - 7 =$

9. $(20) + (-12) - (+15) =$

4. $\dfrac{-96}{-8} =$

10. $(-3)\left(+\frac{2}{3}\right)(-10) =$

5. $(-8) - (12) =$

11. $\dfrac{-14}{21} =$

6. $\dfrac{+1,000}{-10} =$

12. $(-1)\left(+\frac{7}{8}\right) =$

13. Find the difference between 10 degrees below zero and 15 degrees above zero.

14. Inayo has been losing two pounds a day for six days. What is her weight now compared to her weight six days ago?

Answers and solutions start on page 342.

EQUATIONS

An equation is a statement that two amounts are equal. $4m + 2 = 26$ is an equation. In words, this equation says, "Four times some number m plus two is equal to 26." The missing number m is called the **unknown**. The value for m that makes the equation true is called the **solution** of the equation. The solution of the equation above is 6. You can check this by substituting 6 for m as you learned on page 244:

$$4(6) + 2 = 26$$
$$24 + 2 = 26$$
$$26 = 26$$

Since both sides of the equation equal 26, 6 is the correct solution. When both sides of an equation are equal, this means that you have found the correct value for the unknown.

The equal sign (=) makes an equation similar to an old-fashioned balance scale. Imagine a scale with weights on one side and apples on the other side. If you remove an apple from one side, you must remove a corresponding weight from the other side to keep the scale balanced.

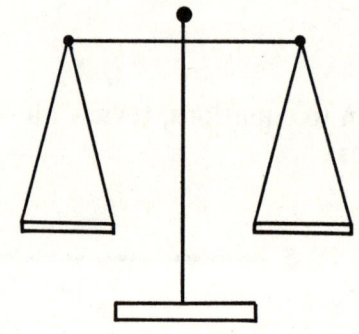

In algebra, we sometimes talk about "balancing" an equation.

To keep an equation balanced, perform the same operation on both sides of the equal sign. You can add 5 to one side of an equation if you also add 5 to the other side. You can divide one side of an equation by 12 if you also divide the other side of the equation by 12.

Solving an Equation

When you solve an equation, your goal is to get the unknown to stand alone in a statement that represents, "Unknown = some value." In an equation with one operation, you can get the unknown standing alone by performing an **inverse** (opposite) operation on the equation.

> The inverse of addition is subtraction.
>
> The inverse of subtraction is addition.
>
> The inverse of multiplication is division.
>
> The inverse of division is multiplication.

Rule for Solving One-Operation Equations

Identify the operation in the equation. Then perform the inverse operation on both sides in order to get a statement that says, "Unknown = value" or "Value = unknown."

EXAMPLE 1: If $6x = 132$, what is the value of x?

Solution: The left side of this equation involves multiplication. Since the inverse of multiplication is division, divide both sides of the equation by 6. On the left side, you have found the value of $1x$. However, you just write x since it is the same as $1x$.

$$\frac{\overset{1}{\cancel{6}}x}{\underset{1}{\cancel{6}}} = \frac{\overset{22}{\cancel{132}}}{\underset{1}{\cancel{6}}}$$

$$x = 22$$

To check, substitute 22 for x.

$$6(22) = 132$$
$$132 = 132$$

EXAMPLE 2: Solve for m in the equation $m + 43 = 74$.

Solution: The left side of this equation involves addition. The inverse of addition is subtraction. Subtract 43 from both sides of the equation. Nothing remains next to the m because $+43 - 43 = 0$. It is not necessary to write the 0.

$$
\begin{array}{rcr}
m + 43 &=& 74 \\
- 43 & & -43 \\
\hline
m &=& 31
\end{array}
$$

To check, substitute 31 for m.

$$31 + 43 = 74$$
$$74 = 74$$

EXAMPLE 3: Find the value of y in the equation $39 = y - 15$.

Solution: The right side of this equation involves subtraction. The inverse of subtraction is addition. Add 15 to both sides.

$$
\begin{array}{rcr}
39 &=& y - 15 \\
+15 & & +15 \\
\hline
54 &=& y
\end{array}
$$

To check, substitute 54 for y.

$$39 = 54 - 15$$
$$39 = 39$$

EXAMPLE 4: Solve for w in $10 = \frac{w}{7}$.

Solution: The right side of this equation involves division. The inverse of division is multiplication. Multiply both sides of the equation by 7.

$$7 \cdot 10 = \frac{w}{7} \cdot 7$$
$$70 = w$$

To check, substitute 70 for w.

$$10 = \frac{70}{7}$$
$$10 = 10$$

Remember that the goal when you solve an equation is to get a statement that says "unknown = value," such as $x = 22$, or "value = unknown," such as $70 = w$. Keep this goal in mind as you work through the next exercise. It may be useful to check your answers by substituting the value of the unknown in the equation.

ALGEBRA EXERCISE 7 ─────────────────────────

Solve each equation.

1. $8y = 96$

2. $f + 20 = 57$

3. $b - 19 = 28$

4. $\frac{x}{3} = 9$

5. $42 = t + 7$

6. $11 = 2y$

7. $n + 36 = 60$

8. $33 = k - 8$

9. $c - 4 = 27$

10. $15p = 75$

11. $9 = \frac{m}{4}$

12. $18 = d - 6$

13. $v - 17 = 0$

14. $25z = 100$

15. $41 = c + 18$

Answers and solutions start on page 342.

Equation of More Than One Operation

The first equation we looked at, $4m + 2 = 26$, has both multiplication and addition. To perform more than one inverse operation, <u>take care of addition or subtraction first</u>. <u>Then take care of multiplication or division.</u>

EXAMPLE 1:   In $4m + 2 = 26$, what is the value of m?

①
$$\begin{array}{r} 4m + 2 = 26 \\ -\ 2 \quad -2 \\ \hline 4m \quad\ = 24 \end{array}$$

②
$$\frac{4m}{4} = \frac{24}{4}$$
$$m = 6$$

Step 1. Subtract 2 from both sides.
Step 2. Divide both sides by 4.

EXAMPLE 2:   Solve for t in $5 = \frac{t}{3} - 7$.

①
$$\begin{array}{r} 5 = \frac{t}{3} - 7 \\ +\ 7 \qquad +7 \\ \hline 12 = \frac{t}{3} \end{array}$$

②
$$3 \cdot 12 = \frac{t}{\overset{}{\underset{1}{3}}} \cdot \overset{1}{3}$$
$$36 = t$$

Step 1. Add 7 to both sides.
Step 2. Multiply both sides by 3.

EXAMPLE 3:   Solve for s in $2s - 3 = -7$.

①
$$\begin{array}{r} 2s - 3 = -7 \\ +\ 3 \quad +3 \\ \hline 2s \quad\ = -4 \end{array}$$

②
$$\frac{\overset{1}{2}s}{\underset{1}{2}} = \frac{-4}{2}$$
$$s = -2$$

Step 1. Add 3 to both sides. Since the signs on the right are different, you must find the difference and give the answer the sign of the larger number.
Step 2. Divide both sides by 2.

EXAMPLE 4:   Find r if $\frac{2}{3}r = 18$.

①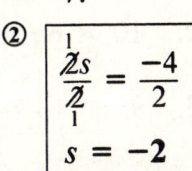

②
$$r = \frac{\overset{9}{18}}{1} \cdot \frac{3}{\underset{1}{2}} = 27$$

Step 1. Since the r is being multiplied by $\frac{2}{3}$, you must divide both sides by $\frac{2}{3}$, which means multiply by the reciprocal, $\frac{3}{2}$.
Step 2. Find $18 \cdot \frac{3}{2}$.

EXAMPLE 5: Solve for c in $12 = -3c + 9$.

$$\boxed{\begin{array}{r} ① \quad 12 = -3c + 9 \\ \underline{-\ 9 \qquad\quad -\ 9} \\ 3 = -3c \end{array}} \qquad \boxed{\begin{array}{c} ② \quad \dfrac{3}{-3} = \dfrac{-3c}{-3} \\ \mathbf{-1} = c \end{array}}$$

Step 1. Subtract 9 from both sides.
Step 2. Divide both sides by -3.

Again, remember that the goal in solving equations is to get a statement that has the unknown standing alone and says "unknown = value" or "value = unknown."

ALGEBRA EXERCISE 8

Solve each equation.

1. $7m - 2 = 54$

2. $\dfrac{a}{3} + 5 = 9$

3. $7 = \dfrac{c}{2} + 3$

4. $82 = 9d + 10$

5. $25c - 17 = 183$

6. $\dfrac{w}{2} - 7 = 3$

7. $-2 = 6x - 8$

8. $\dfrac{1}{3}p + 8 = 11$

9. $40 = 13z + 14$

10. $\dfrac{n}{2} + 3 = 7$

11. $3a + 10 = 1$

12. $\dfrac{3}{4}y - 3 = 12$

13. $39 = 16k - 9$

14. $10 = 6a + 7$

15. $9r + 15 = 18$

Answers and solutions start on page 343.

Equations with Separated Unknowns

Sometimes the unknowns are separated in an equation. You must combine them before you can solve the equation. For the numbers, <u>follow the rules</u> for adding and subtracting signed numbers and keep the unknown (the letter) attached. Look at these examples:

$$\begin{array}{ccc} -5x & 4a & 6c \\ \underline{2x} & \underline{+\ a} & \underline{-5c} \\ -3x & 5a & c \end{array}$$

Notice that in the second example the a was understood to be $1a$. In the third example, the answer c is the same as $1c$.

Rules for Solving Equations with Separated Unknowns

1. Combine the unknowns.

   a. If the unknowns are on the <u>same</u> side of the = sign, combine them according to the rules for adding signed numbers.

   b. If the unknowns are on <u>different</u> sides of the = sign, combine them by using inverse operations.

2. Use inverse operations to solve the equation.

EXAMPLE 1: Solve for x in $5x - 2x + 8 = 26$.

  ①
$$5x - 2x + 8 = 26$$
$$3x + 8 = 26$$

  ②
$$3x + 8 = 26$$
$$\underline{\quad\;\; -8 \quad\;\; -8}$$
$$3x \qquad = 18$$

  ③
$$\frac{3x}{3} = \frac{18}{3}$$
$$x = 6$$

Step 1. Combine the x's: $5x - 2x = 3x$.

Step 2. Subtract 8 from both sides.

Step 3. Divide both sides by 3.

EXAMPLE 2: Solve for a in $9a - 3 = 2a + 11$.

  ①
$$9a - 3 = \quad 2a + 11$$
$$\underline{-2a \qquad\quad -2a}$$
$$7a - 3 = \qquad\; = 11$$

  ②
$$7a - 3 = \quad 11$$
$$\underline{\qquad +3 \quad +3}$$
$$7a \qquad = 14$$

  ③
$$\frac{7a}{7} = \frac{14}{7}$$
$$a = 2$$

Step 1. Since the a's are on different sides of the = sign, use inverse operations to combine them. Subtract $2a$ from both sides.

Step 2. Use inverse operations again to isolate the $7a$ on the left side. Add 3 to both sides.

Step 3. Divide both sides by 7.

EXAMPLE 3: Solve for y in $20 - 2y = 3y$.

  ①
$$20 - 2y = \quad 3y$$
$$\underline{\;\; +2y \qquad +2y}$$
$$20 \qquad = 5y$$

  ②
$$\frac{20}{5} = \frac{5y}{5}$$
$$4 = y$$

Step 1. The unknowns are on different sides of the = sign. We have a choice of moving $3y$ to the left side or $-2y$ to the right side. Since we eventually want all the unknowns on one side and the numbers on the other side, we will move $-2y$ to the right. This leaves the number, 20, on the left side. Add $2y$ to both sides.

Step 2. Divide both sides by 5.

When you are solving equations, your unknown may be negative, as in $-y = 7$. Since $-y$ means the same as $-1y$, there is a simple way to get a value for y. Just divide both sides of the equation by -1. You are performing an inverse operation without changing the values in the equation.

$$-y = 7$$

$$\frac{-1y}{-1} = \frac{7}{-1} = y = -7$$

This means that if you have a negative unknown, you can change the sign of both sides of the equation to get a positive unknown.

EXAMPLE 4: Solve for y in $2y = 3y - 2$

①
$$\begin{array}{r} 2y = 3y - 2 \\ -3y \quad -3y \\ \hline -y = \quad -2 \end{array}$$

②
$$\begin{array}{r} -y = -2 \\ y = +2 \end{array}$$

Step 1. The unknowns are on opposite sides of the equation. Subtract $3y$ from both sides of the equation. This gives $-y$ equal to -2.

Step 2. To get the value for y, divide both sides by -1 or just change the signs on both sides of the equation.

Occasionally, you may need to find the value of two unknowns. In the example below, x and y represent unknown numbers. The unknown y is defined in relation to x.

EXAMPLE 5: If $3y + x = 35$ and $x = 2y$, then what does y equal?

①
$$3y + 2y = 35$$

②
$$\frac{5y}{5} = \frac{35}{5}$$
$$y = 7$$

Step 1. Since $2y$ equals x, substitute $2y$ for x in the first equation.

Step 2. Combine $2y$ and $3y$. Solve for y.

ALGEBRA EXERCISE 9 ━━━━━━━━━━━━━

Solve each equation.

1. $5y - y = 19 + 9$

2. $6t + 8 + 4t = -42$

3. $9c = 44 - 2c$

4. $8m = 2m + 30$

5. $4a + 55 = 9a$

6. $4p = p + 18$

7. $6\frac{1}{2}f = 14 - \frac{1}{2}f$

8. $3 = y + 8y$

9. $8r + 17 = 5r + 32$

10. $7n - 9 = 3n + 7$

11. $6z + 11 = 5z + 20$

12. $5y = 6y - 10$

13. $4y = 7y + 21$

14. If $x = 3y$, solve $x + y = 12$ for y.

15. If $y = \frac{1}{2}x$, solve $x + y = 6$ for x.

Answers and solutions start on page 343.

Equations with Parentheses

Sometimes, part of the information in an equation is contained inside parentheses. Parentheses are used to group together the numbers and letters that are to be multiplied or divided by some other number. In the expression $5(m + 6)$ both m and 6 are to be multiplied by 5.

To simplify an expression with parentheses, multiply each number and letter inside the parentheses by the number outside the parentheses, as you did with the distributive law on page 240.

EXAMPLE 1: $5(m + 6)$ **EXAMPLE 2:** $3(2m - 1)$
 $5 \cdot m + 5 \cdot 6$ $3 \cdot 2m - 3 \cdot 1$
 $5m + 30$ $6m - 3$

When solving an equation that has parentheses, first remove the parentheses by multiplying. Then, follow the rules for solving equations.

EXAMPLE 3: Solve for c in $4(c - 6) = 12$

$$①\ \begin{array}{|l|} \hline 4(c - 6) = 12 \\ 4c - 24\ = 12 \\ \hline \end{array} \quad ②\ \begin{array}{|l|} \hline 4c - 24 =\quad 12 \\ \underline{\ + 24\quad +24} \\ 4c\qquad =\quad 36 \\ \hline \end{array} \quad ③\ \begin{array}{|c|} \hline \frac{4c}{4} = \frac{36}{4} \\ c = 9 \\ \hline \end{array}$$

Step 1.   Remove the parentheses. Multiply both c and -6 by 4.
Step 2.   Add 24 to both sides.
Step 3.   Divide both sides by 4.

EXAMPLE 4: Solve for y in $5(y - 3) = 2(y + 9)$

①
$$5(y - 3) = 2(y + 9)$$
$$5y - 15 = 2y + 18$$

②
$$5y - 15 = 2y + 18$$
$$\underline{-2y \qquad\qquad -2y}$$
$$3y - 15 = \qquad\; + 18$$

③ & ④
$$3y - 15 = +18$$
$$\underline{\quad + 15 \qquad +15}$$
$$\frac{3y}{3} = \frac{33}{3}$$
$$y = 11$$

Step 1. Remove the parentheses. Multiply both y and -3 by 5. Multiply both y and $+9$ by 2.

Step 2. Subtract $2y$ from both sides.

Step 3. Add 15 to both sides.

Step 4. Divide both sides by 3.

ALGEBRA EXERCISE 10

Solve each equation.

1. $3(x + 4) = x + 20$

2. $2a + 9 = 5(a - 3)$

3. $2 - 3c = 4(c + 4)$

4. $2(m - 1) = 7m - 32$

5. $3(y - 2) = y + 4$

6. $80 = 8(s + 7)$

7. $5(t + 4) = 3(t + 10)$

8. $7(m - 2) = 3(m + 6)$

Answers and solutions start on page 344.

INEQUALITIES

You have learned that an equation is a statement that two amounts are equal. **Inequalities** are statements that two amounts are *not* equal. There are four symbols used in writing inequalities:

$<$ means less than

$>$ means greater than

$\leq$ means less than or equal to

$\geq$ means greater than or equal to

For example,

$3 < 4$ means "3 is less than 4"

$4 > 3$ means "4 is greater than 3"

$m \leq 4$ means m is less than or equal to 4, so m could be 4 or any number less than 4 (including negative numbers)

$m \geq 4$ means that m is equal to or greater than 4

Solving inequalities is like solving equations. You can perform inverse operations on both sides of inequalities to find the value of the unknown.

EXAMPLE 1: Solve the inequality $6m + 3 \leq 45$.

$$① \quad \begin{array}{r} 6m + 3 \leq 45 \\ -3 \quad -3 \\ \hline 6m \quad\quad \leq 42 \end{array} \qquad ② \quad \begin{array}{c} \dfrac{6m}{6} \leq \dfrac{42}{6} \\ m \leq 7 \end{array}$$

Step 1. Subtract 3 from both sides.

Step 2. Divide both sides by 6. The solution is $m \leq 7$. This statement is true for the number 7 and for every number less than 7.

EXAMPLE 2: For the inequality $y + 2 > 5$, could y be 2?

$$\begin{array}{r} y + 2 > 5 \\ -2 \quad -2 \\ \hline y \quad > 3 \end{array}$$

Solution: Subtract 2 from both sides. The solution is $y > 3$. This inequality is true for every number greater than 3. Therefore, y could not be 2.

ALGEBRA EXERCISE 11 ────────────────

Solve each inequality.

1. $4c - 3 < 21$

2. $3p + 1 > 7$

3. $9w + 2 \leq 29$

4. $8x - 5 \geq 15$

5. For the inequality $m - 6 > 1$, could m be equal to 7?

6. For $\frac{1}{2} r + 5 \leq 9$, could r equal $\frac{1}{2}$?

7. For $d + 7 \leq 2$, could d be -6?

8. For $2f - 8 < 4$, could f be 4?

Answers and solutions start on page 345.

WRITING ALGEBRAIC EXPRESSIONS

Algebra is a useful tool for solving many word problems. In the next few sections, you will learn to "translate" mathematical ideas into the language of algebra.

Read the following examples, paying close attention to the words that suggest which mathematical symbols to use.

Example	Expression	Notes
A number increased by four	$x + 4$ or $4 + x$	"Increased by" means to add.
Eight less than a number	$y - 8$	"Less" means to subtract.
Eight decreased by a number	$8 - s$	"Decreased by" means to subtract. Compare the order of this example to the last one.
Four times a number a	$4a$	"Times" indicates multiplication.
The sum of a number and one-third of the same number	$c + \frac{1}{3}c$ or $\frac{1}{3}c + c$	"Sum" means to add, and "one-third of" means to multiply.
A number divided by five	$\frac{m}{5}$	Notice that the divisor goes on the bottom.
Five divided by a number	$\frac{5}{n}$	Notice that the unknown is the divisor here.
Twice a number decreased by seven	$2k - 7$	"Twice" means to multiply, and "decreased by" means to subtract.
Three more than half a number	$\frac{1}{2}r + 3$ or $\frac{r}{2} + 3$	"More than" means to add. "Half" can mean to multiply by $\frac{1}{2}$ or to divide by 2.
Three times the sum of six and a number	$3(p + 6)$ or $3(6 + p)$	The parentheses mean to find the sum first. Then multiply by 3.
Five divided into the sum of three and a number	$\frac{x + 3}{5}$ or $\frac{3 + x}{5}$	The fraction bar means to find the sum first. Then divide by 5.

EXAMPLE 1: Write an algebraic expression that stands for "2 times a number added to $\frac{1}{2}$ the number."

① $\boxed{2x}$ ② $\boxed{2x + \frac{1}{2}x}$

Step 1. Call the unknown number x. Write the part of the expression that means 2 times x.

Step 2. Write the rest of the expression. Put in a plus sign for the addition and represent $\frac{1}{2}$ of the number as $\frac{1}{2}x$.

ALGEBRA EXERCISE 12

Write an algebraic expression for each of the following, using x to stand for the unknown number.

1. A number increased by nine

2. The product of seven and a number

3. The sum of two times a number and four times the same number

4. Decrease three times a number by ten

5. Eight more than half a number

6. Subtract one from one-third of a number

7. Two less than five times a number

8. Take three times a number from fifteen

9. The sum of a number and its reciprocal

10. Six more than one-fourth of a number

11. Ten decreased by twice a number

12. The sum of seven and a number, all divided by four

13. Ten times the sum of twelve and a number

14. Four less than a number, all multiplied by nine

Answers and solutions start on page 346.

WRITING EQUATIONS

Now that you can translate words into mathematical language, you are ready for the most important step in solving algebraic word problems: writing an equation for solution.

The first simple word problems are statements of equality that must be translated into an equation and solved. Once you master these, you will work with writing and solving algebra "story problems."

To write equations from words, look for the verb in the statement ("is," "equals," etc.) The verb will tell you that one thing equals another. This will help you decide where to place the equal sign.

EXAMPLE 1:   Eight more than three times a number is twenty. Find the number.

①
$$3x + 8 = 20$$

②
$$
\begin{aligned}
3x + 8 &= 20 \\
-8 \quad &\quad -8 \\
\hline
3x \quad &= 12 \\
\overline{3} \quad &\quad \overline{3} \\
x \quad &= \mathbf{4}
\end{aligned}
$$

Step 1.   Write the equation. Represent the unknown with a letter (x) to write the expression "eight more than three times a number." Use the equal sign to represent that this expression is equal to twenty.

Step 2.   Solve for x.

EXAMPLE 2:   Five times a number equals 12 more than three times the same number. Find the number.

①
$$5x = 12 + 3x$$

②
$$
\begin{aligned}
5x &= 12 + 3x \\
-3x &\quad\quad -3x \\
\hline
\frac{2x}{2} &= \frac{12}{2} \\
x &= \mathbf{6}
\end{aligned}
$$

Step 1.   Write the equation. Here, the verb <u>equals</u> tells where to put the = sign.

Step 2.   Solve for x.

EXAMPLE 3:　One-third of a number decreased by seven is four. Find the number.

①
$$\tfrac{1}{3}x - 7 = 4$$

②
$$
\begin{array}{rcl}
\tfrac{1}{3}x - 7 &=& 4 \\
+7 & & +7 \\
\hline
3 \cdot \tfrac{1}{3}x &=& 11 \cdot 3 \\
x &=& 33
\end{array}
$$

Step 1.　Write the equation. Here, the = sign stands in place of the verb <u>is</u>.

Step 2.　Solve for x.

ALGEBRA EXERCISE 13

For each problem, write an equation. Then solve for the unknown number.

1. Take the product of four and a number and decrease it by two. This equals 18.

2. A number increased by seven equals eight.

3. When twice a number is increased by one, the result is 13.

4. Ten more than six times a number equals 34.

5. Eight less than half a number is 12.

6. A number divided by eight equals six.

7. A number equals five more than half the number.

8. Six times a number decreased by four times the number equals 22.

9. The sum of three times a number and two times the number is 40.

10. Decrease the product of nine times a number by the number. This is equal to 14 more than the number.

Answers and solutions start on page 346.

ALGEBRA WORD PROBLEMS

You can apply your algebra skills to solving many word problems. Algebra word problems may seem difficult. However, if you read a problem carefully and set up an equation based on it, you should be able to solve the equation with ease.

To Solve an Algebra Word Problem:

1. Represent the unknown amount with a letter.
2. Write an equation that shows the relationships expressed in the problem.
3. Solve the equation.
4. Reread the problem to make sure the answer is sensible.

In the first two examples below, you will be working with relationships between numbers. The second two examples require you to understand a situation and to set up an equation based on it.

EXAMPLE 1: One number is twice another. When the larger number is decreased by three, the result is the same as when the smaller number is increased by four. Find the numbers.

①
small number = x
large number = $2x$

② $2x - 3 = x + 4$

③
$$
\begin{aligned}
2x - 3 &= x + 4 \\
- x \qquad & \quad -x \\
\hline
x - 3 &= 4 \\
+ 3 \quad & \quad + 3 \\
\hline
x &= 7 \\
2x &= 14
\end{aligned}
$$

Step 1. We know that there are two numbers. The larger is twice the smaller. Let x = the smaller number and $2x$ = the larger.

Step 2. Write an equation that shows the larger number decreased by 3 is equal to the smaller number increased by 4.

Step 3. Solve the equation.

The smaller number is 7 and the larger number is 14.

When you solve algebraic word problems, you can check your work by substituting the values that you find in the original equation. For example, you found $x = 7$ in the problem above. Substitute 7 for x in the equation that you wrote.

$$
\begin{aligned}
\text{Check:}\quad 2x - 3 &= x + 4 \\
2(7) - 3 &= 7 + 4 \\
14 - 3 &= 11 \\
11 &= 11
\end{aligned}
$$

If both sides of the equation are equal, the value you found for the unknown is correct.

EXAMPLE 2: The sum of two consecutive even integers is 34. Find the integers.

| ① First integer = x
 Next even integer = $x + 2$ | ② $x + x + 2 = 34$ | ③ $x + x + 2 = 34$
 $2x + 2 = 34$
 $\underline{\quad -2 \quad -2}$
 $\dfrac{2x}{2} = \dfrac{32}{2}$
 $x - 16$
 $x + 2 = 18$ |

Step 1. The word **consecutive** means one following another.

The word **integer** means a whole number.

The word **even** means divisible by 2 with no remainder.

Choose a letter to represent the first integer. Then use $x + 2$ to represent the next even integer.

Step 2. Write an equation that shows the sum of the integers is equal to 34

Step 3. Solve the equation.

The first integer is $x = 16$.

The next even integer is $x + 2 = 16 + 2 = 18$.

For some problems, a chart helps organize the information.

EXAMPLE 3. Jon is 3 years older than his brother Pete. Fourteen years ago, Jon was twice as old as Pete. How old are they now?

①	**now**
Pete's age	x
Jon's age	$x + 3$

②	**now**	**14 years ago**
Pete's age	x	$x - 14$
Jon's age	$x + 3$	$x + 3 - 14 = x - 11$

③	④
$x - 11 = 2(x - 14)$	$x - 11 = 2x - 28$ $-11 = 2x - x - 28$ $-11 = x - 28$ $28 - 11 = x$ $17 = x$ $20 = x + 3$

Step 1. You know that Jon is now three years older than his brother. Let Pete = x and Jon = $x + 3$.

Step 2. Fill in the chart for their ages 14 years ago. Since they were both 14 years younger then, you can subtract 14 from both of their current ages. Notice that Jon's age 14 years ago can be simplified.

Step 3. Set up an equation that shows the relationship between their ages 14 years ago as given in the problem. In words, this would be "Jon's age 14 years ago was twice Pete's."

Step 4. Solve the equation for x, Pete's age, and $x + 3$, Jon's age.

Pete is now 17 and Jon is 20.

EXAMPLE 4: The Franks, the Rubens, and the Steins went on a camping trip. The Franks spent $100 more than the Rubens, and the Steins paid twice the Franks' total. If the cost of the trip was $580, how much did each family pay?

①
x = Rubens
$x + 100$ = Franks
$2 (x + 100)$ = Steins

② $x + x + 100 + 2(x + 100) = 580$

③
$$x + x + 100 + 2x + 200 = 580$$
$$4x + 300 = 580$$
$$- 300 - 300$$
$$\frac{4x}{4} = \frac{280}{4}$$
$$x = 70$$
$$x + 100 = 170$$
$$2 (x + 100) = 340$$

Step 1. Since the Franks and the Steins are described in terms of other families, start by assigning x to the Rubens. You can then write algebraic expressions for all three families.

Step 2. Write an equation that expresses the total of the families' expenses as equal to $580.

Step 3. Solve the equation for x. Then use x to find the values of $x + 100$ and $2(x + 100)$.

The Rubens paid $70, the Franks paid $170, and the Steins paid $340.

ALGEBRA EXERCISE 14

Solve each problem.

1. The sum of two consecutive integers is 27. Find the integers.

2. The sum of three consecutive even integers is 60. Find the three integers.

3. A larger number is one more than twice a smaller number. Three times the smaller number is four more than the larger one. Find the numbers.

4. One number is six times as large as another. Five less than the larger number is equal to four times the smaller number increased by three. What are the two numbers?

5. There are three times as many women in a GED class as there are men. When the number of women is decreased by two, the result is the same as when the number of men is increased by ten. Find the number of men and women in the class.

6. Juan and his boss Felipe are house painters. For every dollar that Juan gets, Felipe gets $3. They made $360 for painting a house. How much did Felipe make for the job?

7. For every $4 that Mr. Migliaccio makes, his wife makes $3. The Migliaccios make $294 a week. How much does Mr. Migliaccio make? (Let Mr. Migliaccio = 4x. What will you let Mrs. Migliaccio be equal to?)

8. Paul, Jeff, and Jerry worked together fixing Paul's car. Jeff worked twice as many hours as Jerry, and Paul worked six hours more than Jeff. Altogether they worked 51 hours. How many hours did Paul work?

9. In the last local election in Centerville, three voters stayed home for every voter who went to the polls. There are 22,000 registered voters in Centerville. How many of them stayed home for the last election?

10. Gordon is twice as old as Celeste now. Two years ago, Gordon was three times as old as Celeste. How old are they now?

11. David is four times as old as Laila. In ten years, David will be twice as old as Laila. How old are they now?

12. Mario is now 30 years older than Italo. Eighteen years ago Mario was six times as old as Italo. Find their present ages.

13. Renee is 23 years older than her daughter Sara. In 20 years Renee will be twice as old as her daughter. What are their ages now?

14. Two-thirds of a number minus one-quarter of that number equals 10. What is the number?

Answers and solutions start on page 346.

RECTANGULAR COORDINATES

A special graph called the **rectangular coordinate system** is a tool for showing many algebraic relationships.

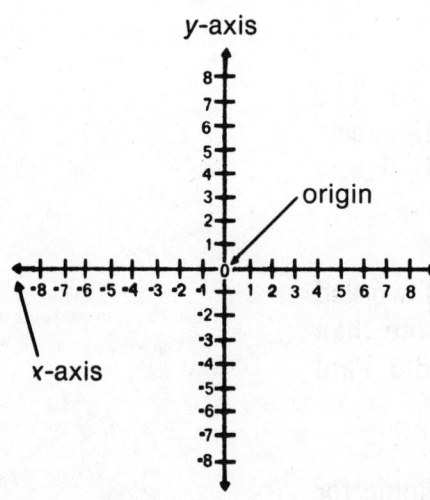

The rectangular coordinate system is made up of a horizontal line called the **x-axis** and a vertical line called the **y-axis.** These two lines intersect at a point marked 0 on the graph. This point is called the **origin.**

A point anywhere on the system can be identified by a pair of numbers called the **coordinates** of the point.

The coordinates of a point are written inside parentheses in the order (*x,y*).

The *x* coordinate indicates a positive value to the right of the *y*-axis or a negative value to the left.

The *y* coordinate indicates a positive value above the *x*-axis or a negative value below it.

The coordinates of the origin are (0,0).

EXAMPLE 1: What are the coordinates of point A in the diagram at the right?

Solution: First, tell how far point A is to the right of the *y*-axis. It is 4 units to the right, or +4. This is the *x*-coordinate. Now, tell how far point A is above the *x*-axis. It is 3 units above the *x*-axis or +3. This is the *y*-coordinate.
The coordinates of point A are **(+4,+3).**

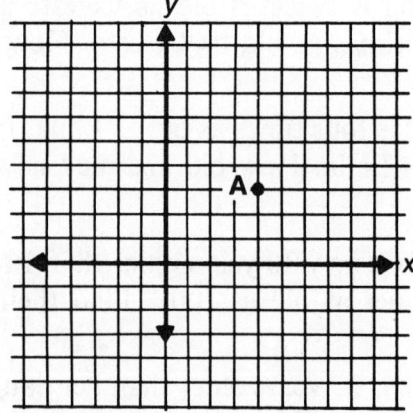

EXAMPLE 2: Point B on the diagram at the right has what coordinates?

Solution: Point B is 2 units to the left of the *y*-axis, or −2, and 1 unit above the *x*-axis, or +1. The coordinates of point B are **(−2,+1).**

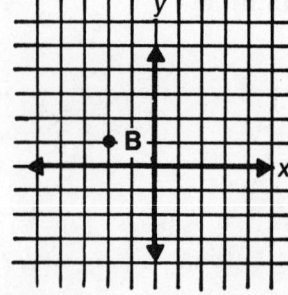

EXAMPLE 3: What are the coordinates of point C in the diagram at the right?

Solution: Point C is 6 units to the left of the y-axis, or −6, and 5 units below the x-axis, or −5. The coordinates of point C are (**−6,−5**).

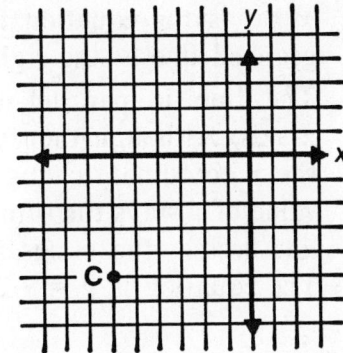

EXAMPLE 4: In the diagram at the right, what are the coordinates of point D?

Solution: Point D is 5 units to the right of the y-axis, or +5, and 3 units below the x-axis, or −3. The coordinates of point D are (**+5,−3**).

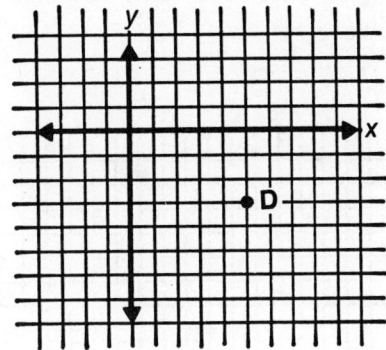

The above examples show how points can be plotted on rectangular coordinates. A line is a connected series of points. If the points are based on the solution to an equation, the points can be connected in a straight line. This is called a **graphed line.**

In Example 4, you identified the coordinates of point D as (+5,−3). These tell you that (+5,−3) are the coordinates of a point and that (+5,−3) is a point on two separate graphed lines; one is parallel to the x-axis, and one is parallel to the y-axis.

Point D lies on the line $x = +5$ shown to the right. The line is drawn parallel to the y-axis through points that have +5 as their x coordinate. For the line $x = +5$, the y coordinates can be −2,−1,0,1,2,3, etc., while the value of the x coordinate for all points is always +5.

Similarly, the point (+5,−3) is a point on a line called $y = −3$ that is parallel to the x-axis. On this line, the x coordinates of points can equal −3,−2,−1,0,1,2,3, etc., while the value of y for all points is always −3.

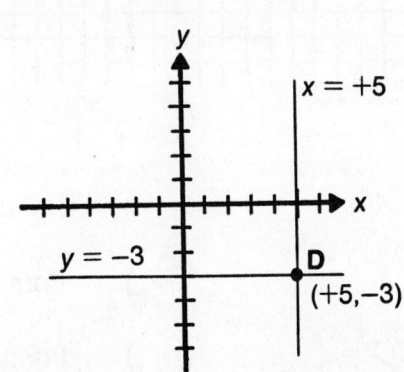

EXAMPLE 5:   What is the equation for the
graphed line at the right?

Solution: The line is parallel to the
x-axis. Although the value of
the x coordinate changes, the
value of y stays the same. The
line crosses the y-axis at -2.
The equation is $y = -2$.

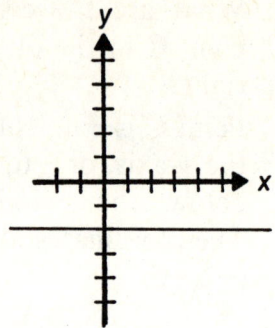

ALGEBRA EXERCISE 15

Write the coordinates for each point shown on the rectangular coordinate
system below.

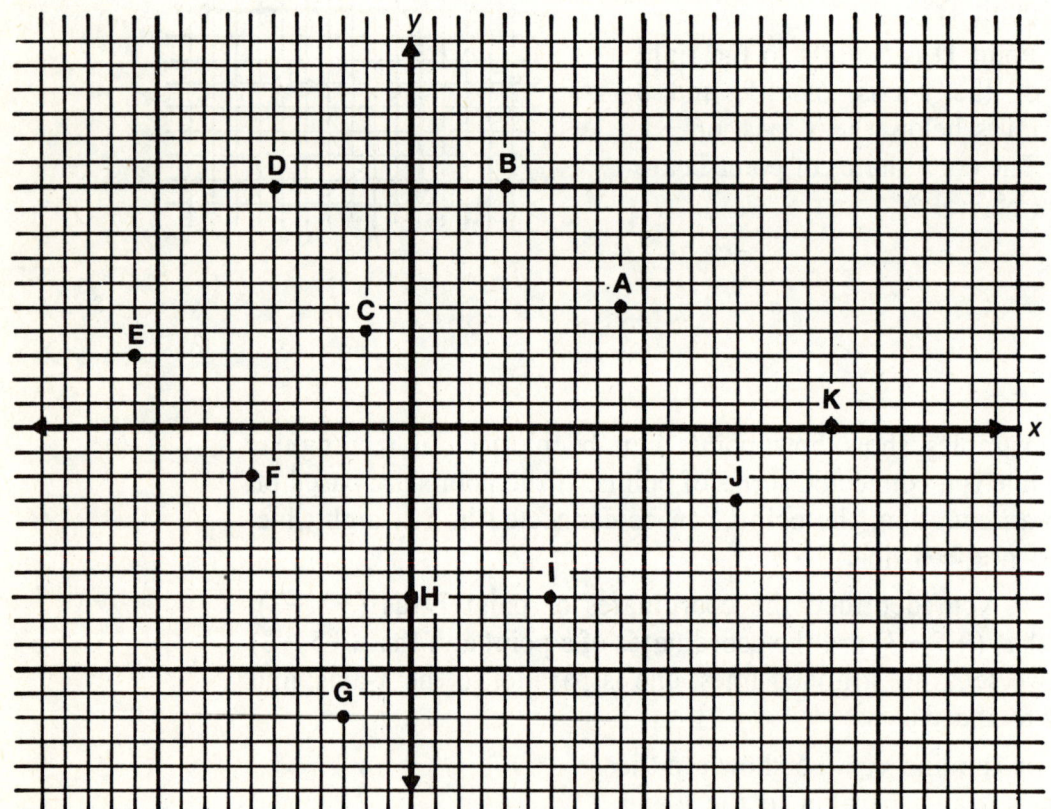

1. Point A = (   )   Point E = (   )   Point I = (

   Point B = (   )   Point F = (   )   Point J = (

   Point C = (   )   Point G = (   )   Point K = (

   Point D = (   )   Point H = (   )

2. What is the equation for the graphed line at the right?

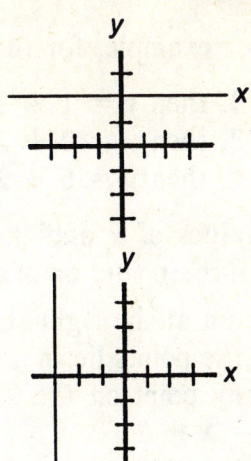

3. Identify the equation for the line shown on this graph.

Answers and solutions start on page 348.

Linear Equations

In addition to drawing a graphed line parallel to the x-axis or the y-axis, you can graph other **linear equations** on rectangular coordinates. This type of equation is called linear because its graph is a straight line.

The equation $y = x$ is one of the simplest linear equations, since in all cases, the x and y coordinates of a point have the same value. You can chart ordered pairs of coordinates for a linear equation, plot the points, and then draw a graphed line through them.

EXAMPLE 1: Graph the line $y = x$.

x	−2	−1	0	1	2	3
y	−2	−1	0	1	2	3

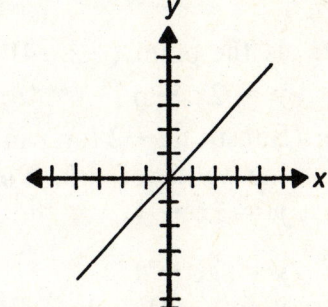

Step 1. Write out some values for x and y. Since $x = y$, the x and y coordinate for a point will have the same value.

Step 2. Plot the points on a graph.

Step 3. Draw a straight line through the points and label the line $y = x$.

The equation $y = x + 2$ has many solutions. If you select values for x, you can find the corresponding values for y by substituting and solving the equation. For example, for the equation: $y = x + 2$

if $x = 1$, then $y = 1 + 2 = 3$
if $x = 3$, then $y = 3 + 2 = 5$
if $x = 6$, then $y = 6 + 2 = 8$

The three values of x and their corresponding values of y are the coordinates for three points on a graphed line: (1,3), (3,5) and (6,8).

The diagram at the right shows these three points. Notice that the points lie in a straight line. The coordinates of any point on the line are solutions to the equation $y = x + 2$.

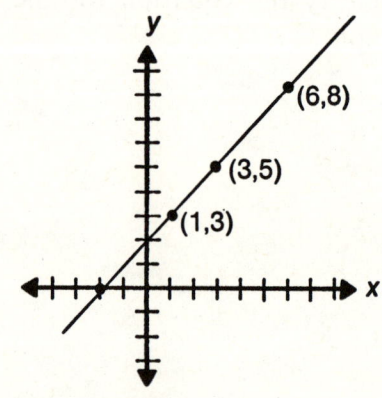

Besides simple linear equations, you can graph more complicated equations. In order to graph such equations, you must solve for at least three pairs of coordinates.

On the GED, you will not graph an equation, but you may need to know whether certain coordinates lie on the graphed line of an equation.

To find whether a point is on the graph of a linear equation, substitute the x coordinate of the point into the equation. If the solution to the equation is the same as the y coordinate, then you know the point is on the graph of that equation.

EXAMPLE 1: Is the point (2,3) on the graph of the equation $y = 2x - 1$?

Solution: Substitute 2 for x in the equation $y = 2x - 1$. This gives y the value of 3. **Point (2,3) is on the graph of $y = 2x - 1$.**

$y = 2x - 1$
$y = 2(2) - 1$
$y = 4 - 1$
$y = 3$

EXAMPLE 2: Is the point (−2,−4) on the graph of the equation $y = 2x - 1$?

Solution: Substitute −2 for x in the equation $y = 2x - 1$. This gives y the value of −5. **Point (−2,−4) is <u>not</u> on the graph of $y = 2x - 1$.**

$y = 2x - 1$
$y = 2(-2) - 1$
$y = -4 - 1$
$y = -5$

For some questions about linear equations, you may only need to look at a graph and count the spaces between points.

EXAMPLE 1: What is the distance from S to U on the graph to the right?

Solution: S is 5 units above the *x*-axis and U is 3 units below.

$$5 + 3 = 8$$

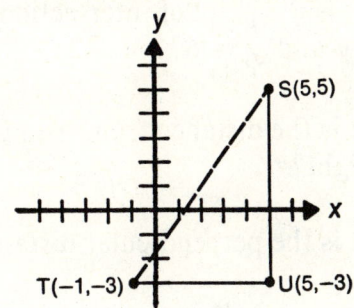

EXAMPLE 2: What is the distance from T to U?

Solution: T is 1 unit to the left of the *y*-axis. U is 5 units to the right.

$$5 + 1 = 6$$

You can also find the distance between two points that lie on a diagonal line. On the graph, a dotted line has been drawn in to connect points T and S. These three points can be connected to form right triangle STU.

Use the Pythagorean Theorem that you learned in geometry to find the length of line ST.

EXAMPLE 3: Find the length of ST.

Solution:
$$c^2 = a^2 + b^2$$
$$ST^2 = SU^2 + TU^2$$
$$ST^2 = 8^2 + 6^2$$
$$ST^2 = 64 + 36$$
$$ST = \sqrt{100}$$
$$ST = 10$$

ALGEBRA EXERCISE 16 ———————————————

Solve each problem.

1. Is the point (6, 7) on the graph of the equation $y = \frac{1}{2}x + 4$?

2. Is the point (3, 4) on the graph of the equation $y = \frac{2}{3}x + 1$?

3. Is the point (8, −5) on the graph of the equation $y = -x + 3$?

Use the graph to the right to answer questions 4–7.

4. What is the point of intersection between the line
   $x = y$ and $y = 6$?

5. What is the distance from A to C on the graph at
   the right?

6. What is the perpendicular distance from B to the
   x-axis?

7. What is the distance from B to C?

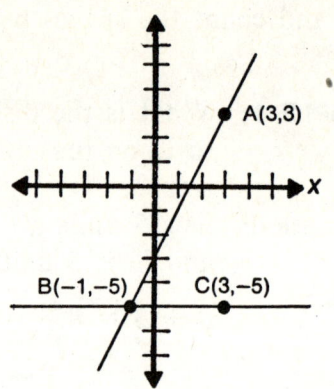

Use the graph to the right to answer questions 8–11.

8. What is the distance from E to F on the graph at
   the right?

9. What is the perpendicular distance from D to the
   x-axis?

10. What is the distance from D to F?

11. Find the length of line DE.

12. The line connecting points R and S is the graph for the equation
    $y = \frac{2}{3}x + 1$. Is the point $(-6, -3)$ on the graph?

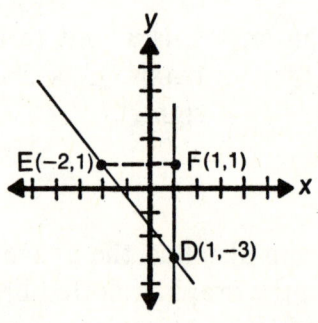

Answers and solutions start on page 348.

Slope and Intercepts

Slope is a measure of how "steep" a line is.

A line that goes up from left to right has a <u>positive</u> slope.

A line that goes down from left to right has a <u>negative</u> slope.

A horizontal line has a <u>zero</u> slope.

A vertical line has an <u>undefined</u> slope.

Look at the four lines labeled A, B, C, and D on the
rectangular coordinate system at the right. Line A has
a positive slope. Line B has a negative slope. Line C
has a zero slope. Line D has an undefined slope.

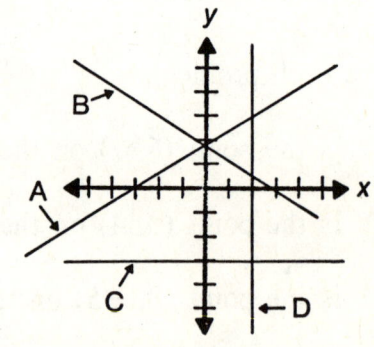

To find the exact slope of a line, substitute values
into the formula:

$$\text{slope} = \frac{y_2 - y_1}{x_2 - x_1}$$

where (x_1, y_1) and (x_2, y_2) are the coordinates of two points on the line of a graph. In words, the formula means: divide the difference between two y-values by the difference between two x-values.

EXAMPLE: What is the slope of a line that passes through the points (2,5) and (6,13)?

Solution: Subtract the y-values: $13 - 5 = 8$, and subtract the x-values: $6 - 2 = 4$. Then divide: $8 \div 4 = +2$.

$$\text{slope} = \frac{13 - 5}{6 - 2} = \frac{8}{4} = +2$$

The **intercept** tells where a line crosses an axis. In the picture at the right, the line labeled m crosses the y-axis at $+3$. The coordinates of the y-intercept are (0,3). The line crosses the x-axis at -2. The coordinates of the x-intercept are $(-2,0)$.

To find the y-intercept of a linear equation, substitute 0 for x and solve for y.

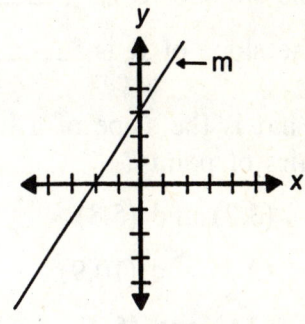

EXAMPLE: What is the y-intercept of the equation $y = 4x - 5$?

Solution: Replace x with 0 and solve for y in $y = 4x - 5$.
The coordinates of the y-intercept are $(0,-5)$.

$$y = 4x - 5$$
$$y = 4 \cdot 0 - 5$$
$$y = 0 - 5$$
$$y = -5$$

To find the x-intercept of a linear equation, substitute 0 for y and solve for x.

EXAMPLE: What is the x-intercept of the equation $y = 3x - 6$?

Solution: Replace y with 0 and solve for x in $y = 3x - 6$.
The coordinates of the x-intercept are (2,0).

$$y = 3x - 6$$
$$0 = 3x - 6$$
$$\underline{+6 \qquad +6}$$
$$\frac{6}{3} = \frac{3x}{3}$$
$$2 = x$$

ALGEBRA EXERCISE 17 ─────────────────

Solve each problem.

1. Identify the slope of each line as positive, negative, zero or undefined.

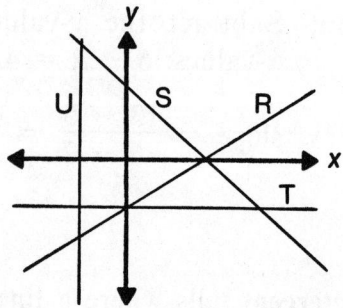

The slope of U is _____

The slope of T is _____

The slope of S is _____

The slope of R is _____

2. What is the slope of a line that passes through each of the following pairs of points?

    a. (3,2) and (5,8)

    b. (2,5) and (10,9)

    c. (3,6) and (5,2)

    d. (2,4) and (−7,10)

    e. (3,−3) and (−1,5)

3. Find the *x*-intercept and the *y*-intercept for each of the following equations.

	x-intercept	*y*-intercept
a. $y = 3x - 9$		
b. $y = x - 6$		
c. $y = \frac{1}{2}x + 5$		
d. $y = \frac{3}{4}x + 12$		
e $y = 2x - 1$		

Answers and solutions start on page 349.

ALGEBRA REVIEW ────────────────────────────

For each problem, fill in the circle that corresponds to the best answer.

1. Which point corresponds to $-\frac{3}{2}$?

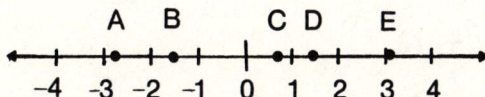

(1) A
(2) B
(3) C
(4) D
(5) E

2. Find the sum of -8, -17, 20, and -9.
   (1) -14
   (2) -4
   (3) -24
   (4) ·34
   (5) -54

3. Simplify $(-12) + (-17) - (+10) - (-6)$.
   (1) -33
   (2) -45
   (3) -13
   (4) $+1$
   (5) -23

4. $(-2)(8)(-\frac{3}{4})(+5)$ is equal to which of the following?
   (1) 120
   (2) -120
   (3) 60
   (4) -60
   (5) 90

5. Simplify $\frac{-35}{42}$.
   (1) -7
   (2) $-\frac{6}{5}$
   (3) $+7$
   (4) $-1\frac{1}{6}$
   (5) $-\frac{5}{6}$

6. Solve for m in $\frac{m}{.05} = 200$.
   (1) 10
   (2) 100
   (3) 1000
   (4) 4
   (5) 40

7. Solve for x in $75 = 9x + 12$.
   (1) 6
   (2) $5\frac{1}{2}$
   (3) $9\frac{2}{3}$
   (4) 7
   (5) 8

8. Solve for a in $\frac{3}{4}a - 8 = -2$.
   (1) $5\frac{1}{2}$
   (2) 6
   (3) 8
   (4) $13\frac{1}{3}$
   (5) -6

9. Solve for n in $7n - 10 = 3n + 14$.
   (1) 40
   (2) 6
   (3) $2\frac{2}{5}$
   (4) -1
   (5) 4

10. Solve for c in $5(c - 4) = 3c + 16$.
    (1) 2
    (2) 4
    (3) 8
    (4) 12
    (5) 18

11. Which of the following could be a value for c in $3c - 8 \leq 7$?
    (1) 4
    (2) $5\frac{1}{2}$
    (3) 6
    (4) 7
    (5) 8

12. Which of the following expresses four times the sum of ten plus a number?

(1) 4(10*n*)
(2) 4*n* + 10
(3) 4(*n* + 10)
(4) 10(*n* + 4)
(5) *n*(4 + 10)

13. Five times a number decreased by six is the same as twice the number increased by 15. Find the number.

(1) 3
(2) 7
(3) 17
(4) $1\frac{4}{5}$
(5) 21

14. Find the second of two consecutive even integers whose sum is 86.

(1) 41
(2) 42
(3) 43
(4) 44
(5) 46

15. At the Central Utility Company, the number of men is 40 less than 3 times the number of women. Altogether the company has 1,260 employees. How many men work there?

(1) 935
(2) 955
(3) 305
(4) 325
(5) 945

16. What are the coordinates of point A in the picture at the right?

(1) (−4,−4)
(2) (−4,−3)
(3) (+3,−4)
(4) (−4,+3)
(5) (−3,+3)

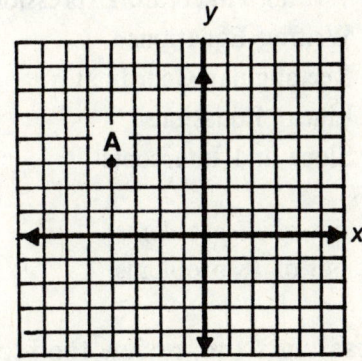

17. What is the distance from A to B on the graph at the right?

(1) 5
(2) 6
(3) 7
(4) 8
(5) 9

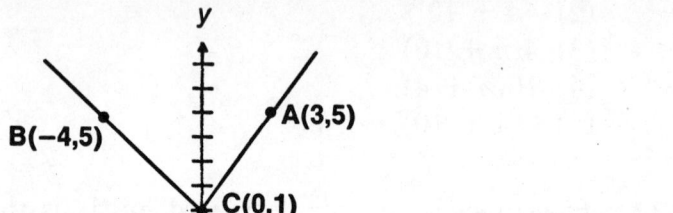

18. What is the slope of the line that passes through points A and C?

(1) +3
(2) +5
(3) $+\frac{3}{4}$
(4) $+\frac{4}{3}$
(5) $+\frac{5}{2}$

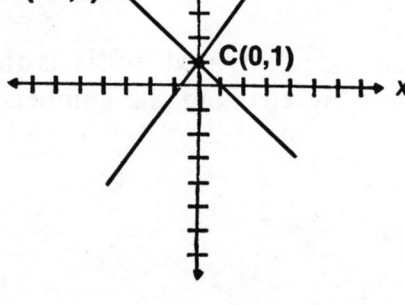

Answers and solutions start on page 350.

ALGEBRA TEST EVALUATION

Problem	Section	Starting Page
1	The Number Line	299
2	Addition of Signed Numbers	300
3	Subtraction of Signed Numbers	304
4	Multiplication of Signed Numbers	306
5	Division of Signed Numbers	308
6	Equations	310
7, 8	Equations of More Than One Operation	313
9, 10	Equations with Separated Unknowns	314
11	Inequalities	318
12	Writing Algebraic Expressions	320
13, 14, 15	Writing Equations	322
16	Rectangular Coordinates	328
17	Linear Equations	331
18	Slope and Intercepts	334

Passing Score: __15__ right out of 18 problems.
Your Score: _____ right out of 18 problems.

If you had less than a passing score, review the sections for the problems you got wrong. Then repeat this test before you go on to the practice GED test.

If you had a passing score, correct any problem you got wrong. Then go to the practice GED test.

ANSWERS AND SOLUTIONS

Algebra Exercise 1

1.	J	**5.**	F			**8.**	H
2.	D	**6.**	G			**9.**	C
3.	B	**7.**	I			**10.**	E
4.	A						

Algebra Exercise 2

1. −12 $\begin{array}{r} -\ 9 \\ -\ 3 \\ \hline -12 \end{array}$

2. −2 $\begin{array}{r} -8 \\ +6 \\ \hline -2 \end{array}$

3. $-8\frac{1}{2}$ $\begin{array}{r} -20\frac{1}{2} \\ +12 \\ \hline -\ 8\frac{1}{2} \end{array}$

4. $-7\frac{1}{2}$ $\begin{array}{r} -20 \\ +12\frac{1}{2} \\ \hline -\ 7\frac{1}{2} \end{array}$

5. +6 $\begin{array}{r} +\ 5 \\ +\ 9 \\ \hline +14 \end{array}$ $\begin{array}{r} +14 \\ -\ 8 \\ \hline +\ 6 \end{array}$

6. −3 $\begin{array}{r} -\ 6 \\ -\ 5 \\ \hline -11 \end{array}$ $\begin{array}{r} -11 \\ +\ 8 \\ \hline -\ 3 \end{array}$

7. −14 $\begin{array}{r} -18 \\ -\ 2 \\ \hline -20 \end{array}$ $\begin{array}{r} -20 \\ +\ 6 \\ \hline -14 \end{array}$

8. +9 $\begin{array}{r} +\ 8 \\ +11 \\ \hline +19 \end{array}$ $\begin{array}{r} -\ 6 \\ -\ 4 \\ \hline -10 \end{array}$ $\begin{array}{r} +19 \\ -10 \\ \hline +\ 9 \end{array}$

9. +13 $\begin{array}{r} -\ 7 \\ -\ 5 \\ -\ 3 \\ -10 \\ -\ 2 \\ \hline -27 \end{array}$ $\begin{array}{r} +40 \\ -27 \\ \hline +13 \end{array}$

10. 0 $\begin{array}{r} +\ 3 \\ +\ 7 \\ \hline +10 \end{array}$ $\begin{array}{r} +10 \\ -10 \\ \hline 0 \end{array}$

11. −17 $\begin{array}{r} -10 \\ -\ 3 \\ -\ 9 \\ \hline -22 \end{array}$ $\begin{array}{r} -22 \\ +\ 5 \\ \hline -17 \end{array}$

12. 0 $\begin{array}{r} +\ 8 \\ +\ 7 \\ \hline +15 \end{array}$ $\begin{array}{r} -\ 9 \\ -\ 4 \\ -\ 2 \\ \hline -15 \end{array}$ $\begin{array}{r} +15 \\ -15 \\ \hline 0 \end{array}$

13. −14° $\begin{array}{r} -10 \\ -\ 4 \\ \hline -14 \end{array}$

14. −22 yards $\begin{array}{r} -37 \\ +15 \\ \hline -22 \end{array}$

15. no $\begin{array}{r} +235 \\ +\ 55 \\ \hline +290 \text{ in the bank} \end{array}$ $\begin{array}{r} -150 \\ -120 \\ -\ 30 \\ \hline -300 \text{ in checks} \end{array}$

Algebra Exercise 3

1. +2 $(+6) - (+4) =$
$+6 - 4 = +2$

2. −11 $(-8) - (+3) =$
$-8 - 3 = -11$

3. −1 $(-9) - (-8) =$
$-9 + 8 = -1$

4. +19 $(+10) - (-9) =$
$+10 + 9 = 19$

5. +1 $(+8) - (7) =$
$8 - 7 = 1$

6. 0 $(-9) - (-9) =$
$-9 + 9 = 0$

7. −22 $-10 - (12) =$
$-10 - 12 = -22$

8. +7 $(+6) - (-3) + (-2) =$
$+6 + 3 - 2 = 7$

9. −23 $(-9) - (+4) - (+10) =$
$-9 - 4 - 10 = -23$

10. −29 $(-15) - (20) + (+6) =$
$-15 - 20 + 6 = -29$

Exercise 3 cont'd.

11. −27 $(-8) + (-13) - (+6) =$
$-8 - 13 - 6 = -27$

12. 4 $(-3) + (-4) - (-5) - (-6) =$
$-3 - 4 + 5 + 6 = 4$

13. 65 $(+22) - (-43) =$
$+22 + 43 = 65$

14. 24° drop in temperature $(+7) - (-17) =$
$7 + 17 = 24$

Algebra Exercise 4

1. −18
2. +36
3. −45
4. +24
5. −9

6. $-\frac{1}{2}$
7. +96
8. +721
9. −40

10. +5
11. −192
12. −28
13. −28

14. 9 accounts $(-2)\,(4) = -8$
$17 - 8 = 9$

15. 18 fewer children $(+3)\,(-6) = -18$

Algebra Exercise 5

1. +2
2. −2
3. −8
4. $-\frac{2}{3}$
5. −3

6. $-\frac{5}{6}$
7. +8
8. $-\frac{1}{2}$
9. −6
10. +12

11. $+\frac{4}{5}$
12. −13
13. $+\frac{3}{2}$ or $1\frac{1}{2}$
14. $-\frac{9}{5}$ or $-1\frac{4}{5}$
15. $-\frac{3}{4}$

Algebra Exercise 6

1. +24 $(15) - (-9) = +15 + 9 = +24$

2. −120 $(-6)(+20) = -120$

3. −8 $-4 + 3 - 7 = -11 + 3 = -8$

4. +12 $\frac{-96}{-8} = +12$

5. −20 $(-8) - (12) = -8 - 12 = -20$

6. −100 $\frac{+1,000}{-10} = -100$

7. −43 $-29 - 14 = -43$

8. +1 $(-20) - (-21) = -20 + 21 = +1$

9. −7 $(20) + (-12) - (+15) =$
$20 - 12 - 15 = -7$

10. +20 $(-3)(+\frac{2}{3})(-10) = -\overset{1}{\cancel{3}} \cdot \frac{2}{\underset{1}{\cancel{3}}} \cdot -10 = +20$

11. $-\frac{2}{3}$ $\frac{-14 \div 7}{21 \div 7} = -\frac{2}{3}$

12. $-\frac{7}{8}$ $(-1)(+\frac{7}{8}) = -\frac{7}{8}$

13. 25° $15 - (-10) = 15 + 10 = 25$

14. 12 pounds less $(-2)(+6) = -12$

Algebra Exercise 7

1. 12 $\frac{8y}{8} = \frac{96}{8}$
$y = 12$

2. 37 $\begin{array}{r} f + 20 = 57 \\ \underline{-20 } \quad \underline{-20} \\ f = 37 \end{array}$

3. 47 $\begin{array}{r} b - 19 = 28 \\ \underline{+19 } \quad \underline{+19} \\ b = 47 \end{array}$

4. 27 $3 \cdot \frac{x}{3} = 9 \cdot 3$
$x = 27$

5. 35 $\begin{array}{r} 42 = t + 7 \\ \underline{-7} \quad \underline{-7} \\ 35 = t \end{array}$

6. $5\frac{1}{2}$ $\frac{11}{2} = \frac{2y}{2}$
$5\frac{1}{2} = y$

7. 24 $\begin{array}{r} n + 36 = 60 \\ \underline{-36 } \quad \underline{-36} \\ n = 24 \end{array}$

8. 41 $\begin{array}{r} 33 = k - 8 \\ \underline{+8} \quad \underline{+8} \\ 41 = k \end{array}$

9. 31
$$c - 4 = 27$$
$$\underline{+4} \quad \underline{+4}$$
$$c = 31$$

10. 5
$$\frac{15p}{15} = \frac{75}{15}$$
$$p = 5$$

11. 36
$$4 \cdot 9 = \frac{m}{4} \cdot 4$$
$$36 = m$$

12. 24
$$18 = d - 6$$
$$\underline{+6} \quad \underline{+6}$$
$$24 = d$$

13. 17
$$v - 17 = 0$$
$$\underline{+17} \quad \underline{+17}$$
$$v = 17$$

14. 4
$$\frac{25z}{25} = \frac{100}{25}$$
$$z = 4$$

15. 23
$$41 = c + 18$$
$$\underline{-18} \quad \underline{-18}$$
$$23 = c$$

Algebra Exercise 8

1. 8
$$7m - 2 = 54$$
$$\underline{+2} \quad \underline{+2}$$
$$\frac{7m}{7} = \frac{56}{7}$$
$$m = 8$$

2. 12
$$\frac{a}{3} + 5 = 9$$
$$\underline{-5} \quad \underline{-5}$$
$$3 \cdot \frac{a}{3} = 4 \cdot 3$$
$$a = 12$$

3. 8
$$7 = \frac{c}{2} + 3$$
$$\underline{-3} \quad \underline{-3}$$
$$2 \cdot 4 = \frac{c}{2} \cdot 2$$
$$8 = c$$

4. 8
$$82 = 9d + 10$$
$$\underline{-10} \quad \underline{-10}$$
$$\frac{72}{9} = \frac{9d}{9}$$
$$8 = d$$

5. 8
$$25c - 17 = 183$$
$$\underline{+17} \quad \underline{+17}$$
$$\frac{25c}{25} = \frac{200}{25}$$
$$c = 8$$

6. 20
$$\frac{w}{2} - 7 = 3$$
$$\underline{+7} \quad \underline{+7}$$
$$2 \cdot \frac{w}{2} = 10 \cdot 2$$
$$w = 20$$

7. 1
$$-2 = 6x - 8$$
$$\underline{+8} \quad \underline{+8}$$
$$\frac{+6}{6} = \frac{6x}{6}$$
$$1 = x$$

8. 9
$$\frac{1}{3}p + 8 = 11$$
$$\underline{-8} \quad \underline{-8}$$
$$3 \cdot \frac{1}{3}p = 3 \cdot 3$$
$$p = 9$$

9. 2
$$40 = 13z + 14$$
$$\underline{-14} \quad \underline{-14}$$
$$\frac{26}{13} = \frac{13z}{13}$$
$$2 = z$$

10. 8
$$\frac{n}{2} + 3 = 7$$
$$\underline{-3} \quad \underline{-3}$$
$$2 \cdot \frac{n}{2} = 4 \cdot 2$$
$$n = 8$$

11. -3
$$3a + 10 = 1$$
$$\underline{-10} \quad \underline{-10}$$
$$\frac{3a}{3} = \frac{-9}{3}$$
$$a = -3$$

12. 20
$$\frac{3}{4}y - 3 = 12$$
$$\underline{+3} \quad \underline{+3}$$
$$\frac{4}{3} \cdot \frac{3}{4}y = 15 \cdot \frac{4}{3}$$
$$y = 20$$

13. 3
$$39 = 16k - 9$$
$$\underline{+9} \quad \underline{+9}$$
$$\frac{48}{16} = \frac{16k}{16}$$
$$3 = k$$

14. $\frac{1}{2}$
$$10 = 6a + 7$$
$$\underline{-7} \quad \underline{-7}$$
$$\frac{3}{6} = \frac{6a}{6}$$
$$\frac{1}{2} = a$$

15. $\frac{1}{3}$
$$9r + 15 = 18$$
$$\underline{-15} \quad \underline{-15}$$
$$\frac{9r}{9} = \frac{3}{9}$$
$$r = \frac{1}{3}$$

Algebra Exercise 9

1. 7
$$5y - y = 19 + 9$$
$$\frac{4y}{4} = \frac{28}{4}$$
$$y = 7$$

2. -5
$$6t + 8 + 4t = -42$$
$$+ 8 + 10t = -42$$
$$\underline{-8} \quad \underline{-8}$$
$$\frac{10t}{10} = \frac{-50}{10}$$
$$t = -5$$

Exercise 9 cont'd.

3. 4
$$9c = 44 - 2c$$
$$\underline{+2c \qquad + 2c}$$
$$\frac{11c}{11} = \frac{44}{11}$$
$$c = 4$$

4. 5
$$8m = 2m + 30$$
$$\underline{-2m \quad -2m}$$
$$\frac{6m}{6} = \frac{30}{6}$$
$$m = 5$$

5. 11
$$4a + 55 = 9a$$
$$\underline{-4a \qquad -4a}$$
$$\frac{55}{5} = \frac{5a}{5}$$
$$11 = a$$

6. 6
$$4p = p + 18$$
$$\underline{-p \quad -p}$$
$$\frac{3p}{3} = \frac{18}{3}$$
$$p = 6$$

7. 2
$$6\tfrac{1}{2}f = 14 - \tfrac{1}{2}f$$
$$\underline{+\tfrac{1}{2}f \qquad + \tfrac{1}{2}f}$$
$$\frac{7f}{7} = \frac{14}{7}$$
$$f = 2$$

8. $\tfrac{1}{3}$
$$3 = y + 8y$$
$$\frac{3}{9} = \frac{9y}{9}$$
$$\frac{1}{3} = y$$

9. 5
$$8r + 17 = 5r + 32$$
$$\underline{-5r \qquad -5r}$$
$$3r + 17 = 32$$
$$\underline{\quad - 17 \qquad -17}$$
$$\frac{3r}{3} = \frac{15}{3}$$
$$r = 5$$

10. 4
$$7n - 9 = 3n + 7$$
$$\underline{-3n \quad = -3n}$$
$$4n - 9 = \quad + 7$$
$$\underline{\quad + 9 \qquad + 9}$$
$$\frac{4n}{4} = \frac{16}{4}$$
$$n = 4$$

11. 9
$$6z + 11 = 5z + 20$$
$$\underline{-5z \quad = -5z}$$
$$z + 11 = \quad 20$$
$$\underline{\quad - 11 \qquad -11}$$
$$z = 9$$

12. 10
$$5y = 6y - 10$$
$$\underline{-6y \quad -6y}$$
$$-y = -10$$
$$y = +10$$

13. -7
$$4y = 7y + 21$$
$$\underline{-7y \quad -7y}$$
$$\frac{-3y}{-3} = \frac{21}{-3}$$
$$y = -7$$

14. 3
$$x + y = 12$$
$$3y + y = 12$$
$$\frac{4y}{4} = \frac{12}{4}$$
$$y = 3$$

15. 4
$$x + y = 6$$
$$x + \tfrac{1}{2}x = 6$$
$$1\tfrac{1}{2}x = 6$$
$$\tfrac{2}{3} \cdot \tfrac{3}{2}x = 6 \cdot \tfrac{2}{3}$$
$$x = 4$$

Algebra Exercise 10

1. 4
$$3(x + 4) = x + 20$$
$$3x + 12 = x + 20$$
$$\underline{- x \qquad -x}$$
$$2x + 12 = \quad + 20$$
$$\underline{\quad - 12 \qquad - 12}$$
$$\frac{2x}{2} = \frac{8}{2}$$
$$x = 4$$

2. 8
$$2a + 9 = 5(a - 3)$$
$$2a + 9 = 5a - 15$$
$$\underline{-2a \qquad -2a}$$
$$\quad + 9 = 3a - 15$$
$$\underline{\quad + 15 \qquad + 15}$$
$$\frac{+ 24}{3} = \frac{3a}{3}$$
$$8 = a$$

3. −2

$$2 - 3c = 4(c + 4)$$

$$
\begin{array}{rcl}
2 - 3c &=& 4c + 16 \\
\underline{+\,3c} & & \underline{+3c} \\
2 &=& 7c + 16 \\
\underline{-16} & & \underline{-\,16} \\
\underline{-14} &=& \underline{7c} \\
7 & & 7 \\
-2 &=& c
\end{array}
$$

4. 6

$$2(m - 1) = 7m - 32$$

$$
\begin{array}{rcl}
2m - 2 &=& 7m - 32 \\
\underline{-2m} & & \underline{-2m} \\
-2 &=& 5m - 32 \\
\underline{+\,32} & & \underline{+\,32} \\
\underline{+\,30} &=& \underline{5m} \\
5 & & 5 \\
6 &=& m
\end{array}
$$

5. 5

$$3(y - 2) = y + 4$$

$$
\begin{array}{rcl}
3y - 6 &=& y + 4 \\
\underline{-\,y} & & \underline{-y} \\
2y - 6 &=& +\,4 \\
\underline{+\,6} & & \underline{+\,6} \\
\underline{2y} &=& \underline{+10} \\
2 & & 2 \\
y &=& 5
\end{array}
$$

6. 3

$$80 = 8(s + 7)$$

$$
\begin{array}{rcl}
80 &=& 8s + 56 \\
\underline{-56} & & \underline{-\,56} \\
\underline{24} &=& \underline{8s} \\
8 & & 8 \\
3 &=& s
\end{array}
$$

7. 5

$$5(t + 4) = 3(t + 10)$$

$$
\begin{array}{rcl}
5t + 20 &=& 3t + 30 \\
\underline{-3t} & & \underline{-3t} \\
2t + 20 &=& +\,30 \\
\underline{-\,20} & & \underline{-\,20} \\
\underline{2t} &=& \underline{10} \\
2 & & 2 \\
t &=& 5
\end{array}
$$

8. 8

$$7(m - 2) = 3(m + 6)$$

$$
\begin{array}{rcl}
7m - 14 &=& 3m + 18 \\
\underline{-3m} & & \underline{-3m} \\
4m - 14 &=& +\,18 \\
\underline{+\,14} & & \underline{+\,14} \\
\underline{4m} &=& \underline{32} \\
4 & & 4 \\
m &=& 8
\end{array}
$$

Algebra Exercise 11

1. $c < 6$

$$
\begin{array}{rcl}
4c - 3 &<& 21 \\
\underline{+\,3} & & \underline{+\,3} \\
\underline{4c} &<& \underline{24} \\
4 & & 4 \\
c &<& 6
\end{array}
$$

2. $p > 2$

$$
\begin{array}{rcl}
3p + 1 &>& 7 \\
\underline{-\,1} & & \underline{-1} \\
\underline{3p} &>& \underline{6} \\
3 & & 3 \\
p &>& 2
\end{array}
$$

3. $w \le 3$

$$
\begin{array}{rcl}
9w + 2 &\le& 29 \\
\underline{-\,2} & & \underline{-\,2} \\
\underline{9w} &\le& \underline{27} \\
9 & & 9 \\
w &\le& 3
\end{array}
$$

4. $x \ge 2\frac{1}{2}$

$$
\begin{array}{rcl}
8x - 5 &\ge& 15 \\
\underline{+\,5} & & \underline{+\,5} \\
\underline{8x} &\ge& \underline{20} \\
8 & & 8 \\
x &\ge& 2\tfrac{1}{2}
\end{array}
$$

5. no

$$
\begin{array}{rcl}
m - 6 &>& 1 \\
\underline{+\,6} & & \underline{+6} \\
m &>& 7
\end{array}
$$

Since m is greater than 7, it cannot equal 7.

6. yes

$$
\begin{array}{rcl}
\tfrac{1}{2}r + 5 &\le& 9 \\
\underline{-\,5} & & \underline{-5} \\
2 \cdot \tfrac{1}{2}r &\le& 4 \cdot 2 \\
r &\le& 8
\end{array}
$$

Since r is less than or equal to 8, r could be $\frac{1}{2}$.

7. yes

$$
\begin{array}{rcl}
d + 7 &\le& 2 \\
\underline{-\,7} & & \underline{-7} \\
d &\le& -5
\end{array}
$$

Since d is less than or equal to −5, it could be −6.

Exercise 11 cont'd.

8. yes

$$2f - 8 < 4$$
$$\underline{+8 \qquad +8}$$
$$\frac{2f}{2} < \frac{12}{2}$$
$$f < 6$$

Since f is less than 6,
f could be 4.

Algebra Exercise 12

1. $x + 9$ or $9 + x$
2. $7x$
3. $2x + 4x$ or $4x + 2x$
4. $3x - 10$
5. $\frac{1}{2}x + 8$ or $8 + \frac{1}{2}x$
6. $\frac{1}{3}x - 1$
7. $5x - 2$

8. $15 - 3x$
9. $x + \frac{1}{x}$ or $\frac{1}{x} + x$
10. $\frac{1}{4}x + 6$ or $6 + \frac{1}{4}x$
11. $10 - 2x$
12. $\frac{x + 7}{4}$ or $\frac{7 + x}{4}$
13. $10(x + 12)$ or $10(12 + x)$
14. $9(x - 4)$

Algebra Exercise 13

1. 5
$$4x - 2 = 18$$
$$\underline{+2 \qquad +2}$$
$$\frac{4x}{4} = \frac{20}{4}$$
$$x = 5$$

2. 1
$$x + 7 = 8$$
$$\underline{-7 \qquad -7}$$
$$x = 1$$

3. 6
$$2x + 1 = 13$$
$$\underline{-1 \qquad -1}$$
$$\frac{2x}{2} = \frac{12}{2}$$
$$x = 6$$

4. 4
$$6x + 10 = 34$$
$$\underline{-10 \qquad -10}$$
$$\frac{6x}{6} = \frac{24}{6}$$
$$x = 4$$

5. 40
$$\frac{1}{2}x - 8 = 12$$
$$\underline{\phantom{\frac{1}{2}x}+8 \qquad +8}$$
$$2 \cdot \frac{1}{2}x = 20 \cdot 2$$
$$x = 40$$

6. 48
$$8 \cdot \frac{x}{8} = 6 \cdot 8$$
$$x = 48$$

7. 10
$$x = \frac{1}{2}x + 5$$
$$\underline{-\frac{1}{2}x \qquad -\frac{1}{2}x}$$
$$2 \cdot \frac{1}{2}x = 5 \cdot 2$$
$$x = 10$$

8. 11
$$6x - 4x = 22$$
$$\frac{2x}{2} = \frac{22}{2}$$
$$x = 11$$

9. 8
$$3x + 2x = 40$$
$$\frac{5x}{5} = \frac{40}{5}$$
$$x = 8$$

10. 2
$$9x - x = 14 + x$$
$$8x = 14 + x$$
$$\underline{-x \qquad \qquad -x}$$
$$\frac{7x}{7} = \frac{14}{7}$$
$$x = 2$$

Algebra Exercise 14

1. 13 and 14

The first integer = x
The next integer = $x +$

$$x + x + 1 = 27$$

$$2x + 1 = 27$$
$$\underline{-1 \quad -1}$$
$$\frac{2x}{2} = \frac{26}{2}$$

$$x = 13$$
$$x + 1 = 13 + 1 = 14$$

2. 18, 20, and 22

The first even integer $= x$

The second even integer $= x + 2$

The third even integer $= x + 4$

$$x + x + 2 + x + 4 = 60$$

$$3x + 6 = 60$$

$$\frac{-6 \quad -6}{\frac{3x}{3} = \frac{54}{3}}$$

$$x = 18$$

$$x + 2 = 18 + 2 = 20$$

$$x + 4 = 18 + 4 = 22$$

3. 5 and 11

The small number $= x$

The large number $= 2x + 1$

$$3x = 4 + 2x + 1$$

$$3x = 5 + 2x$$

$$\frac{-2x \qquad -2x}{x = 5}$$

$$2x + 1 = 2(5) + 1 = 10 + 1 = 11$$

4. 4 and 24

The small number $= x$

The large number $= 6x$

$$6x - 5 = 4x + 3$$

$$\frac{-4x \qquad -4x}{2x - 5 = \qquad 3}$$

$$\frac{+5 \qquad +5}{\frac{2x}{2} = \frac{8}{2}}$$

$$x = 4$$

$$6x = 24$$

5. 6 men and 18 women

The number of men $= x$

The number of women $= 3x$

$$3x - 2 = x + 10$$

$$\frac{-x \qquad -x}{2x - 2 = \qquad 10}$$

$$\frac{+2 \qquad +2}{\frac{2x}{2} = \frac{12}{2}}$$

$$x = 6$$

$$3x = 18$$

6. $270

Juan's wage $= x$

Felipe's wage $= 3x$

$$x + 3x = 360 \qquad\qquad x = 90$$

$$\frac{4x}{4} = \frac{360}{4} \qquad\qquad 3x = 270$$

7. $168

Mr. Migliaccio's wages $= 4x$

Mrs. Migliaccio's wages $= 3x$

$$4x + 3x = 294$$

$$\frac{7x}{7} = \frac{294}{7}$$

$$x = 42$$

$$3x = 126$$

$$4x = 168$$

8. 24 hours

Jerry's hours $= x$

Jeff's hours $= 2x$

Paul's hours $= 2x + 6$

$$x + 2x + 2x + 6 = 51$$

$$5x + 6 = 51$$

$$\frac{-6 \qquad -6}{\frac{5x}{5} = \frac{45}{5}}$$

$$x = 9$$

$$2x = 18$$

$$2x + 6 = 24$$

9. 16,500 voters

Voters who voted $= x$

Voters who stayed home $= 3x$

$$x + 3x = 22,000$$

$$\frac{4x}{4} = \frac{22,000}{4}$$

$$x = 5,500$$

$$3x = 16,500$$

10. Celeste is 4. Gordon is 8.

	now	2 years ago
Celeste's age	x	$x - 2$
Gordon's age	$2x$	$2x - 2$

$$2x - 2 = 3(x - 2)$$

$$2x - 2 = 3x - 6$$

$$\frac{-2x \qquad\qquad -2x}{-2 = x - 6}$$

$$\frac{+6 \qquad\qquad +6}{4 = x}$$

$$8 = 2x$$

Exercise 14 cont'd.

11.   Laila is 5. David is 20.

	now	in 10 years
Laila's age	x	$x + 10$
David's age	$4x$	$4x + 10$

$$4x + 10 = 2(x + 10)$$
$$4x + 10 = 2x + 20$$
$$\underline{-2x \qquad = -2x}$$
$$2x + 10 = \quad + 20$$
$$\underline{\quad - 10 \qquad - 10}$$
$$\frac{2x}{2} = \frac{10}{2}$$
$$x = 5$$
$$4x = 20$$

12.   Mario is 54. Italo is 24.

	now	18 years ago
Mario's age	$x + 30$	$x + 30 - 18 = x + 12$
Italo's age	x	$x - 18$

$$x + 12 = 6(x - 18)$$
$$x + 12 = 6x - 108$$
$$\underline{-x \qquad - x}$$
$$+ 12 = 5x - 108$$
$$\underline{+108 \qquad + 108}$$
$$\frac{120}{5} = \frac{5x}{5}$$
$$24 = x$$
$$54 = x + 30$$

13.   Sara is 3. Renee is 26.

	now	in 20 years
Sara's age	x	$x + 20$
Renee's age	$x + 23$	$x + 23 + 20 = x + 43$

$$x + 43 = 2(x + 20)$$
$$x + 43 = 2x + 40$$
$$\underline{-x \qquad - x}$$
$$43 = x + 40$$
$$\underline{-40 \qquad - 40}$$
$$3 = x$$
$$26 = x + 23$$

14.   24

Unknown number $= x$.
$$\frac{2}{3}x - \frac{1}{4}x = 10$$
$$\frac{8}{12}x - \frac{3}{12}x = 10$$
$$\frac{12}{5} \cdot \frac{5}{12}x = 10 \cdot \frac{12}{5}$$
$$x = \frac{120}{5}$$
$$x = 24$$

Algebra Exercise 15

1. $A = (+9, +5)$ $G = (-3, -12)$
$B = (+4, +10)$ $H = (0, -7)$
$C = (-2, +4)$ $I = (+6, -7)$
$D = (-6, +10)$ $J = (+14, -3)$
$E = (-11, +3)$ $K = (+18, 0)$
$F = (-7, -2)$

2. $y = +2$

3. $x = -3$

Algebra Exercise 16

1.   yes $y = \frac{1}{2}x + 4$
$y = \frac{1}{2} \cdot 6 + 4$
$y = 3 + 4$
$y = 7$
(6,7) is on the graph.

2.   no $y = \frac{2}{3}x + 1$
$y = \frac{2}{3} \cdot 3 + 1$
$y = 2 + 1$
$y = 3$
(3,4) is not on the graph.

3.   yes $y = -x + 3$
$y = -8 + 3$
$y = -5$
(8,−5) is on the graph.

4.   (6,6)   *$y = 6$ and $x = y$*
Therefore, the point of intersection is (6,6).

5.   8   *A is 3 units above the x-axis.*
C is 5 units below the x-axis.
The distance from A to C is $3 + 5 = 8$.

6.   5   *B is 5 units below the x-axis.*

7.   4   *B is 1 unit left of the y-axis.*
C is 3 units right of the y-axis.
The distance from B to C is $1 + 3 = 4$.

8.   3   *E is 2 units to the left of the y-axis.*
F is 1 unit to the right of the y-axis.
The distance from E to F is $2 + 1 = 3$.

9.   3   *D is 3 units below the x-axis.*

10. 4   *D is 3 units below the x-axis.*
F is 1 unit above the x-axis.
The distance from D to F is 1 + 3 = 4.

11. 5 $DE^2 = EF^2 + DF^2$
$DE^2 = 3^2 + 4^2$
$DE^2 = 9 + 16$
$DE^2 = 25$
$DE = \sqrt{25} = 5$

12. yes $y = \frac{2}{3}x + 1$
$y = \frac{2}{3}(-6) + 1$
$y = -4 + 1$
$y = -3$
(−6,−3) is on the graph.

Algebra Exercise 17

1. The slope of U is undefined.
The slope of T is zero.
The slope of S is negative.
The slope of R is positive.

2. a. +3
$slope = \frac{8-2}{5-3} = \frac{6}{2} = +3$

b. $+\frac{1}{2}$
$slope = \frac{9-5}{10-2} = \frac{4}{8} = +\frac{1}{2}$

c. −2
$slope = \frac{2-6}{5-3} = \frac{-4}{2} = -2$

d. $-\frac{2}{3}$
$slope = \frac{10-4}{-7-2} = \frac{6}{-9} = -\frac{2}{3}$

e. −2
$slope = \frac{5-(-3)}{-1-3} = \frac{5+3}{-1-3} = \frac{8}{-4} = -2$

3. a. x-intercept = (3,0)   y-intercept = (0,−9)
$y = 3x - 9$ $y = 3x - 9$
$0 = 3x - 9$ $y = 3\cdot 0 - 9$
$\underline{+9 \qquad\quad +9}$ $y = -9$
$\frac{9}{3} = \frac{3x}{3}$
$3 = x$

b. x-intercept = (6,0)   y-intercept = (0,−6)
$y = x - 6$ $y = x - 6$
$0 = x - 6$ $y = 0 - 6$
$\underline{+6 \quad\; +6}$ $y = -6$
$6 = x$

c. x-intercept = (−10,0)   y-intercept = (0,5)
$y = \frac{1}{2}x + 5$ $y = \frac{1}{2}x + 5$
$0 = \frac{1}{2}x + 5$ $y = \frac{1}{2}\cdot 0 + 5$
$\underline{-5 \qquad -5}$ $y = +5$
$2\cdot -5 = 2\cdot\frac{1}{2}x$
$-10 = x$

d. x-intercept = (−16,0)   y-intercept = (0,12)
$y = \frac{3}{4}x + 12$ $y = \frac{3}{4}x + 12$
$0 = \frac{3}{4}x + 12$ $y = \frac{3}{4}\cdot 0 + 12$
$\underline{-12 = \qquad -12}$ $y = +12$
$\frac{4}{3}\cdot -12 = \frac{4}{3}\cdot\frac{3}{4}x$
$-16 = x$

e. x-intercept = $(\frac{1}{2},0)$ y-intercept = (0,−1)
$y = 2x - 1$ $y = 2x - 1$
$0 = 2x - 1$ $y = 2\cdot 0 - 1$
$\underline{+1 \qquad +1}$ $y = -1$
$\frac{1}{2} = \frac{2x}{2}$
$\frac{1}{2} = x$

Algebra Review

1. (2) B

2. (1) −14

$$
\begin{array}{cc}
-8 & -34 \\
-17 & +20 \\
\underline{-9} & \underline{-14} \\
-34 &
\end{array}
$$

3. (1) −33

$(-12) + (-17) - (+10) - (-6) =$

$-12 - 17 - 10 + 6 =$

$$
\begin{array}{cc}
-12 & -39 \\
-17 & +\ 6 \\
\underline{-10} & \overline{-33} \\
-39 &
\end{array}
$$

4. (4) 60

$$\frac{-2}{1} \times \frac{\overset{2}{\cancel{8}}}{1} \times \frac{-3}{\underset{1}{\cancel{4}}} \times \frac{5}{1} = 60$$

Two negative numbers make the answer positive.

5. (5) $-\dfrac{5}{6}$

Reduce by 7: $\dfrac{-35}{42} = -\dfrac{5}{6}$

6. (1) 10

$$.05 \cdot \frac{m}{.05} = 200 \cdot .05$$

$$m = 10$$

7. (4) 7

$$
\begin{array}{l}
75 = 9x + 12 \\
\underline{-12 \quad\quad -\ 12} \\
\dfrac{63}{9} = \dfrac{9x}{9} \\
7 = x
\end{array}
$$

8. (3) 8

$$
\frac{3}{4}a - 8 = -2
$$
$$
\underline{+8 \quad\quad +8}
$$
$$
\frac{4}{3} \cdot \frac{3}{4}a = +6 \cdot \frac{4}{3}
$$
$$
a = 8
$$

9. (2) 6

$$
\begin{array}{ll}
7n - 10 = & 3n + 14 \\
\underline{-3n} & \underline{-3n} \\
4n - 10 = & +\ 14 \\
\underline{+\ 10} & \underline{+\ 10} \\
\dfrac{4n}{4} = & \dfrac{24}{4} \\
n = & 6
\end{array}
$$

10. (5) 18

$$
\begin{array}{ll}
5(c - 4) = & 3c + 16 \\
5c - 20 = & 3c + 16 \\
\underline{-3c} & \underline{-3c} \\
2c - 20 = & 16 \\
\underline{+\ 20} & \underline{+\ 20} \\
\dfrac{2c}{2} = & \dfrac{36}{2} \\
c = & 18
\end{array}
$$

11. (1) 4

$$
\begin{array}{ll}
3c - 8 \leq & 7 \\
\underline{+\ 8} & \underline{+8} \\
\dfrac{3c}{3} \leq & \dfrac{15}{3} \\
c \leq & 5
\end{array}
$$

Therefore, c must be less than or equal to 5. Answer (1) is the only possibility.

12. (3) $4(n + 10)$

13. (2) 7

$$
\begin{array}{ll}
5x - 6 = & 2x + 15 \\
\underline{-2x} & \underline{-2x} \\
3x - 6 = & +\ 15 \\
\underline{+\ 6} & \underline{+\ 6} \\
\dfrac{3x}{3} = & \dfrac{21}{3} \\
x = & 7
\end{array}
$$

14. (4) 44

first integer = x

next even integer = x + 2

$$
\begin{array}{ll}
x + x + 2 = & 86 \\
2x + 2 = & 86 \\
\underline{-\ 2} & \underline{-\ 2} \\
\dfrac{2x}{2} = & \dfrac{84}{2} \\
x = & 42
\end{array}
$$

second integer = x + 2 = 44

15. (1) 935

number of women = x

number of men = 3x − 40

$$
\begin{array}{ll}
x + 3x - 40 = & 1260 \\
4x - 40 = & 1260 \\
\underline{+\ 40} & \underline{+\ 40} \\
\dfrac{4x}{4} = & \dfrac{1300}{4} \\
x = & 325
\end{array}
$$

men = 3x − 40 = 3(325)   40 = 935

16. (4) (−4,+3)

17. (3) 7

A is 3 units right of the y-axis.
B is 4 units left of the y-axis.
The distance between A and B
is 3 + 4 = 7.

18. (4) $+\frac{4}{3}$

slope $= \frac{5-1}{3-0} = \frac{4}{3}$

Mathematics Post-Test

The math test that follows will give you an opportunity to see if you are prepared for the GED math test. Like the GED test, it consists of 50 multiple-choice questions and should take 90 minutes to complete.

Work carefully, answering every question. If you are done in less than 90 minutes, use the remaining time to go over your work. If you are not finished, mark where you are at 90 minutes and then finish the test. In this way, you can see two things: whether you can finish a test in the time allotted and whether you have mastered the material in this book and are ready for the test.

Fill in your answers on the separate answer sheet on page 363. If you like, you can tear that sheet out of the book and use it along with the test. Only mark one answer per question; multiple answers marked will count as a wrong answer. A blank also counts as a wrong answer on the GED test, so answer every question.

When you are finished, check your answers in the answer key that follows. For any wrong answers, study the solution carefully to learn from your mistakes. Fill in the evaluation chart at the end of the test to see what your weak areas are and if you are ready to take the GED math test.

1. Tina makes $4.25 an hour and $136 per week. How many hours does she work per week?

    (1) 16   (2) 17   (3) 32   (4) 34   (5) 40

2. To finish her bookshelf, Niah bought $4\frac{1}{4}$ feet of wood. She used $2\frac{1}{2}$ feet for one shelf. How much wood did she have left?

    (1) $\frac{3}{4}$ ft.   (2) 1 ft.   (3) $1\frac{3}{4}$ ft.
    (4) 2 ft.   (5) $2\frac{1}{4}$ ft.

3. Which of the following gives the area of the shaded figure at the right?

    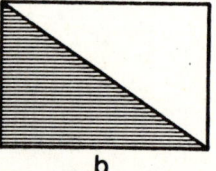

    (1) $A = \frac{1}{2}bc$
    (2) $A = 2b + 2c$
    (3) $A = bc$
    (4) $A = b^2 + c^2$
    (5) $A = b + c$

4. Yolanda paid $150 down and $48 a month over 30 months for new furniture. What total amount did she pay for the furniture?

    (1) $5,940   (2) $1,440   (3) $1,290
    (4) $1,390   (5) $1,590

5. If coffee costs $3.40 a pound, what is the price of $\frac{3}{4}$ pound of coffee?

    (1) $4.53   (2) $2.45   (3) $2.27
    (4) $2.40   (5) $2.55

Use the circle graph below to answer questions 6 through 8.

WHERE THE U.S. BUDGET DOLLAR COMES FROM

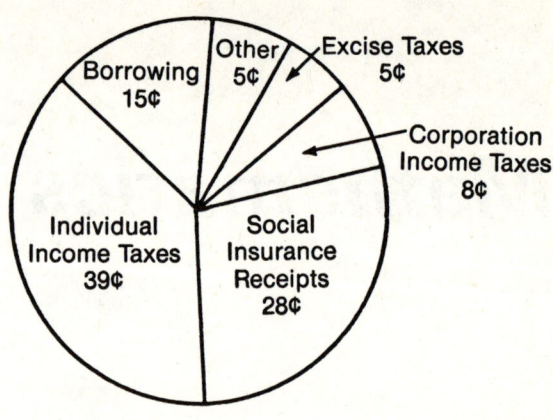

6. Together, borrowing and excise taxes account for what percent of the U.S. budget dollar?

    (1) 5% (2) 13% (3) 15% (4) 20% (5) 28%

7. The budget amount from corporation income taxes represents about what fraction of the amount from individual income taxes?

    (1) $\frac{1}{8}$ (2) $\frac{1}{5}$ (3) $\frac{1}{4}$ (4) $\frac{1}{3}$ (5) $\frac{1}{2}$

8. Which of the following statements is <u>false</u>?

    (1) Individual income taxes are the largest source of money in the U.S. budget.
    (2) Corporation income taxes produce more money for the U.S. budget than excise taxes.
    (3) Together, individual income taxes and social insurance receipts make up more than 75% of the U.S. budget.
    (4) The U.S. borrows almost twice as much as it receives from corporation income taxes.
    (5) Social insurance receipts account for more than $\frac{1}{4}$ of the U.S. budget.

9. A field is twice as long as it is wide. The field is 50 feet long. In square feet, what is its size?

    (1) 75   (2) 100   (3) 110
    (4) 150   (5) 1,250

10. What is the value of m in $5 : m = 9 : 35$?

    (1) $19\frac{4}{9}$ (2) $\frac{7}{9}$ (3) $1\frac{2}{7}$
    (4) 63   (5) 36

11. One mile is equal to 5,280 feet. How many feet are there in .75 mile?

    (1) 2,640   (2) 4,500   (3) 3,960
    (4) 2,220   (5) 7,040

12. A stereo originally selling for $450 was on sale for $360. What is the discount rate on the stereo?

    (1) 10%   (2) 15%   (3) 20%
    (4) 80%   (5) 90%

13. For the expression $4t < 36$, what is a possible value of t?

    (1) 8   (2) 9   (3) 10   (4) 36   (5) 360

14. In 1970, 6,546,817 American-made cars were sold. In 1981, 6,255,340 cars were sold, and in 1982, 6,266,313 cars were sold. How many more cars were sold in 1970 than in 1981?

    (1) 290,487   (2) 291,577   (3) 290,577
    (4) 291,477   (5) 391,477

15.

    Which point on the number line above represents $-\frac{7}{4}$?

    (1) A   (2) B   (3) C   (4) D   (5) E

16. Fred drove 205 miles on 14 gallons of gasoline. To the nearest tenth, how many miles did Fred drive on one gallon of gasoline?

    (1) 16.1   (2) 15.3   (3) 14.6
    (4) 13.2   (5) 12.8

17. Adrienne makes $14,000 a year. Her employer withholds $2,800 from her salary. What is the ratio of the amount withheld to her take-home pay?

    (1) 1:6   (2) 5:1   (3) 4:1
    (4) 1:5   (5) 1:4

18. Which of the following is equal to $(9 + 8) + 12$?

    (1) $9(8 + 12)$
    (2) $(9 + 8)12$
    (3) $9 \cdot 12 + 8 \cdot 12$
    (4) $9 + 12 + 8 + 12$
    (5) $9 + (8 + 12)$

Use the table below for questions 19 and 20

Population and Unemployment in Selected Cities		
	Population (in thousands)	Unemployment (%)
	1970 1980	1980
Atlanta	417 425	7.9
Chicago	3,063 3,005	9.1
Cleveland	609 578	10.2
Dallas	845 904	4.8
Detroit	1,290 1,203	16.5

19. Which cities shown in the table had a population increase from 1970 to 1980?

    (1) Chicago & Detroit
    (2) Atlanta & Chicago
    (3) Atlanta & Dallas
    (4) Dallas & Detroit
    (5) Cleveland & Dallas

20. According to the table, which of the following statements is true?

    (1) The population of each city increased by about the same percent.
    (2) The smaller cities had higher unemployment rates.
    (3) The 1980 unemployment rate in Dallas was more than three times the unemployment rate in Cleveland.
    (4) The cities with population increases from 1970 to 1980 had lower 1980 unemployment rates than the cities with population decreases.
    (5) Cleveland had the highest unemployment rate.

GO ON TO THE NEXT PAGE

21. Carla has 18 feet of picture frame. She has to frame pictures that are 5 inches by 7 inches. How many pictures can she frame if she uses all the wood?

(1) 3   (2) 6   (3) 9   (4) 12   (5) 30

22. In the picture at the right, ∠BOC = 47°. What is the measurement of ∠AOB?

(1) 43°
(2) 133°
(3) 143°
(4) 153°
(5) 47°

23.

stock price	$9\frac{1}{8}$	9	$8\frac{7}{8}$	$8\frac{3}{4}$	—
date	May 5	May 6	May 7	May 8	May 9

If the stock price continues in the same pattern, what will be the price on May 9?

(1) $8\frac{7}{8}$ (2) 9   (3) $8\frac{3}{8}$ (4) $8\frac{5}{8}$ (5) $8\frac{1}{2}$

24. Pat can type 92 words per minute. Which of the following will tell how many minutes Pat needs to type a 1,500-word letter?

(1) Divide 1,500 by 92.
(2) Multiply 92 by 1,500.
(3) Divide 92 by 1,500.
(4) Multiply 92 by 60, then divide by 1,000.
(5) Divide 1,500 by 60, then multiply by 92.

25. Over nine months, Pat repaid a $1,200 loan borrowed at a $12\frac{1}{2}$% interest rate. How much did Pat pay back to the bank?

(1) $112.50   (2) $150   (3) $1,200
(4) $1,312.50   (5) $1,500

26. What is the area of a lot that is 13.2 meters long and 8.35 meters wide?

(1) 22 sq.m
(2) 26.4 sq.m
(3) 43 sq.m
(4) 110.22 sq.m
(5) 1,100 sq.m

27. Leo's recipe for barbecued chicken calls for 3 pounds of chicken for six people. How many pounds of chicken would he use for nine people?

(1) 3   (2) $4\frac{1}{2}$ (3) 6   (4) 9   (5) 21

28. How many cubic feet of concrete are needed to finish a driveway section that is 12 feet long, 10 feet wide, and $\frac{3}{4}$ foot deep?

(1) $12\frac{3}{4}$ (2) 45   (3) 90
(4) 120   (5) 180

29. Which of the following is equal to $9^2 - 4^3$?

(1) $(9 - 4)(9 - 4)$
(2) $(9 \times 9) - (4 \times 4 \times 4)$
(3) $(9 - 4)^3$
(4) $(9 \times 2) - (4 \times 3)$
(5) $(9 + 9) - (4 + 4 + 4)$

30. Fumio shipped three cartons weighing 19 lb. 6 oz., 15 lb. 8 oz., and 12 lb. 10 oz., respectively. What was the total weight of the three cartons?

(1) 47 lb. 8 oz.
(2) 46 lb. 4 oz.
(3) 48 lb.
(4) 48 lb. 4 oz.
(5) 48 lb. 8 oz.

31. Pilar paid $15.30 for $4\frac{1}{2}$ yards of material. What was the price of one yard?

(1) $3.40   (2) $7.00   (3) $1.70
(4) $5.10   (5) $6.89

32. If $6x - 7 + 2x = 3x + 13$, what is the value of x?

(1) 20   (2) 5   (3) 3   (4) 4   (5) 13

33. What is the distance between points A and B on the graph at the right?

(1) 2
(2) 4
(3) 6
(4) 8
(5) 10

34 M₁ and Mrs. Granata bought a car in 1982 for $5,500. By 1983, the value of the car had depreciated by 40%. What was the value of the car in 1983?

(1) $5,100   (2) $4,400   (3) $2,200
(4) $2,800   (5) $3,300

35. A house stands next to a tree. The house is 24 feet tall and its shadow is 18 feet long. The tree is 12 feet tall. How long is its shadow?

(1) 6 ft.   (2) 9 ft.   (3) 12 ft.
(4) 16 ft.   (5) 54 ft.

36. When four times a number is increased by 16, the result is the same as when seven times the same number is decreased by 2. Find the number.

(1) $4\frac{2}{3}$ (2) 6   (3) $3\frac{1}{2}$ (4) 4   (5) $10\frac{1}{2}$

37. Sandy used to be able to throw a football 36 yards 2 feet. Now she can throw it 3 yards 1 foot farther. How far can she now throw the ball?

(1) 39 ft.   (2) 40 ft.   (3) 33 yds. 1 ft.
(4) 34 yds.   (5) 40 yds.

Use the graph below to answer questions 38 through 40.

MONTHLY SALES TOTALS FOR BETTER BOILERS, INC.

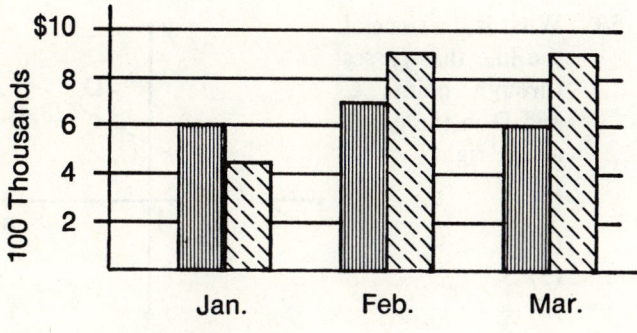

Key: ▦ 1982
      ▧ 1983

38. For which month shown on the graph were the sales for Better Boilers the lowest?

(1) Jan. '82
(2) Jan. '83
(3) Feb. '82
(4) Feb. '83
(5) Mar. '82

39. The ratio of January 1983 sales to February 1983 sales was about:

(1) 2:1   (2) 4:3   (3) 3:4   (4) 1:2   (5) 1:3

40. According to the graph, which of the following statements is true?

(1) For the first three months of 1982, sales rose steadily.
(2) For the first three months of 1982, sales dropped steadily.
(3) For the first three months of 1983, sales remained about the same.
(4) The largest one-month jump in sales occurred between January and February 1982.
(5) The month that shows the greatest difference in sales between 1982 and 1983 is March.

41. Solve for w in $3(w - 4) = 5(w - 7)$.

(1) $1\frac{1}{2}$ (2) $4\frac{1}{2}$ (3) $5\frac{1}{2}$ (4) $2\frac{7}{8}$ (5) $11\frac{1}{2}$

42. In a recent poll, 520 people were against storing nuclear waste near their communities. This represents 65% of the people who were polled. How many of the people polled were *not* against storing nuclear waste near their communities?

(1) 462   (2) 320   (3) 520
(4) 338   (5) 280

GO ON TO THE NEXT PAGE

43. On the map at the right, the distance along I-80 from A to B is 40 miles. The distance along I-95 from A to C is 30 miles. Find the distance along Route 3 from B to C.

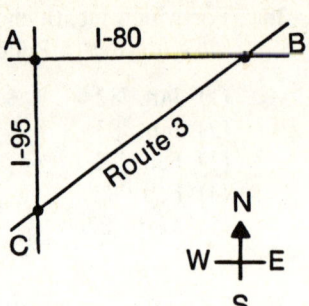

   (1) 35 mi.
   (2) 40 mi.
   (3) 45 mi.
   (4) 50 mi.
   (5) 70 mi.

44. One number is twice as big as another. Three times the smaller number increased by one is equal to the larger number increased by nine. Find the numbers.

   (1) 3 and 6
   (2) $4\frac{1}{2}$ and 9
   (3) 8 and 16
   (4) 9 and 18
   (5) 12 and 24

45. What is one side of a square tablecloth that covers 16 square feet?

   (1) 4 ft.   (2) 4 ft. 6 in.   (3) 8 ft.
   (4) 64 ft.   (5) 256 ft.

46. The formula for converting Fahrenheit to Celsius temperature is $C = \frac{5}{9}(F - 32)$ where F is the Fahrenheit temperature. Find the Celsius temperature that corresponds to 104° Fahrenheit.

   (1) 20° (2) 40° (3) 60° (4) 72° (5) 27°

47. During a two-week holiday rush, Rosa worked 12 hours overtime one week and 15 hours overtime the next week. She made $7 an hour for overtime. Which expression tells how much Rosa made working overtime those two weeks?

   (1) 15(12 + 7)
   (2) 12(15 + 7)
   (3) 7(12 + 15)
   (4) (15 × 12) + 7
   (5) (15 × 12) + (7 × 12)

48. Sol, Kevin, and Hiro worked together building a garage. Kevin worked twice as many hours as Sol, and Hiro worked 10 hours more than Kevin. Altogether they worked 100 hours. How many hours did Hiro work?

   (1) 46 (2) 33 (3) 50 (4) 36 (5) 26

49. The shaded area in the picture at the right shows the paved area around a circular pool. The diameter of the pool is 14 feet. Find the amount of square feet of paved area. Use $\pi = \frac{22}{7}$.

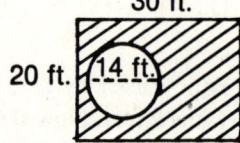

   (1) 551
   (2) 586
   (3) 404
   (4) 502
   (5) 446

50. What is the slope of the line that passes through points C and D in the graph at the right?

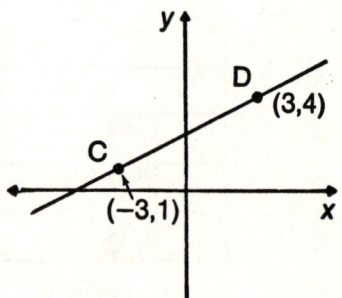

   (1) 2
   (2) $\frac{1}{2}$
   (3) 3
   (4) $\frac{1}{3}$
   (5) −2

ANSWERS AND SOLUTIONS

1. (3) 32

$$4.25.\overline{)136.00.}$$
$$\begin{array}{r} 32 \\ \hline 127\ 5 \\ \hline 8\ 50 \\ 8\ 50 \end{array}$$

2. (3) $1\frac{3}{4}$ ft.

$$\begin{array}{r} 4\frac{1}{4} = 3\frac{5}{4} \\ -2\frac{1}{2} = -2\frac{2}{4} \\ \hline 1\frac{3}{4} \end{array}$$

3. (1) $A = \frac{1}{2}bc$

The triangle is $\frac{1}{2}$ the rectangle with a base b and a height c.

4. (5) $1,590

$$\begin{array}{r} \$48 \\ \times\ \ 30\ wk. \\ \hline \$1,440 \end{array} \qquad \begin{array}{r} \$1,440 \\ +\ 150\ down \\ \hline \$1,590 \end{array}$$

5. (5) $2.55

$$\frac{3}{\cancel{4}_{1}} \times \frac{\$\overset{.85}{\cancel{3.40}}}{1} = \$2.55$$

6. (4) 20%

$$\begin{array}{lll} borrowing & = & 15¢ \\ excise\ taxes & = & +\ 5¢ \\ \hline total & = & 20¢ \end{array}$$

20¢ out of a dollar = 20%.

7. (2) $\frac{1}{5}$

Corporation taxes = 8¢
Individual taxes = 39¢
(which is close to 40¢)

$$\frac{8¢}{40¢} = \frac{1}{5}$$

8. (3) Together, individual income taxes and social insurance receipts make up over 75% of the U.S. budget.

The statement is false.

$$\begin{array}{lll} individual\ taxes & = & 39¢ \\ social\ insurance & = & 28¢ \\ \hline total & = & 67¢ \end{array}$$

67¢ out of a dollar is less than 75%.

9. (5) 1,250

The problem is asking you to find the number of square feet in a field. You have to find the area.

l = 50 feet
$w = \frac{1}{2}$ length = 25 feet

$A = lw$
$A = 50 \times 25 = 1,250$ square feet

10. (1) $19\frac{4}{9}$

$$\frac{5}{m} = \frac{9}{35} \qquad \begin{array}{r} 35 \\ \times\ 5 \\ \hline 175 \end{array} \qquad 9\overline{)175}\ \ 19\frac{4}{9}$$
$$\begin{array}{r} \underline{9} \\ 85 \\ \underline{81} \\ 4 \end{array}$$

11. (3) 3,960

$$\begin{array}{r} 5,280 \\ \times\ .75 \\ \hline 264\ 00 \\ 3\ 696\ 0 \\ \hline 3,960.00 \end{array}$$

12. (3) 20%

$$\begin{array}{ll} \$\ 450 & original\ price \\ -360 & sale\ price \\ \hline \$\ 90 & discount \end{array}$$

90	%
450	100

$$\begin{array}{r} 90 \\ \times 100 \\ \hline 9,000 \end{array} \qquad 450\overline{)9,000}\ \ 20\%$$

13. (1) 8

$$\frac{4t}{4} < \frac{36}{4}$$
$$t < 9$$

If t is less than 9, then the only possible value given is choice (1) 8.

14. (4) 291,477

$$\begin{array}{r} 6,546,817\ (1970) \\ -6,255,340\ (1981) \\ \hline 291,477 \end{array}$$

15. (2) B

$$-\frac{7}{4} = -1\frac{3}{4}$$

16. (3) 14.6

$$14\overline{)205.00}\ \ \begin{array}{l} 14.64\ \ to\ the\ nearest \\ \ \ \ \ \ \ \ tenth = 14.6 \end{array}$$
$$\begin{array}{r} \underline{14} \\ 65 \\ \underline{56} \\ 9\ 0 \\ \underline{8\ 4} \\ 60 \\ \underline{56} \end{array}$$

17. (5) 1:4

$$\begin{array}{r} \$14,000 \\ -\ 2,800 \\ \hline \$11,200 \end{array}$$

withheld : take-home
2,800 : 11,200
1 : 4

18. (5) $9 + (8 + 12)$

19. (3) Atlanta & Dallas

20. (4) The cities with population increases from 1970 to 1980 had lower 1980 unemployment rates than the cities with population decreases.

21. (3) 9

$$P = 2l + 2w$$
$$P = 2(7) + 2(5)$$
$$P = 14 + 10 = 24 \ \ inches$$

$$\begin{array}{r} 2 \ feet \ per \ picture \\ 12\overline{)24} \end{array}$$

$$\begin{array}{r} 9 \ pictures \\ 2\overline{)18} \end{array}$$

22. (2) 133°

$$\begin{array}{r} 180° \\ -47° \\ \hline 133° \end{array}$$

23. (4) $8\frac{5}{8}$

$$9\frac{1}{8} \searrow \ 9 \searrow \ 8\frac{7}{8} \searrow \ 8\frac{3}{4} \searrow \ 8\frac{5}{8}$$
$$-\frac{1}{8} \quad -\frac{1}{8} \quad -\frac{1}{8} \quad -\frac{1}{8}$$

24. (1) Divide 1,500 by 92

25. (4) 1,312.50

$$I = prt$$
$$(9 \ months = \frac{9}{12} = \frac{3}{4})$$

$$I = 1,200 \times \frac{12\frac{1}{2}}{100} \times \frac{3}{4}$$

$$(\frac{12\frac{1}{2}}{100} = \frac{25}{2} \div 100$$

$$= \frac{\overset{1}{25}}{2} \times \frac{1}{\underset{4}{100}} = \frac{1}{8})$$

$$\overset{75}{\cancel{1,200}} \times \frac{1}{\underset{1}{8}} \times \frac{3}{\underset{2}{4}} = \frac{225}{2} = 112.50$$

$$1,200 + 112.50 = 1,312.50$$

26. (4) 110.22 sq. m

$$A = lw$$
$$A = 13.2 \times 8.35$$
$$A = 110.22 \ sq. \ m$$

27. (2) $4\frac{1}{2}$

$$\frac{people}{pounds} \ \frac{6}{3} = \frac{9}{x}$$

$$\begin{array}{r} 9 \\ \times 3 \\ \hline 27 \end{array} \quad \begin{array}{r} 4\frac{1}{2} \\ 6\overline{)27} \\ \underline{24} \\ 3 \end{array}$$

28. (3) 90

$$V = lwh$$

$$V = \cancel{12}^{3} \times 10 \times \frac{3}{\cancel{4}_{1}} = 90$$

29. (2) $(9 \times 9) - (4 \times 4 \times 4)$

30. (1) 47 lb. 8 oz

$$\begin{array}{r} 19 \ lb. \ \ 6 \ oz. \\ 15 \ lb. \ \ 8 \ oz. \\ +12 \ lb. \ 10 \ oz. \\ \hline 46 \ lb. \ 24 \ oz. \ = \ 47 \ lb. \ 8 \ oz. \end{array}$$

31. (1) $3.40

$$\$15.30 \div 4\frac{1}{2} = \frac{15.30}{1} \div \frac{9}{2} =$$

$$\frac{\overset{1.70}{\cancel{15.30}}}{1} \times \frac{2}{\cancel{9}} = \$3.40$$

32. (4) 4

$$\begin{array}{rcl} 6x - 7 + 2x & = & 3x + 13 \\ 8x - 7 & = & 3x + 13 \\ \underline{-3x} & & \underline{-3x} \\ 5x - 7 & = & 13 \\ \underline{+7} & & \underline{+7} \\ \dfrac{5x}{5} & = & \dfrac{20}{5} \\ x & = & 4 \end{array}$$

33. (3) 6

A is 2 units left of the y-axis.
B is 4 units right of the y-axis.
4 units + 2 units = 6 units

34. (5) $3,300

part	40
5,500	100

$$\begin{array}{r} 5,500 \\ \times \ \ \ 40 \\ \hline 220,000 \end{array} \quad \begin{array}{r} 2,200 \ depreciation \\ 100\overline{)220,000} \end{array}$$

$$\begin{array}{lcl} original \ value & = & \$5,500 \\ depreciation & = & -2,200 \\ 1983 \ value & = & \$3,300 \end{array}$$

35. (2) 9 ft.

$$\frac{building}{shadow} \ \frac{24}{18} = \frac{12}{x}$$

$$\begin{array}{r} 18 \\ \times 12 \\ \hline 36 \\ 18 \\ \hline 216 \end{array} \quad \begin{array}{r} 9 \\ 24\overline{)216} \\ \underline{216} \end{array}$$

36. (2) 6

$$\begin{array}{rcl} 4x + 16 & = & 7x - 2 \\ \underline{-4x} & & \underline{-4x} \\ 16 & = & 3x - 2 \\ \underline{+ \ 2} & & \underline{+ \ 2} \\ \dfrac{18}{3} & = & \dfrac{3x}{3} \\ 6 & = & x \end{array}$$

37. (5) 40 yds.

$$36 \text{ yds. 2 ft.}$$
$$\underline{+\ 3 \text{ yds. 1 ft.}}$$
$$39 \text{ yds. 3 ft.} = 40 \text{ yds.}$$

38. (2) Jan. '83

39. (4) 1:2

Jan. '83 is about $450,000.
Feb. '83 is about $900,000.
450,000:900,000 = 1:2

40. (5) The month that shows the greatest difference in sales between 1982 and 1983 is March.

41. (5) $11\frac{1}{2}$

$$3(w - 4) = 5(w - 7)$$
$$3w - 12 = 5w - 35$$
$$\underline{-3w \qquad\quad -3w}$$
$$-12 = 2w - 35$$
$$\underline{+ 35 \qquad\quad + 35}$$
$$\frac{23}{2} = \frac{2w}{2}$$
$$11\frac{1}{2} = w$$

42. (5) 280

520	65
whole	100

$$\begin{array}{r} 520 \\ \times\ 100 \\ \hline 52,000 \end{array} \qquad \begin{array}{r} 800 \\ 65)\overline{52,000} \\ \underline{52\ 0} \\ 000 \end{array}$$

number polled = 800
number against = −520
number not against = 280

43. (4) 50 mi.

$$c^2 = a^2 + b^2$$
$$c^2 = 30^2 + 40^2$$
$$c^2 = 900 + 1,600$$
$$c^2 = 2,500$$
$$c = \sqrt{2,500}$$
$$c = 50 \text{ mi.}$$

44. (3) 8 and 16

small number = x
large number = 2x

$$3x + 1 = 2x + 9$$
$$\underline{-2x \qquad\ -2x}$$
$$x + 1 = 9$$
$$\underline{-1 \qquad\ -1}$$
$$x = 8$$
$$2x = 16$$

45. (1) 4 ft.

The area of a square is represented by the formula $A = s^2$. To find one side, find the square root of the area.

$$\sqrt{16} = 4$$

46. (2) 40°

$$C = \frac{5}{9}(F - 32) = \frac{5}{9}(104 - 32)$$
$$= \frac{5}{\cancel{9}_1} \times \frac{\cancel{72}^{8}}{1} = 40°$$

47. (3) 7(12 + 15)

48. (1) 46

Sol's hours = x
Kevin's hours = 2x
Hiro's hours = 2x + 10

$$x + 2x + 2x + 10 = 100$$
$$5x + 10 = 100$$
$$\underline{-10 \qquad -10}$$
$$\frac{5x}{5} = \frac{90}{5}$$
$$x = 18$$

Hiro worked 2x + 10
$$= 2(18) + 10$$
$$= 46$$

49. (5) 446

area of rectangle:
A = lw
$$= 20 \times 30 = 600 \text{ sq. ft}$$
area of pool: radius = 7 ft.
$A = \pi r^2$

$$= \frac{22}{\cancel{7}_1} \times \frac{\cancel{7}^{1}}{1} \times \frac{7}{1} = 154 \text{ sq. ft.}$$

paved area:
$$\begin{array}{r} 600 \text{ sq. ft.} \\ \underline{-154} \\ 446 \text{ sq. ft.} \end{array}$$

50. (2) $\frac{1}{2}$

$$slope = \frac{y_2 - y_1}{x_2 - x_1}$$
$$= \frac{4 - 1}{3 - (-3)}$$
$$= \frac{3}{3 + 3} = \frac{3}{6} = \frac{1}{2}$$

POST-TEST EVALUATION CHART

On the chart below, circle the number of any questions that you missed. If any section has half or more questions wrong, you should review that material before going on to the GED test.

Section	Problems
Whole Numbers Pages 23-53	4, 14, 24
Decimals Pages 54-86	1, 11, 16
Fractions Pages 87-137	2, 5, 7, 23, 31
Percents Pages 138-174	12, 25, 34, 42
Ratio & Proportion Pages 176-183	10, 17, 27, 35, 39
Graphs & Tables Pages 184-208	6, 8, 19, 20, 21, 38, 40
Measurement Pages 209-232	30, 37
Intro to Alg. & Geom. Pages 234-253	18, 29, 46, 47
Geometry Pages 254-298	3, 9, 21, 22, 26, 28, 43, 45, 49
Algebra Pages 299-351	13, 15, 32, 33, 36, 41, 44, 48, 50

Scoring Chart

The chart below shows you two scores similar to the scores on the GED test. The first column of scores shows the number of problems you got correct. The second column shows the score this would result in. Remember, for the GED test, you must have a minimum score to pass the test. You must have at least half of each test correct to get the minimum average score needed to get a certificate. Check with your teacher or local GED testing center to find what these scores are. If you have a score less than 45, you should do some more work before going on to the actual GED test.

Problems Right (out of 50)	Score (approximate—varies from test to test)
13 or less	less than 35
19	40
24	45
29	50
35	55
39 or more	60 or more

Answer Grid

1 ① ② ③ ④ ⑤	13 ① ② ③ ④ ⑤	26 ① ② ③ ④ ⑤	38 ① ② ③ ④ ⑤
2 ① ② ③ ④ ⑤	14 ① ② ③ ④ ⑤	27 ① ② ③ ④ ⑤	39 ① ② ③ ④ ⑤
3 ① ② ③ ④ ⑤	15 ① ② ③ ④ ⑤	28 ① ② ③ ④ ⑤	40 ① ② ③ ④ ⑤
4 ① ② ③ ④ ⑤	16 ① ② ③ ④ ⑤	29 ① ② ③ ④ ⑤	41 ① ② ③ ④ ⑤
5 ① ② ③ ④ ⑤	17 ① ② ③ ④ ⑤	30 ① ② ③ ④ ⑤	42 ① ② ③ ④ ⑤
6 ① ② ③ ④ ⑤	18 ① ② ③ ④ ⑤	31 ① ② ③ ④ ⑤	43 ① ② ③ ④ ⑤
7 ① ② ③ ④ ⑤	19 ① ② ③ ④ ⑤	32 ① ② ③ ④ ⑤	44 ① ② ③ ④ ⑤
8 ① ② ③ ④ ⑤	20 ① ② ③ ④ ⑤	33 ① ② ③ ④ ⑤	45 ① ② ③ ④ ⑤
9 ① ② ③ ④ ⑤	21 ① ② ③ ④ ⑤	34 ① ② ③ ④ ⑤	46 ① ② ③ ④ ⑤
10 ① ② ③ ④ ⑤	22 ① ② ③ ④ ⑤	35 ① ② ③ ④ ⑤	47 ① ② ③ ④ ⑤
11 ① ② ③ ④ ⑤	23 ① ② ③ ④ ⑤	36 ① ② ③ ④ ⑤	48 ① ② ③ ④ ⑤
12 ① ② ③ ④ ⑤	24 ① ② ③ ④ ⑤	37 ① ② ③ ④ ⑤	49 ① ② ③ ④ ⑤
	25 ① ② ③ ④ ⑤		50 ① ② ③ ④ ⑤